ANALYTICAL DYNAMICS

ANALYTICAL DYNAMICS

A New Approach

Firdaus E. Udwadia and Robert E. Kalaba

University of Southern California

CAMBRIDGE UNIVERSITY PRESS
Cambridge, New York, Melbourne, Madrid, Cape Town, Singapore, São Paulo

Cambridge University Press
The Edinburgh Building, Cambridge CB2 8RU, UK

Published in the United States of America by Cambridge University Press, New York

www.cambridge.org
Information on this title: www.cambridge.org/9780521482172

First published 1996
This digitally printed version 2008

A catalogue record for this publication is available from the British Library

Library of Congress Cataloguing in Publication data

Udwadia, F. E.
Analytical dynamics : a new approach / Firdaus E. Udwadia, Robert
E. Kalaba.
p. cm.
Includes bibliographical references and index.
ISBN 0-521-48217-8
1. Dynamics. I. Kalaba, Robert E. II. Title.
QA845.U39 1996
531'.11–dc20 95-13004
CIP

ISBN 978-0-521-48217-2 hardback
ISBN 978-0-521-04833-0 paperback

CONTENTS

PREFACE

There are many treatises in the field of analytical mechanics that have been written in this century. This book is different in that it presents a new and fresh approach to the central problem of the motion of discrete mechanical systems. A system of point masses differs from a set of point masses in that the masses of a system satisfy certain constraints. This book primarily deals with the statement and analytical resolution of the problem of constrained motion and we provide the explicit equations of motion that govern large classes of constrained mechanical systems. The simplicity of the results has encouraged us to write a text which we hope will be well within the grasp of the average college senior in science and engineering.

We assume that the student has had an elementary level course dealing with statics and dynamics, and some exposure to elementary linear algebra, though the latter is not essential, because most of what is needed is contained in Chapter 2 of this book. Being pitched at the junior/senior undergraduate level, we have tried to take pains in introducing concepts slowly, gradually building them up in depth through a continual process of revisitation. We have also restricted our "For Further Reading List" at the end of each chapter principally to two books (those by Pars and Rosenberg), though there are also many other excellent treatises on analytical dynamics. Rather than deluge the student with a plethora of reference material we have selected these two excellent texts which share between them many common threads of presentation.

We make no claim to cover the whole of analytical dynamics. If our book made any approach to completeness it would be many times larger than it is. Our main aim is to provide a new and different approach to the subject, an approach that brings basic concepts to the fore and is often simpler to use in many practical situations.

The book is divided into eight chapters. The first chapter deals with an introduction to the problem of constrained motion of a system of particles, and, looking ahead, provides the solution to the problem. The second chapter is a brief review of matrix algebra with special emphasis on the tools of linear algebra which we will be needing. The third chapter introduces Gauss's principle in rectangular coordinates and obtains the solution to the problem of constrained motion in closed form for constraints which are defined by functions which depend on the velocities of the particles, their displacements and time. The fourth chapter provides a more constructive proof for the equations of motion of constrained systems that were presented in the previous chapter and shows computational results utilizing MATLAB and Mathematica, along with sample programs used to generate these results.

Chapter 5 of the book shows how these results can be extended to Lagrangian coordinates and explains in depth the various types of constraints usually met with in Lagrangian mechanics. We introduce the principle of virtual work and define more explicitly what we mean by an unconstrained system, and a constrained system. We show that there are various alternative formulations that can be used to describe the constrained motion of mechanical systems. Chapter 6 provides another proof for the constrained equations of motion using the elementary principle of virtual work which was introduced in the previous chapter. We go on to provide an elementary account of rigid body motion, and illustrate the use of our approach to such situations. The seventh chapter is a revisitation of Gauss's principle albeit at greater depth along with concepts such as virtual displacements and virtual work. The last chapter builds connections between the various key concepts which were previously introduced and weaves together several ideas related to the determination of the equation of motion for constrained mechanical systems. We end with what we began with – Gauss's principle.

Several illustrative examples have been provided in the book to firm up the concepts introduced; we hope that the problems at the end of each chapter will be helpful in further clarification.

We wish to express our appreciation to all our students and colleagues who have encouraged us in writing this book. We have received many suggestions, advice and comments from numerous colleagues during the preparation of this book. We thank them all. We hope that it will provide a new impetus to the field of analytical mechanics, particularly to the understanding of the motion of constrained systems, a long-standing problem of some importance.

1 Introduction

Consider a point-particle of mass m. By a point-particle we mean that the particle has negligible dimensions and can be imagined as occupying a point in space. Let the particle at any given time t, occupy a point whose rectangular coordinates are, say, $\{x(t), y(t), z(t)\}$. Our right-handed X-Y-Z coordinate frame will always be "inertial." By this we mean that the coordinate frame is either fixed relative to the stars, or that it is moving with constant velocity. Let the components of the net force acting on the particle at time t be $\{F_x(t), F_y(t), F_z(t)\}$.

Newton's second law of motion informs us that, at each instant of time t, the mass of the particle times its acceleration is a measure of the force acting on it. With increasing time, the particle will trace out a trajectory. The X-component of the velocity, $\dot{x}(t)$, of the particle at time t is the derivative of $x(t)$ with respect to time; the X-component of the acceleration, $\ddot{x}(t)$, of the particle at time t, is the second derivative of $x(t)$ with respect to time. Likewise for the Y- and

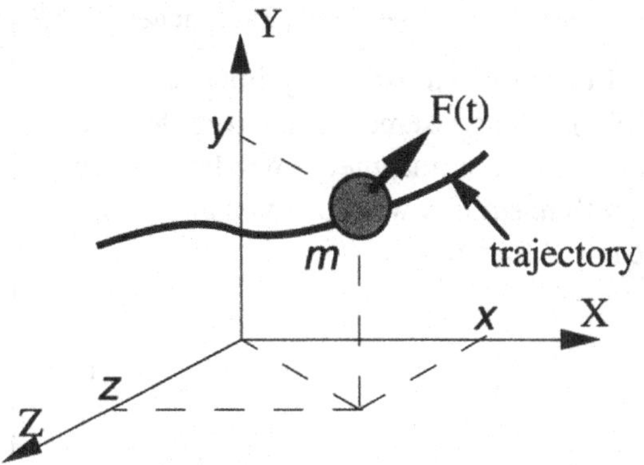

Figure 1.1: A point-particle under the action of a force $F(t)$

Z-components. We then have three scalar equations describing the motion of the particle given by

$$\begin{aligned} m\ddot{x}(t) &= F_x(t), \\ m\ddot{y}(t) &= F_y(t), \\ m\ddot{z}(t) &= F_z(t). \end{aligned} \tag{1.1}$$

Notice that Newton's law relates the acceleration *at each instant of time* to the force on the particle at that instant. Were we to be given the position, $\{(x(t_0), y(t_0), z(t_0)\}$, of the particle and its velocity, $\{(\dot{x}(t_0), \dot{y}(t_0), \dot{z}(t_0)\}$, at some time t_0, we could integrate equation (1.1) (at least numerically) to obtain the trajectory of the particle shown in Figure 1.1. The position and velocity of the particle at time t_0 are often called the "initial" position and velocity of the particle, and the time t_0 is often referred to as the "initial" time.

Equation (1.1) can be rewritten as

$$\begin{aligned} \ddot{x}(t) &= \frac{1}{m}F_x(t) = a_x(t), \\ \ddot{y}(t) &= \frac{1}{m}F_y(t) = a_y(t), \\ \ddot{z}(t) &= \frac{1}{m}F_z(t) = a_z(t), \end{aligned} \tag{1.2}$$

where $a_x(t)$ is the force acting on the particle in the X-direction divided by its mass. Note that this quantity equals the acceleration of the particle in the X-direction. Likewise for the Y- and Z-directions. We now illustrate what we mean by constrained motion by two simple examples.

Example 1.1

Consider a particle subjected to the given externally impressed force components $F_x(t)$, $F_y(t)$ and $F_z(t)$. Let us assume that the particle is constrained to move on the X-axis. One can imagine a small bead moving (without friction) along a very thin, straight wire, oriented along the X-direction. Assume that at the initial time t_0,

$$\begin{aligned} x(t_0) &= 5, \\ y(t_0) &= 0, \\ z(t_0) &= 0, \end{aligned} \tag{1.3}$$

and that

$$\dot{x}(t_0) = 10,$$
$$\dot{y}(t_0) = 0, \tag{1.4}$$
$$\dot{z}(t_0) = 0.$$

Note that these initial conditions satisfy the constraint at time $t = t_0$, namely that the particle lies along the X-axis and has a nonzero velocity only along the X-direction.

By an "impressed" force we mean a force, which is often externally applied, and whose components are known (or given) functions of time, position and velocity of the particle. Sometimes such a force is also termed as a "given" force.

Were there no constraints on the motion of the particle and the particle were free to move in the 3-dimensional X–Y–Z coordinate space under the action of the given force components, its equations of motion would be those given by equation set (1.1).

However, the particle *is* constrained. It can move only in one dimension, along the X-axis. Hence additional forces, beyond those impressed upon it, need to be applied to it so that the particle remains on the X-axis. These forces are applied by the thin wire on our bead, hence keeping the bead on the wire at all times. The equations of motion describing the constrained motion of the particle then are

$$m\ddot{x}(t) = F_x(t) + F_x^c(t),$$
$$m\ddot{y}(t) = F_y(t) + F_y^c(t), \tag{1.5}$$
$$m\ddot{z}(t) = F_z(t) + F_z^c(t),$$

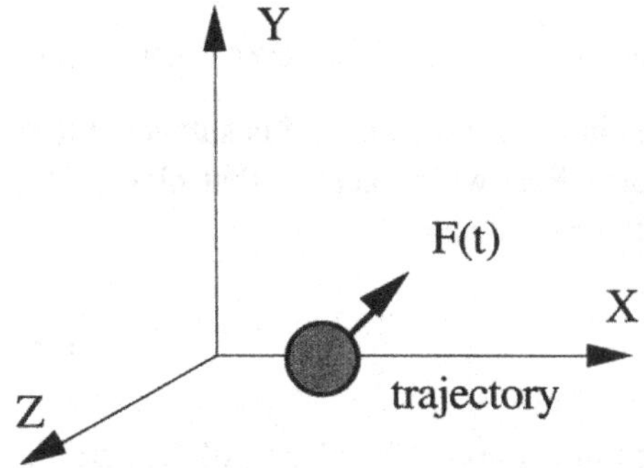

Figure 1.2. Constrained motion

Example 1.1 (continued)

where the X-, Y-, and Z-components of the force of constraint are denoted by $F_x^c(t)$, $F_y^c(t)$ and $F_z^c(t)$, respectively. We will show, in this book, how we can explicitly determine the forces of constraint in problems such as these which involve constrained motion.

That the motion is constrained to the X-axis, can be expressed by the relations

$$\begin{aligned} y(t) &= 0, \\ z(t) &= 0, \end{aligned} \tag{1.6}$$

for all time t or equivalently by the equations

$$\begin{aligned} \dot{y}(t) &= 0, \\ \dot{z}(t) &= 0, \end{aligned} \tag{1.7}$$

for all time t or by

$$\begin{aligned} \ddot{y}(t) &= 0, \\ \ddot{z}(t) &= 0, \end{aligned} \tag{1.8}$$

for all time t. Note that the latter two sets of equations are equivalent to (1.6) when considered along with equations (1.3) and (1.4). Also, the particle's configuration is completely described by its position $x(t)$ at time t. The components of the force of constraint $F_x^c(t)$, $F_y^c(t)$ and $F_z^c(t)$, are not yet known, and need to be determined so that in conjunction with the given externally impressed (applied) forces $F_x(t)$, $F_y(t)$ and $F_z(t)$ the particle remains on the X-axis.

Example 1.2

Consider a particle which moves in the XY-plane, and is subjected to a known, externally impressed force $\mathbf{F}(t)$ which acts in that plane. The equations of motion describing the particle are

$$\begin{aligned} m\ddot{x}(t) &= F_x(t), \\ m\ddot{y}(t) &= F_y(t). \end{aligned} \tag{1.9}$$

At each time t, the acceleration components $a_x(t)$ and $a_y(t)$ of the particle under the influence of the known, impressed force $\mathbf{F}(t)$ are then simply given by

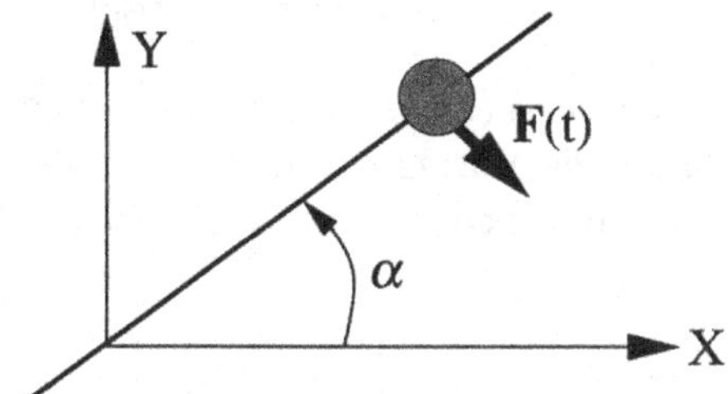

Figure 1.3. Motion in two dimensions constrained to a line

$$a_x(t) = \ddot{x}(t) = F_x(t)/m,$$
$$a_y(t) = \ddot{y}(t) = F_y(t)/m. \quad (1.10)$$

The quantities $a_x(t)$ and $a_y(t)$ can be thought of as the acceleration components of the unconstrained particle subjected to the impressed force $\mathbf{F}(t)$.

Now let the particle be constrained to lie along a line of constant inclination, described by

$$y(t) = x(t) \tan \alpha. \quad (1.11)$$

At the initial time t_0, the particle's position $\{x(t_0), y(t_0)\}$ and initial velocity $\{\dot{x}(t_0), \dot{y}(t_0)\}$ are therefore such that the equations

$$y(t_0) = x(t_0) \tan \alpha \quad (1.12)$$

and

$$\dot{y}(t_0) = \dot{x}(t_0) \tan \alpha \quad (1.13)$$

are satisfied. Then the equations of motion describing the constrained motion of the particle may be written as

$$m\ddot{x}(t) = F_x(t) + F_x^c(t),$$
$$m\ddot{y}(t) = F_y(t) + F_y^c(t), \quad (1.14)$$

where the additional forces of constraint that need to be applied to the particle to ensure that it moves along the line are indicated by the superscript "c." Thus we see that the acceleration components $\ddot{x}(t)$ and $\ddot{y}(t)$ of the particle which is constrained to lie on the line defined by equation

Example 1.2 (continued)

(1.11) are no longer those given by equation (1.10). In fact, by dividing each of the equations (1.14) by m and then using equation (1.10), the acceleration components $\ddot{x}(t)$ and $\ddot{y}(t)$ of the constrained particle can be written as

$$\begin{aligned} \ddot{x}(t) &= a_x(t) + \frac{F_x^c(t)}{m}, \\ \ddot{y}(t) &= a_y(t) + \frac{F_y^c(t)}{m}. \end{aligned} \tag{1.15}$$

Equation (1.15) shows that, in general, the acceleration components of the unconstrained particle differ from those of the constrained particle and that this difference is created by the additional force components $F_x^c(t)$ and $F_y^c(t)$ which are brought into play by the presence of the constraint.

We note that the equation of constraint (1.11) can also be expressed, as before, after two differentiations with respect to time, as

$$\ddot{y}(t) = \ddot{x}(t) \tan \alpha . \tag{1.16}$$

Furthermore, because of the constraint, the particle's position can be determined by its X-coordinate or its Y-coordinate: e.g., given the X-coordinate at any time t, its Y-coordinate can be obtained from equation (1.11).

From a purely algebraic viewpoint, we observe from equation (1.10) that a knowledge of the impressed force components $F_x(t)$ and $F_y(t)$ allows us to uniquely determine the accelerations of the unconstrained system, since there are two equations for the two unknown accelerations $a_x(t)$ and $a_y(t)$. The introduction of our constraint changes this situation radically. Equation (1.14) involves two algebraic equations involving the four unknowns $\ddot{x}(t)$, $\ddot{y}(t)$, $F_x^c(t)$ and $F_y^c(t)$. This equation, along with equation (1.16) provides a total of three equations, for the four unknowns. Clearly an additional equation is required if we are to determine these four unknowns uniquely. We shall show that this additional equation arises in analytical mechanics from the acceptance of a principle set forth by Gauss, called Gauss's principle.

Now instead of a single particle, consider a set of particles of masses m_1, $m_2, \ldots, m_3$, whose coordinates at any time t are described by the $3n$ numbers

$\{x_1(t),y_1(t),z_1(t)\},\{x_2(t),y_2(t),z_2(t)\},\ldots,\{x_n(t),y_n(t),z_n(t)\}$. We assume that the rectangular coordinate frame is, as usual, "inertial," and that the forces impressed on each of these point-masses are given and denoted by $\{F_{1_x}(t), F_{1_y}(t), F_{1_z}(t)\}$, $\{F_{2_x}(t), F_{2_y}(t), F_{2_z}(t)\}, \ldots, \{F_{n_x}(t), F_{n_y}(t), F_{n_z}(t)\}$ respectively. Then Newton's second law informs us that if each particle is free to move independently of the motion of any of the other particles, then the equations of motion for the set of particles can be described by the $N = 3n$ equations

$$\left.\begin{aligned} m_i\ddot{x}_i &= F_{i_x} \\ m_i\ddot{y}_i &= F_{i_y} \\ m_i\ddot{z}_i &= F_{i_z} \end{aligned}\right\}; \; i = 1,\ 2, \ldots, n, \tag{1.17}$$

or, as before,

$$\left.\begin{aligned} \ddot{x}_i &= \frac{1}{m_i}F_{i_x} = a_{i_x} \\ \ddot{y}_i &= \frac{1}{m_i}F_{i_y} = a_{i_y} \\ \ddot{z}_i &= \frac{1}{m_i}F_{i_z} = a_{i_z} \end{aligned}\right\}; \; i = 1,\ 2, \ldots, n, \tag{1.18}$$

where a_{i_x} is the X-component of the force acting on the ith mass divided by the corresponding mass m_i. We note that we preclude the possibility of two particles occupying the same location at any one given time. This condition is often referred to as the condition of impenetrability of the point-particles of our set.

Physical objects are finite in their dimensions, and they are deformable. Therefore, point-particles are clearly abstractions of the real world. A set of masses may be represented by such a set of particles when their dimensions are negligibly small compared to the characteristic dimensions that govern the kind of dynamical behavior which is of interest to us. When the deformation of these masses is considered negligible, again relative to some characteristic dimension which governs the dynamics and our interest, the masses are assumed to be "rigid" or undeformable.

Example 1.3

To fix our ideas, let us consider a pair of particles; hence $n = 2$. Then the components of the accelerations of the particles are described by the 6 equations,

Example 1.3 (continued)

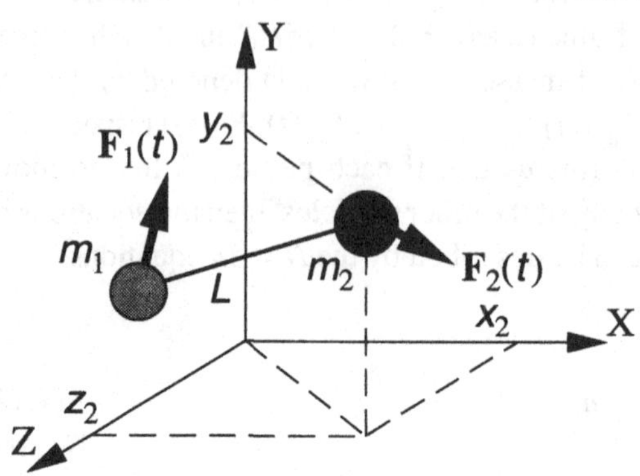

Figure 1.4. Constrained motion of two point-particles

$$\begin{aligned}
m_1\ddot{x}_1 &= F_{1_x}(t),\ m_2\ddot{x}_2 = F_{2_x}(t),\\
m_1\ddot{y}_1 &= F_{1_y}(t),\ m_2\ddot{y}_2 = F_{2_y}(t),\\
m_1\ddot{z}_1 &= F_{1_z}(t),\ m_2\ddot{z}_2 = F_{2_z}(t),
\end{aligned} \tag{1.19}$$

where F_{1_x} and F_{2_x} are the X-components of the "impressed" or "given" forces acting on the two masses m_1 and m_2 respectively.

Now let us suppose that this pair of particles forms a system so that the motion of each particle is possibly constrained by the motion of the other. Let us say that the distance between the particles is constrained to be a fixed quantity – the two particles being, say, at the two ends of a weightless rod of constant length, L.

Now the two particles cannot move independently of each other; for at any instant of time the distance between them must be fixed and equal to L. This requires that

$$(x_2 - x_1)^2 + (y_2 - y_1)^2 + (z_2 - z_1)^2 = L^2. \tag{1.20}$$

Furthermore, the two particles constrain one another's motion by applying forces on each other in addition to the forces that are externally impressed upon them. Thus the equations of motion for the two masses now become

$$\begin{aligned}
m_1\ddot{x}_1 &= F_{1_x}(t) + F_{1_x}^c(t),\\
m_1\ddot{y}_1 &= F_{1_y}(t) + F_{1_y}^c(t),\\
m_1\ddot{z}_1 &= F_{1_z}(t) + F_{1_z}^c(t),
\end{aligned} \tag{1.21}$$

and

$$\begin{aligned} m_2\ddot{x}_2 &= F_{2_x}(t) + F^c_{2_x}(t), \\ m_2\ddot{y}_1 &= F_{2_y}(t) + F^c_{2_y}(t), \\ m_2\ddot{z}_2 &= F_{2_z}(t) + F^c_{2_z}(t), \end{aligned} \tag{1.22}$$

where the terms $F^c_{1_x}(t)$ and $F^c_{2_x}(t)$ are the additional forces on the particles in the X-direction caused by the fact that they have to satisfy the constraint equation (1.20). Again given that the positions and velocities of the two particles at the initial time t_0 satisfy the constraint (1.20), the constraint can be expressed by differentiating it twice as

$$\begin{aligned} &(x_2 - x_1)(\ddot{x}_2 - \ddot{x}_1) + (y_2 - y_1)(\ddot{y}_2 - \ddot{y}_1) + (z_2 - z_1)(\ddot{z}_2 - \ddot{z}_1) \\ &= -(\dot{x}_2 - \dot{x}_1)^2 - (\dot{y}_2 - \dot{y}_1)^2 - (\dot{z}_2 - \dot{z}_1)^2. \end{aligned} \tag{1.23}$$

Observe that the equations obtained by differentiating the constraint equations in Examples 1.1, 1.2 and 1.3 are all linear in the acceleration components.

The Vector–Matrix Notation

It would be convenient at this point to introduce a more compact way of representing equations (1.17) and (1.18). We do this by employing the vector–matrix notation, a notation which will be very useful throughout this book.

Let us denote by the column vector, $\mathbf{x}_1(t)$, the three numbers $x_1(t)$, $y_1(t)$, $z_1(t)$ at time t, so that

$$\mathbf{x}_1(t) = \begin{bmatrix} x_1(t) \\ y_1(t) \\ z_1(t) \end{bmatrix}. \tag{1.24}$$

The quantities $x_1(t)$, $y_1(t)$, $z_1(t)$ are called the components of the vector $\mathbf{x}_1(t)$. The vector $\mathbf{x}_1(t)$ can be thought of as a matrix with 3 rows and 1 column and can be referred to as a 3 by 1 matrix, or more simply as a 3-vector. Similarly, we can define the vector $\mathbf{F}_1(t)$ as

$$\mathbf{F}_1(t) = \begin{bmatrix} F_{1_x}(t) \\ F_{1_y}(t) \\ F_{1_z}(t) \end{bmatrix}. \tag{1.25}$$

The derivative with respect to time of the vector $\mathbf{x}_1(t)$ is defined as the derivative of each of its components so that

$$\dot{\mathbf{x}}_1(t) = \frac{d}{dt}\mathbf{x}_1(t) = \frac{d}{dt}\begin{bmatrix} x_1(t) \\ y_1(t) \\ z_1(t) \end{bmatrix} = \begin{bmatrix} \frac{d}{dt}x_1(t) \\ \frac{d}{dt}y_1(t) \\ \frac{d}{dt}z_1(t) \end{bmatrix}. \tag{1.26}$$

The second derivative $\ddot{\mathbf{x}}_1(t)$ with respect to time can be defined similarly as the derivative of the column vector $\dot{\mathbf{x}}_1(t)$.

Hence Newton's law of motion can be expressed for the mass m_1 by the vector equation

$$m_1\ddot{\mathbf{x}}_1(t) = \mathbf{F}_1(t) \tag{1.27}$$

where we define the product of the scalar m_1 by the vector $\ddot{\mathbf{x}}_1(t)$ as the product of each of the (three) components of the vector $\ddot{\mathbf{x}}_1(t)$ with the scalar m_1. Notice that the vector equation (1.27) implies three scalar equations. Furthermore, we can write equation (1.17) in vector–matrix notation as

$$\begin{bmatrix} m_1 & 0 & \cdot & \cdot & \cdot & \cdot & \cdot & 0 \\ 0 & m_1 & 0 & \cdot & \cdot & \cdot & \cdot & 0 \\ \cdot & 0 & m_1 & 0 & \cdot & \cdot & \cdot & \cdot \\ \cdot & \cdot & \cdot & \cdot & \cdot & \cdot & \cdot & \cdot \\ \cdot & \cdot & \cdot & \cdot & \cdot & \cdot & \cdot & \cdot \\ \cdot & \cdot & \cdot & \cdot & \cdot & \cdot & \cdot & \cdot \\ \cdot & \cdot & \cdot & \cdot & 0 & m_n & 0 & 0 \\ \cdot & \cdot & \cdot & \cdot & \cdot & 0 & m_n & 0 \\ \cdot & \cdot & \cdot & \cdot & \cdot & \cdot & 0 & m_n \end{bmatrix} \begin{bmatrix} \ddot{x}_1 \\ \ddot{y}_1 \\ \ddot{z}_1 \\ \cdot \\ \cdot \\ \cdot \\ \ddot{x}_n \\ \ddot{y}_n \\ \ddot{z}_n \end{bmatrix} = \begin{bmatrix} F_{1_x} \\ F_{1_y} \\ F_{1_z} \\ \cdot \\ \cdot \\ \cdot \\ F_{n_x} \\ F_{n_y} \\ F_{n_z} \end{bmatrix} \tag{1.28}$$

or more compactly as

$$M\ddot{\mathbf{x}}(t) = \mathbf{F}(t), \tag{1.29}$$

where $\ddot{\mathbf{x}}(t) = [\ddot{\mathbf{x}}_1^T(t)\ \ddot{\mathbf{x}}_1^T(t) \cdots \ddot{\mathbf{x}}_n^T(t)]^T$, $\mathbf{F}(t) = [\mathbf{F}_1^T(t)\ \mathbf{F}_2^T(t) \cdots \mathbf{F}_n^T(t)]^T$ and the superscript T denotes the operation (called transpose) of rendering a column (row) vector into a row (column) vector. Each of these vectors has $N = 3n$ components; each vector $\mathbf{x}_i(t), i = 1, 2, \ldots, n$, has 3 components. The quantity M is an N by N matrix with zeros everywhere except along its diagonal, as shown in equation (1.28). A matrix of the type M is called a diagonal matrix, since its nonzero entries are only along its diagonal. It is often written by showing its diagonal entries, as $M = Diag\{m_1, m_1, m_1, \ldots, m_n, m_n, m_n\}$. Notice that the masses of the particles occur in sets of three along the diagonal of the matrix M.

In general, a matrix may have all its entries nonzero. We define the product of an m by n matrix A (whose ith row, jth column element is denoted a_{ij}) with an n by 1 column vector **b** as an m by 1 column vector **c**, whose components are obtained as

$$c_i = \sum_{k=1}^{k=n} a_{ik} b_k, \quad i = 1, 2, \ldots, m. \tag{1.30}$$

or, more simply, as

$$\mathbf{c} = A\mathbf{b}. \tag{1.31}$$

Notice that the column vector **c** can also be thought of as an m *by* 1 matrix and its transpose $\mathbf{c}^T$ as a 1 by m matrix.

Using the notation provided in equations (1.31) and (1.30), the reader may now verify that equation (1.17) can indeed be written as equation (1.28), and more compactly as equation (1.29).

More generally, the product of an m by r matrix A and an r by n matrix B is defined as the m by n matrix C, whose i–jth element (i.e., the ith row, jth column element) is given by

$$c_{ij} = \sum_{k=1}^{k=r} a_{ik} b_{kj}, \quad i = 1, 2, \ldots, m; \; j = 1, 2, \ldots, n, \tag{1.32}$$

or, more simply, written as

$$C = AB. \tag{1.33}$$

The matrix counterpart of equation (1.18) can be written as

$$\ddot{\mathbf{x}} = M^{-1}\mathbf{F}(t) = \mathbf{a}(t) \tag{1.34}$$

where the matrix M^{-1} is called the inverse of the matrix M and is simply a diagonal matrix described by $M^{-1} = Diag\{m_1^{-1}, m_1^{-1}, m_1^{-1}, \ldots, m_n^{-1}, m_n^{-1}, m_n^{-1}\}$. Note that since the masses are all positive numbers, the matrix M^{-1} is defined.

Furthermore, each component of the vector on the right hand side of equation (1.28) could be dependent on the components of the vectors $\mathbf{x}(t)$ and $\dot{\mathbf{x}}(t)$ and hence equation (1.29) could be written more explicitly as

$$M\ddot{\mathbf{x}}(t) = \mathbf{F}(\mathbf{x}(t), \dot{\mathbf{x}}(t), t). \tag{1.35}$$

Example 1.4

Consider the equations of motion of the constrained system represented by equation (1.5) for the particle described in Example 1.1. The three scalar equations of motion can be written more compactly by using the vector–matrix notation. Let us use the following notation:

$$\begin{aligned} M &= Diag\{m, m, m\}, \\ \ddot{\mathbf{x}}(t) &= [\ddot{x}(t) \quad \ddot{y}(t) \quad \ddot{z}(t)]^T, \\ \mathbf{F}(t) &= [F_x(t) \quad F_y(t) \quad F_z(t)]^T, \\ \mathbf{F}^c(t) &= [F_x^c(t) \quad F_y^c(t) \quad F_z^c(t)]^T. \end{aligned} \tag{1.36}$$

Then the equation set (1.5) can be expressed as:

$$M\ddot{\mathbf{x}} = \mathbf{F}(t) + \mathbf{F}^c(t), \tag{1.37}$$

where, by the sum of the two 3 by 1 vectors $\mathbf{F}$ and $\mathbf{F}^c$, we mean a new 3 by 1 vector each of whose components is a sum of the corresponding components of the vectors $\mathbf{F}$ and $\mathbf{F}^c$, i.e., we define

$$\mathbf{F}(t) + \mathbf{F}^c(t) = [F_x(t) + F_x^c(t) \quad F_y(t) + F_y^c(t) \quad F_z(t) + F_z^c(t)]^T. \tag{1.38}$$

Example 1.5

Throughout this book we shall be employing the vector–matrix notation. As a simple prelude to this nomenclature, consider the linear set of equations

$$\begin{aligned} 2x + 3y &= 5, \\ 4x + 6y &= 10. \end{aligned} \tag{1.39}$$

They can be expressed as

$$\begin{bmatrix} 2 & 3 \\ 4 & 6 \end{bmatrix} \begin{bmatrix} x \\ y \end{bmatrix} = \begin{bmatrix} 5 \\ 10 \end{bmatrix}. \tag{1.40}$$

Example 1.6

An important consequence of our definition of the product of a matrix and a column vector is that if $\mathbf{c} = A\mathbf{b}$, then $\mathbf{c}^T = \mathbf{b}^T A^T$. This follows directly from the definition of a product of a matrix and a vector.

Example 1.7

Often in mechanics we will come across a column vector (or a matrix) each of whose components (or elements) is a function of the components of another vector. Consider the $3n$ by 1 vectors $\mathbf{x}(t) = [x_1(t) \; x_2(t) \; x_3(t) \ldots x_{3n}(t)]^T$, and $\mathbf{F} = [F_1 \; F_2 \; F_3 \ldots F_{3n}]^T$. If each component of the vector $\mathbf{F}$ is dependent on the $3n$ components of the vector $\mathbf{x}$, i.e., on $x_1, x_2, x_3, \ldots, x_{3n}$, so that

$$F_i = f_i(x_1(t), x_2(t), x_3(t), \ldots, x_{3n}(t)), \quad i = 1, 2, \ldots, 3n, \tag{1.41}$$

then we denote this functional dependence by writing $\mathbf{F}(\mathbf{x}(t))$.

Similarly, if each element of a matrix A is a function of the components of a vector $\mathbf{x}(t)$, we indicate this functional dependence by writing $A(\mathbf{x}(t))$.

Example 1.8

a. We can write Equation (1.16) in vector–matrix notation as

$$A\mathbf{w} = 0, \tag{1.42}$$

where the 1 by 2 matrix A is in this case a row vector given by

$$A = [\tan\alpha \quad -1], \tag{1.43}$$

and

$$\mathbf{w} = [\ddot{x} \quad \ddot{y}]^T. \tag{1.44}$$

b. We can write Equation (1.23) in vector–matrix notation as

$$A\mathbf{w} = \mathbf{b}$$

where the 1 by 6 matrix A is given by

$$\begin{aligned} A = [&-(x_2 - x_1) \quad (x_2 - x_1) \quad -(y_2 - y_1) \quad (y_2 - y_1) \\ &-(z_2 - z_1) \quad (z_2 - z_1)] \end{aligned} \tag{1.45}$$

and

$$\mathbf{w} = [\ddot{x}_1 \quad \ddot{x}_2 \quad \ddot{y}_1 \quad \ddot{y}_2 \quad \ddot{z}_1 \quad \ddot{z}_2]^T. \tag{1.46}$$

The scalar $\mathbf{b} = -(\dot{x}_2 - \dot{x}_1)^2 - (\dot{y}_2 - \dot{y}_1)^2 - (\dot{z}_2 - \dot{z}_1)^2$. Note that we can write $A(\mathbf{x}(t))$.

Example 1.9

Consider a pendulum bob suspended from a massless rod, moving in the XY-plane. The mass of the bob is m. The motion of the bob may be described by the coordinates (x, y), as shown in Figure 1.5. The bob is constrained to move so that the distance of the bob from its point of suspension is a constant L. The acceleration due to gravity is downwards, and of magnitude g.

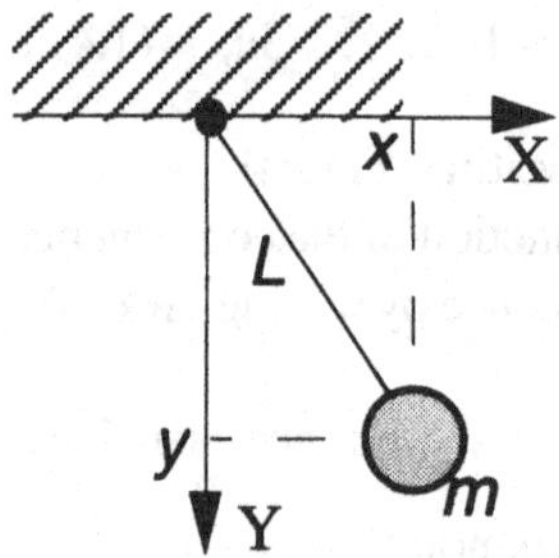

Figure 1.5. A simple pendulum

Were the rod nonexistent, the motion of the bob would be unconstrained. This unconstrained motion of the bob can then be expressed as

$$\begin{aligned} m\ddot{x} &= 0, \\ m\ddot{y} &= mg. \end{aligned} \tag{1.47}$$

Thus, were the bob not constrained to be at a constant distance from the point of suspension, the acceleration of the bob would be vertically downwards, under the force of gravity. It is because the bob is constrained to satisfy the equation

$$x^2+y^2 = L^2 \tag{1.48}$$

that its motion is the familiar swinging motion of a pendulum. Equations (1.47) can be expressed as

$$M\ddot{\mathbf{x}} = \mathbf{F}, \tag{1.49}$$

where

$$M = \begin{bmatrix} m & 0 \\ 0 & m \end{bmatrix}, \ddot{\mathbf{x}} = \begin{bmatrix} \ddot{x} \\ \ddot{y} \end{bmatrix}, \text{and } \mathbf{F} = \begin{bmatrix} 0 \\ mg \end{bmatrix}. \tag{1.50}$$

Also, differentiating the constraint equation (1.48) twice, we get

$$A(\mathbf{x}(t))\,\ddot{\mathbf{x}} = \mathbf{b}, \tag{1.51}$$

where

$$A = [x(t) \qquad y(t)], \text{ and the scalar } \mathbf{b} = [-\dot{x}^2 - \dot{y}^2]. \tag{1.52}$$

Observe that equation set (1.47), which describes the unconstrained motion, and equation (1.51), which describes the constraint, are both linear in the accelerations. These two sets of equations taken together with the initial conditions completely describe the motion of the pendulum bob, as we shall now see.

Because the bob is required to move so that it satisfies the constraint, additional forces, over and above those present in the unconstrained motion, will be exerted on the bob, causing it to move so that at every instant of time, equation (1.48) is satisfied. Hence the constrained motion of the system will be described by the equation

$$M\ddot{\mathbf{x}}(t) = \mathbf{F}(t) + \mathbf{F}^c(t), \tag{1.53}$$

where the 2 by 1 vector $\mathbf{F}^c(t)$ represents the force of constraint. We assume, as before, that the initial position and velocity of the bob are compatible with the constraint.

The determination of the equations of motion of the constrained system is equivalent to determining this additional constraint vector $\mathbf{F}^c(t)$.

Example 1.10

Consider a point-particle of mass m moving in three dimensions subjected to the given externally impressed force $\mathbf{F} = [F_x(t), F_y(t), F_z(t)]^T$. Let the position of the particle be denoted by the vector $\mathbf{x} = [x(t) \quad y(t) \quad z(t)]^T$. The motion of the particle is described by the equation

$$M\ddot{\mathbf{x}} = \mathbf{F} \tag{1.54}$$

where the 3 by 1 vector $\ddot{\mathbf{x}} = [\ddot{x} \quad \ddot{y} \quad \ddot{z}]^T$, and the matrix $M = Diag\,\{m, m, m\}$. Now if the particle is, in addition, required to satisfy the relation

$$\dot{y} = z\dot{x}, \tag{1.55}$$

Example 1.10 (continued)

additional forces would need to be applied to the particle so that it satisfies this constraint. We assume that the initial conditions are given and that they are compatible with the constraint equation (1.55). Once again the constrained motion of the particle would be governed by the equation

$$M\ddot{\mathbf{x}} = \mathbf{F}(t) + \mathbf{F}^{c}(t), \tag{1.56}$$

where $\mathbf{F}^{c}(t)$ is the 3 by 1 vector of constraint forces which needs to be applied so that the constraint equation (1.55) is satisfied.

The central issue in the determination of the time evolution of the constrained system is the determination of this constraint force vector $\mathbf{F}^{c}(t)$. This vector must be determined in accordance with the accepted principles of analytical dynamics. Once this is known at each time t, using the given initial conditions, equation (1.56) could be integrated (in principle, or at least numerically) to obtain the trajectory, $\mathbf{x}(t)$, whose three components at any time are then $x(t)$, $y(t)$ and $z(t)$.

We note that the constraint equation (1.55) is fundamentally different from the types of constraint equations that we have considered so far. Our previous constraint equations (e.g., equations (1.6), (1.11) and (1.20)) were essentially equations which interrelated the components of the position of the particle(s) at each instant of time; these equations could, of course, be differentiated to obtain relations between the velocities (or accelerations) as we have done on several occasions before, i.e., in obtaining, for example, equations (1.7), (1.16) and (1.23). However, equation (1.55) *cannot* be integrated to obtain a relation between the components of the position of the particle at each instant of time t; for to do that, we would need to obtain $\dot{x}$ and z as functions of the time t; but this is not possible unless we know how the system evolves in time, and this evolution is dependent on our knowledge of the equations of motion that the system satisfies. But at the present time, though these equations of motion may be known in form to look like equation (1.56), they are unknown in content because we have yet to find the constraint force vector $\mathbf{F}^{c}(t)$. As always, this constraint force must be such that under its action in combination with the given impressed force $\mathbf{F}(t)$, the motion of the particle must satisfy the constraint equation (1.55).

Such constraints as equation (1.55) are called *nonintegrable* or *nonholonomic* constraints, precisely because we cannot integrate them to obtain relations between the components of the position(s) of the particle(s) without knowing the time-evolution of the system.

It should be noted that though the constraint equation (1.55) cannot be integrated, it *can* be differentiated, to obtain a relation, which is again linear in the components of the accelerations. Differentiating equation (1.55) we get the linear relation

$$A(\mathbf{x}(t))\ddot{\mathbf{x}} = \mathbf{b}, \tag{1.57}$$

where, the 1 by 3 matrix $A = [-z \quad 1 \quad 0]$, and the 1 by 1 matrix (scalar) $\mathbf{b} = \dot{z}\dot{x}$.

On the Nature of Constraints

While we will be looking at the different types of constraints in some detail in later chapters, we want here to present a brief overview of the types of constraints so that the reader is acquainted with some of the fundamental ideas, which, incidentally, are very simple.

Through the examples provided in previous sections, we have illustrated the central problem of constrained motion, namely that of finding the forces of constraint that engender motion which is compatible with the constraint, while being cognizant of the impressed forces applied to the system.

We have noted that there appear to be two different types of constraints which are of importance in practical situations that arise in mechanics – integrable and nonintegrable constraints. This classification of constraints has been a crucial issue in the development of mechanics since the time of Lagrange. Systems with nonintegrable constraints, being more difficult to handle, have engendered special methods for their analysis. Over the last 200 years or so, a considerable body of literature has developed around the determination of the equations of motion for such systems. Many of these contributions have come from physicists and mathematicians like Lagrange, Euler, Gauss, Volterra, Gibbs, Appell, Boltzman and Dirac.

In this book we shall show that both these constraints can be handled with equal ease by the methods that we develop. Contrary to accepted practice, we do not require any new concepts to handle nonintegrable as opposed to integrable constraints. The methods provided in this book are applicable to nearly all of the situations that one encounters in mechanics and, as we will see, they encompass the entirety of Lagrangian mechanics. This general applicability of our approach to both integrable and nonintegrable constraints hinges, in part, on an observation that we have already made – that, for most situations of practical significance, irrespective of whether the constraint is integrable or not, the differentiated form of the constraint involving the accelerations is linear in the accelerations.

The Central Problem of Analytical Dynamics

We are now in a suitable position to state the central problem of constrained motion. Consider a system of n point-particles, whose motion is described by the coordinates of each particle with respect to an inertial Cartesian coordinate frame of reference. Let the initial position and velocity of each particle be known. Denoting the vector of displacements by $\mathbf{x} = [x_1\ x_2\ \ldots x_{3n}]^T$, this means that we are given each of the $3n$ components of the vectors $\mathbf{x}(t_0)$ and $\dot{\mathbf{x}}(t_0)$. We have denoted, for simplicity, the components of the positions of the particles successively as $x_1, x_2, x_3, \ldots,$ etc. The forces impressed on the particles are given, and are denoted similarly by the $3n$-vector $\mathbf{F}(t) = [F_1(t)\ F_2(t)\ldots F_{3n}(t)]^T$. The unconstrained motion of the system can be expressed by the $3n$ by $3n$ matrix equation

$$M\ddot{\mathbf{x}}(t) = \mathbf{F}(\mathbf{x}(t), \dot{\mathbf{x}}(t), t) \tag{1.58}$$

or, as before,

$$\ddot{\mathbf{x}} = M^{-1}\mathbf{F}(\mathbf{x}(t), \dot{\mathbf{x}}(t), t) = \mathbf{a}(t) \tag{1.59}$$

where the matrix M is diagonal and the masses appear in sets of threes along the diagonal as described before. For the moment we will consider that this system of equations can be obtained through the use of Newton's laws as we did in the previous sections of this chapter. (In later chapters in the book we will take excursions into Lagrange's formulation and use generalized coordinates, but these are not essential to the basic understanding of the problem here.) Note that equation (1.58) can be thought of as a "recipe" for determining the accelerations of the unconstrained system at any time t, given the impressed forces acting on it, as functions of time! Also, $\mathbf{a}(t)$ is the acceleration at the time t of the unconstrained system were it to have an "initial" position of $\mathbf{x}(t)$ and a velocity of $\dot{\mathbf{x}}(t)$, prescribed at time t.

Now we constrain the system through the use of a given set of consistent constraint equations of the form

$$D(\mathbf{x}(t), t)\dot{\mathbf{x}} = \mathbf{g}(\mathbf{x}(t), t) \tag{1.60}$$

where the matrix D is an m by $3n$ matrix and $\mathbf{g}$ is an m by 1 vector. Initial conditions, at time t, are provided on the positions and velocities of the particles which are compatible with (that is, they satisfy the equations describing) this set of constraints.

The central issue of constrained motion can now be stated as follows: Given $\mathbf{x}(t)$, $\dot{\mathbf{x}}(t)$ and the impressed force $\mathbf{F}(t)$ determine the instantaneous acceleration $\ddot{\mathbf{x}}(t)$ of the system at time t, in the presence of these constraints. Put alternatively, can we determine how, and by how much, the accelerations of the constrained system deviate, or differ, from those of the unconstrained system?

As we know, the presence of the constraint (1.60) causes additional forces of constraint to be applied to the particles and the equation of motion of the constrained system of particles now becomes

$$M\ddot{\mathbf{x}}(t) = \mathbf{F}(\mathbf{x}(t), \dot{\mathbf{x}}(t), t) + \mathbf{F}^c. \tag{1.61}$$

Our main concern then is the determination of this constraint force vector $\mathbf{F}^c(t)$. This is an alternate statement of the central problem of constrained motion. Note that this constraint force vector must be such that: (1) under the *combined action* of itself and the given impressed force vector $\mathbf{F}(t)$, the system must satisfy the constraints, and (2) it must be in accordance with the accepted principles of analytical dynamics. An example of such an accepted principle of analytical dynamics is a principle which was first enunciated by Gauss, and is nowadays referred to as Gauss's principle. We shall consider it in detail in Chapter 3.

We can also differentiate with respect to time the constraint equation (1.60), to obtain

$$A(\mathbf{x}, t)\ddot{\mathbf{x}} = \mathbf{b}(\mathbf{x}, \dot{\mathbf{x}}, t). \tag{1.62}$$

We will assume throughout this book that the functions involved in equation (1.60) are sufficiently smooth so that such differentiation is possible. This equation, along with the initial conditions on the positions and velocities of the particles at the initial time t_0, can be used to describe the constraints on the system. The reader will note that though the vector $\mathbf{b}$ in equation (1.62) differs from the vector $\mathbf{g}$ of equation (1.60), the matrices D and A in the two equations are the same.

The reader may well wonder why the constraint equation is chosen to have the special form described by equation (1.60) or (1.62), and whether such a form is general enough to encompass most situations of practical interest. We saw in our examples that the constraints did fit the form of equation (1.62).

In fact, in this book we will use a more general form than (1.62) of the constraint equation, namely,

$$\boxed{A(\mathbf{x}, \dot{\mathbf{x}}, t)\ddot{\mathbf{x}}(t) = \mathrm{b}(\mathbf{x}, \dot{\mathbf{x}}, t)} \tag{1.63}$$

where the matrix A is an m by $3n$ matrix. This form of the constraint equation may be thought of as arising from a set of m constraints of the form

$$\boxed{\varphi_i(\mathbf{x}, \dot{\mathbf{x}}, t) = 0, \ i = 1, 2, \ldots, m}. \tag{1.64}$$

By appropriately differentiating these m equations with respect to time, we obtain the more general form of the constraint equation given by (1.63). We will always assume that the functions φ_i are sufficiently smooth to allow such differentiations. We will think of equation (1.63) as being the standard form of the constraint equation.

Thus the central problem of constrained motion can be succinctly stated as follows:

1. given the equation of motion (1.58) which describes the **unconstrained** motion of a system at time t,
2. given the position $\mathbf{x}(t)$ and the velocity $\dot{\mathbf{x}}(t)$ of the constrained system at time t, and
3. given the constraints which are described by the equation (1.63) (or equivalently by equations (1.64)),

find the equations of motion (the acceleration) for the **constrained** system at time t, in accordance with the agreed upon principles of analytical mechanics. It is the solution of this central problem with which this book deals.

A Look Ahead

To give a flavor of the simplicity of the result that we will obtain in this book, we provide the reader with a "peek" at the final outcome. We realize that while many of the readers may be unfamiliar with the type of linear algebra used to describe this result, it is instructive to see its form and its astonishing simplicity. In the chapters to follow we shall not only derive this result but will amplify on its deeper ramifications and the parsimony and aesthetic with which Nature appears to work.

The equation of motion of the constrained system described in the previous section, at each instant of time, is

$$\boxed{M\ddot{\mathbf{x}} = \mathbf{F} + M^{1/2}(AM^{-1/2})^{+}(\mathbf{b} - AM^{-1}\mathbf{F})} \tag{1.65}$$

where, $(AM^{-1/2})^{+}$ is the Moore–Penrose generalized inverse of the matrix $(AM^{-1/2})$.

We shall refer to equation (1.65) as the fundamental equation. Notice that the product of the two matrices A and $M^{-1/2}$ appears conspicuously in our result. We shall call the matrix $AM^{-1/2}$ as the Constraint Matrix. It remains to be explained what we mean by the $3n$ by $3n$ matrix $M^{1/2}$ and what we mean by the Moore–Penrose generalized inverse (which we denote by the superscript +) of a matrix. We will shortly be taking this up in the next chapter.

One can consider equation (1.65) as an extension of Newton's equation which is now applicable to constrained systems. The effect of the constraint described by equation (1.63) is to add another term on the right hand side of equation (1.65) in addition to the given impressed force **F**.

We observe that the constrained motion is obtained in a simple manner, no matter how nonlinear the mechanical system being considered is (we assume of course the usual differentiability conditions for the nonlinear system). Furthermore, comparing equations (1.61) and (1.65), we obtain the explicit relation for the constraint force vector given by

$$\boxed{\mathbf{F}^c(t) = M^{1/2}(AM^{-1/2})^+(\mathbf{b} - AM^{-1}\mathbf{F})}. \tag{1.66}$$

One may gain greater insight into the motion of constrained systems by using equation (1.59) in equations (1.65) and (1.66). We then have

$$\boxed{M\ddot{\mathbf{x}} = \mathbf{F} + M^{1/2}(AM^{-1/2})^+(\mathbf{b} - A\mathbf{a})}, \tag{1.67}$$

and

$$\boxed{\mathbf{F}^c(t) = M^{1/2}(AM^{-1/2})^+(\mathbf{b} - A\mathbf{a})}. \tag{1.68}$$

However, the interpretation and use of the last two equations needs to be done with some care. Let us say that at any time t, the position and velocity of various point-particles of the **constrained** system are known, i.e., $\mathbf{x}(t)$ and $\dot{\mathbf{x}}(t)$ are known at time t. As mentioned earlier, the central issue in the study of constrained motion is the determination of the acceleration $\ddot{\mathbf{x}}(t)$ of the constrained system at that instant of time t. We notice that because the system has to satisfy the constraints, its acceleration $\ddot{\mathbf{x}}(t)$ at time t will differ from what it would have been had there been no constraints on it. In fact its acceleration in the absence of constraints would be exactly $\mathbf{a}(t)$, given by equation (1.59). Note that $\mathbf{a}(t)$ is determined by equation (1.59) using our knowledge of $\mathbf{x}(t)$ and $\dot{\mathbf{x}}(t)$, which may be required to determine **F** at time t. However, this acceleration $\mathbf{a}(t)$ so obtained will not, in general, satisfy the constraint equation (1.63) when $\ddot{\mathbf{x}}(t)$ is replaced by $\mathbf{a}(t)$ in it. This amounts to saying that were we to use the acceleration corresponding to the unconstrained system $\mathbf{a}(t)$, in equation (1.63), the left hand side of equation (1.63) would naturally not equal the right hand side. Thus the difference between the left hand side and the right hand side of (1.63) would then naturally be the extent to which the constraint equation (1.63) remains dissatisfied. We can denote this difference vector at each instant of time t by

$$\mathbf{e}(t) = (\mathbf{b} - A\mathbf{a}), \tag{1.69}$$

and equation (1.68) can be rewritten now as

$$\mathbf{F}^c(t) = M^{1/2}(AM^{-1/2})^+\mathbf{e} = K\mathbf{e} \tag{1.70}$$

where the matrix $K = M^{1/2}(AM^{-1/2})^+$.

Now equation (1.68) informs us that Nature creates a constraint force vector, $\mathbf{F}^c(t)$, which is directly proportional to this difference vector, $\mathbf{e}(t)$, closing the gap as it were, and thereby causing the constraint equation (1.63) to be satisfied. Since we are dealing with vectors, we have a matrix of proportionality (and not just a constant of proportionality), and this matrix of proportionality is provided by the matrix $K = M^{1/2}(AM^{-1/2})^+$. While we have not yet discussed the Moore–Penrose generalized inverse which enters here, suffice it to say that it exists (and is unique) for any given matrix.

As mentioned earlier, to understand in depth the relations stated in this section and to derive them, we need greater familiarity with some elements of linear algebra. Hence our next goal is to present some of the fundamentals of linear algebra and, to explain to those who are unfamiliar, the meaning of the generalized inverse of a matrix, a concept we have to deal with if we are to understand the matrix of proportionality which appears in these results.

PROBLEMS

1.1 In Example 1.1 assume that the particle is constrained to move along the X-axis such that $x(t) = p\sin(qt)$, where p and q are constants. Express the constraint in the form of equation (1.63). Are the initial conditions $x(\frac{\pi}{2q}) = 0$, $\dot{x}(\frac{\pi}{2q}) = 1$ compatible with the constraint?

1.2 In Example 1.2 consider the particle to move along a line whose inclination α depends on the time t. Let $\alpha(t) = p\sin(qt)$, where p and q are constants. Write the constraint equation in terms of the particle's accelerations in the vector–matrix form as $A\ddot{\mathbf{x}} = \mathbf{b}$.

1.3 In Example 1.3 the distance between the particles L changes with time according to the law $L(t) = [1 - 0.5\sin(qt)]$, where q is a constant. How does this alter the constraint on the system? Express this constraint again in the form $A\ddot{\mathbf{x}} = \mathbf{b}$.

1.4 Prove the result in Example 1.6.

1.5 Write the equations of motion for the pendulum shown in Example 1.9 when the point of suspension is accelerating downwards with an acceleration $g/2$, given that $x(0) = 0$, $y(0) = L$, $\dot{x}(0) = a$ and $\dot{y}(0) = 0$. What would the motion be if the point of suspension accelerated downwards with an acceleration g instead? Assume, in each case, that the point of suspension starts from rest.

1.6 Verify that in the notation of equation (1.58), the X-component of the displacement of the jth particle is x_{3j-2}, the Y-component is x_{3j-1} and the Z-component is x_{3j}. Similarly, the X-component of the impressed force on the jth particle is F_{3j-2}, the Y-component is F_{3j-2} and the Z-component is F_{3j}.

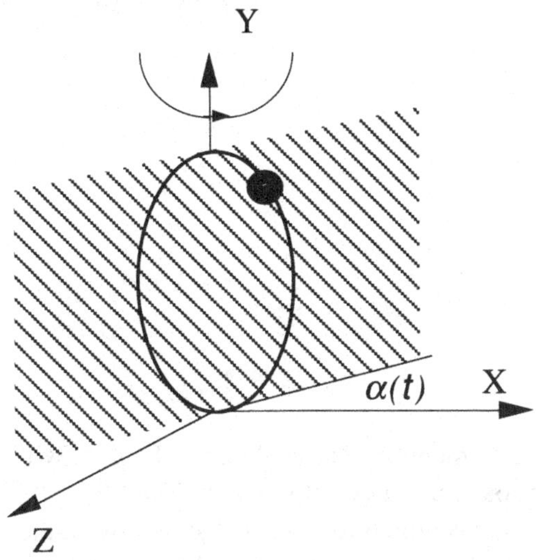

Figure 1.6. A bead on a rotating ring

1.7 A particle-bead of mass m is constrained to move on a ring of radius r which rotates about a vertical axis passing through its center at a constant angular speed $\omega = \dot{\alpha}(t)$. See Figure 1.6. Write the constraint equation in the form $A\ddot{\mathbf{x}} = \mathbf{b}$.

1.8 In Example 1.10 how many of the quantities x, y, z, $\dot{x}$, $\dot{y}$, $\dot{z}$ can be independently prescribed at the initial time, t_0, so that the constraint equation (1.55) is satisfied?

1.9 Differentiate equation (1.60) with respect to time and show that the matrix A in equation (1.62) is the same as the matrix D in equation (1.60).

For Further Reading

1.R1 This is an excellent introductory book on analytical mechanics. The student might wonder if the given impressed force $\mathbf{F}$ at time t, defined in equation (1.1), can be a function of the particle's acceleration. That this is not, in general, permissible is pointed out on pages 15 and 16 of this book.

2.P1 This is an excellent book on analytical mechanics though pitched at a slightly more advanced level. The discussion of the unconstrained motion of a particle in one dimension is beautifully and exhaustively presented on pages 3 to 11. A good discussion on why the force acting on a particle cannot be a function of its acceleration is provided on page 11.

2 Matrix Algebra

In this chapter we will deal with the elements of matrix algebra which we will use throughout the rest of this book. The material presented here is self-contained. However, the student who has little prior knowledge of most of the topics in this chapter may find it helpful to refer to some of the books indicated at the end in our "For Further Reading" list. In classical mechanics we will be mainly dealing with vectors (and matrices) whose elements will be real. Hence unless explicitly stated, we shall deal in this book with real vectors and real matrices, as opposed to vectors and matrices whose elements may be complex numbers.

The chapter is divided into two main sections. The first is entitled Preliminaries, the second is called Generalized Inverse of a Matrix. We recommend that the reader begin by going to the second section after skimming over the two subsections that precede it. The details will become clearer as greater familiarity with the topics is gained. It is then that (s)he may go back to the section entitled Preliminaries, so that a better groundwork for the chapters to follow will be laid.

Preliminaries

This section contains some of the preliminary material related to vector spaces. The reader is welcome to skip the material here if (s)he is familiar with it. For the beginner, it may be better to simply skim over this section lightly at the first reading. Various definitions and the use of the concepts developed here will become clearer as we proceed to later chapters, at which time the reader may refer to this section for specific topics as and when they are encountered.

Vector Spaces, Subspaces and Basis

Let $a_1, a_2, \ldots, a_n$ be a set of n real numbers (or more generally, elements from a field). Then we define the n by 1 vector $\mathbf{a}$ to be

$$\mathbf{a} = \begin{bmatrix} a_1 \\ a_2 \\ \cdot \\ \cdot \\ a_n \end{bmatrix} = \begin{bmatrix} a_1 & a_2 & \ldots & a_n \end{bmatrix}^T = \begin{bmatrix} a_1, a_2, \ldots, a_n \end{bmatrix}^T, \tag{2.1}$$

where the superscript T denotes the transpose. The numbers a_i, $i = 1, 2, \ldots, n$, are called the components (or elements) of the n-vector $\mathbf{a}$. Since the vector $\mathbf{a}$ has n components, it is referred to as an n-vector. If every component of an n-vector $\mathbf{a}$ is zero, it is called a zero vector. If any one component of an n-vector $\mathbf{a}$ is nonzero, it is called a nonzero vector.

We define the rule of addition of two vectors and the rule of multiplication of a real scalar with a vector in the following way. Consider two n by 1 vectors $\mathbf{a}$ and $\mathbf{b}$. Then for any two scalars α and β, we have

$$\alpha\,\mathbf{a} + \beta\,\mathbf{b} = \alpha \begin{bmatrix} a_1 \\ a_2 \\ \cdot \\ \cdot \\ a_n \end{bmatrix} + \beta \begin{bmatrix} b_1 \\ b_2 \\ \cdot \\ \cdot \\ b_n \end{bmatrix} = \begin{bmatrix} \alpha\, a_1 + \beta\, b_1 \\ \alpha\, a_2 + \beta\, b_2 \\ \cdot \\ \cdot \\ \alpha\, a_n + \beta\, b_n \end{bmatrix}. \tag{2.2}$$

Note that for two vectors to be added together, they must have the same number of components. Two n-vectors are equal when each component of one vector equals the corresponding component of the other vector.

VECTOR SPACE Denote by V_n a set of n-vectors such that (1) for any two vectors in the set V_n, their sum is a vector which also belongs to the set V_n and (2) for any vector in V_n, its product with any real number is a vector which also belongs to the set V_n. Then the set V_n is called a vector space.

The set of all possible n-vectors (for a fixed positive integer n) is then a vector space. We will denote this special vector space by R_n.

VECTOR SUBSPACE Let S_n be a subset of the vectors in the vector space V_n. If the set S_n is itself a vector space, then S_n is called a subspace of the vector space V_n.

The set $\{\mathbf{0}\}$ is a subspace of every vector space V_n. The vector $\mathbf{0}$ is defined as the n-vector all of whose components equal zero; it is also called the null vector.

Example 2.1

Consider the set of all possible vectors of the form $\alpha\mathbf{e}_1 + \beta\mathbf{e}_2$, where $\mathbf{e}_1$ and $\mathbf{e}_2$ are the two fixed 3-vectors $[1\ 0\ 0]^T$ and $[0\ 1\ 0]^T$ respectively. The numbers α and β are any two real numbers. The set of vectors, denoted by S_3, in which each vector has the form

$$\mathbf{x} = \alpha\begin{bmatrix}1\\0\\0\end{bmatrix} + \beta\begin{bmatrix}0\\1\\0\end{bmatrix} \tag{2.3}$$

obtained by taking different values of α and β, forms a subspace. Such a subspace is referred to as the subspace spanned by the two vectors $\mathbf{e}_1$ and $\mathbf{e}_2$. Note that this subspace does not include the entire space R_3. This is because this set of vectors, S_3, does not include, say, the vector $\mathbf{e}_3 = [0\ 0\ 1]^T$. That is, no matter how we choose α and β, we cannot have $\mathbf{x}$ equal the vector $\mathbf{e}_3$.

LINEAR INDEPENDENCE Consider the set of vectors $\{\mathbf{v}_1, \mathbf{v}_2, \ldots, \mathbf{v}_m\}$ where each vector is an n-vector belonging to R_n. This set of vectors is defined to be linearly dependent if and only if there exist a set of scalars $\{c_1, c_2, \ldots, c_m\}$, at least one of which is not equal to zero, such that

$$\sum_{i=1}^{i=m} c_i\mathbf{v}_i = \mathbf{0}. \tag{2.4}$$

If the only set of scalars such that (2.4) is satisfied is the set $\{0, 0, \ldots, 0\}$, then the set of vectors is defined to be linearly independent.

Example 2.2

a. Consider the three vectors $\mathbf{e}_1 = [1 \quad 0 \quad 0]^T$, $\mathbf{e}_2 = [0 \quad 1 \quad 0]^T$, and $\mathbf{e}_3 = [0 \quad 0 \quad 1]^T$. To determine if these vectors are linearly dependent, we need to find three scalars (not all equal to zero) such that

$$\sum_{i=1}^{i=3} c_i\mathbf{e}_i = \mathbf{0}. \tag{2.5}$$

Note that this set of three equations can be written as

$$\begin{bmatrix} 1 & 0 & 0 \\ 0 & 1 & 0 \\ 0 & 0 & 1 \end{bmatrix} \begin{bmatrix} c_1 \\ c_2 \\ c_3 \end{bmatrix} = \begin{bmatrix} 0 \\ 0 \\ 0 \end{bmatrix}. \qquad (2.6)$$

But this implies that $c_1 = c_2 = c_3 = 0$, and hence the three vectors are linearly independent. The 3 by 3 diagonal matrix with +1's along its diagonal is called an Identity Matrix and is denoted by I_3

b. Consider the vectors $\mathbf{e}_1$, $\mathbf{e}_2$ and $\mathbf{e}_3$ as defined in part (a) above along with the vector $\mathbf{x} = [a \quad b \quad c]^T$. Let us find if these four vectors are linearly independent or not. If they are linearly dependent, then we must be able to find numbers c_1, c_2, c_3, c_4, not all zero, so that $c_1\mathbf{e}_1 + c_2\mathbf{e}_2 + c_3\mathbf{e}_3 + c_4\,\mathbf{x} = \mathbf{0}$. But this is possible if we choose $c_1 = a$, $c_2 = b$, $c_3 = c$, and $c_4 = -1$. The reader may want to spend a minute to verify this. Hence the four vectors are linearly dependent.

c. Consider the three linearly independent vectors $\mathbf{e}_1$, $\mathbf{e}_2$ and $\mathbf{e}_3$ as defined in part (a) above along with the vector $\mathbf{0} = [0 \quad 0 \quad 0]^T$. These four vectors are now linearly dependent because $c_1\mathbf{e}_1 + c_2\mathbf{e}_2 + c_3\mathbf{e}_3 + c_4\mathbf{0} = \mathbf{0}$, when $c_4 = 1$, and $c_1 = c_2 = c_3 = 0$. Thus adding the vector $\mathbf{0}$ to any set of vectors always makes the enlarged set of vectors linearly dependent.

d. Consider the 2-vector $\mathbf{a} = [a_1 \ a_2]$. Let us create an n-vector by appending $(n-2)$ components which are each zero to this vector $\mathbf{a}$, so that we obtain the n-vector $\mathbf{a}' = [a_1 \ a_2 \ 0 \ 0 \cdot\cdot 0]$. Consider the 2-vector $\mathbf{b} = [b_1 \ b_2]$; similarly create the n-vector $\mathbf{b}' = [b_1 \ b_2 \ 0 \ 0 \cdot\cdot 0]$. If the vectors $\mathbf{a}$ and $\mathbf{b}$ are linearly independent, then the vectors $\mathbf{a}'$ and $\mathbf{b}'$ are also linearly independent. This follows directly from the definition of linear independence.

SET OF VECTORS SPANNING A VECTOR SPACE Let V_n be a vector space. If every vector which belongs to V_n, can be expressed as a linear combination of a set of vectors $\{\mathbf{v}_1, \mathbf{v}_2, \ldots, \mathbf{v}_m\}$, then this set of vectors is defined as spanning the vector space V_n. Note that this set of vectors need not be linearly independent.

Example 2.3

Consider the set of vectors $\{\mathbf{e}_1, \mathbf{e}_2, \mathbf{e}_3, \mathbf{x}\}$ defined in Example 2.2, part (b). These vectors are linearly dependent and they span the entire space R_3.

Example 2.3 (continued)

This is because any vector $\mathbf{y} = [p \quad q \quad r]^T$ in R_3 can be expressed as $\mathbf{y} = p\mathbf{e}_1 + q\mathbf{e}_2 + r\mathbf{e}_3 + 0\mathbf{x}$. Note that the set of vectors $\{\mathbf{e}_1, \mathbf{e}_2, \mathbf{e}_3\}$, which is linearly independent, also spans the entire space R_3. Thus including the vector $\mathbf{x}$ in the set of vectors that span R_3 is really unnecessary.

BASIS OF A VECTOR SPACE Let V_n be a vector space. If every vector in V_n can be expressed as a linear combination of the set of linearly independent vectors $\{\mathbf{v}_1, \mathbf{v}_2, \ldots, \mathbf{v}_m\}$, then the set of vectors $\{\mathbf{v}_1, \mathbf{v}_2, \ldots, \mathbf{v}_m\}$ is called a basis set for V_n. In general, a basis set for a given vector space V_n is not unique. There may be many different basis sets for V_n. However, the number of vectors, m, in any basis set of V_n is unique and is defined as the dimension of V_n.

Example 2.4

a. The vectors $\mathbf{e}_1$, $\mathbf{e}_2$ and $\mathbf{e}_3$ of Example 2.2, part (a), form a basis for R_3.

We have shown in Example 2.2 that they are linearly independent, and in Example 2.3 that they span the space R_3. Since there are 3 basis vectors for R_3, the dimension of R_3 is 3.

b. Another basis set for R_3 is formed by the vectors $[2 \quad 0 \quad 0]^T$, $[0 \quad 3 \quad 0]^T$ and $[0 \quad 1 \quad 1]^T$. Note that though this basis set of vectors is different from that of part (a), the number of vectors in the basis set equals three, the dimension of the space R_3.

c. The vector space spanned by the 3-vectors $[1 \; 1 \; 0]^T$ and $[1 \quad -1 \quad 0]^T$ has dimension 2.

Example 2.5

Consider the vector space R_n formed of all possible real n-vectors. Any n-vector $\mathbf{a}$ in this space can be expressed as

$$\mathbf{a} = \begin{bmatrix} a_1 \\ a_2 \\ \cdot \\ \cdot \\ a_n \end{bmatrix} = a_1 \begin{bmatrix} 1 \\ 0 \\ 0 \\ \cdot \\ 0 \end{bmatrix} + a_2 \begin{bmatrix} 0 \\ 1 \\ 0 \\ \cdot \\ 0 \end{bmatrix} + \cdot \cdot \cdot + a_n \begin{bmatrix} 0 \\ 0 \\ \cdot \\ 0 \\ 1 \end{bmatrix} \qquad (2.7)$$

Denoting by $\mathbf{e}_j$ the vector $[0 \quad 0 \quad . \quad . \quad 1 \quad 0 \quad . \quad .]^T$ which has all its components equal to zero except for the jth component which is 1, we find that the vectors $\mathbf{e}_j$, $j = 1, 2, \ldots, n$ form a basis set for the space R_n because they are linearly independent and span R_n. Note that though we can find another set of linearly independent vectors to span R_n, the number of such vectors will always be n, the dimension of the space R_n.

INNER PRODUCTS AND ORTHOGONALITY Given two vectors **a** and **b** in a vector space V_n. The inner product of **a** and **b** is defined by the scalar $\mathbf{a}^T\mathbf{b}$. If this number equals zero, the vectors **a** and **b** are defined to be orthogonal. The quantity $\|\mathbf{a}\| = +\sqrt{\mathbf{a}^T\mathbf{a}}$ is called the Euclidean length of the vector **a**.

Example 2.6

The set of vectors $\mathbf{e}_j$, $j = 1, 2, \ldots, n$, defined in Example 2.5 are pairwise orthogonal, $\mathbf{e}_i^T\mathbf{e}_j = 0$, for $i \neq j$. Note also that $\mathbf{e}_i^T\mathbf{e}_i = 1$, for all i. The Euclidean length of the vector $\mathbf{e}_i$ is therefore unity.

Example 2.7

a. For any vector **a**, $\mathbf{a}^T\mathbf{a} \geq 0$, the equality being valid if and only if $\mathbf{a} = \mathbf{0}$.

This is simple to show using the definition of the product $\mathbf{a}^T\mathbf{a}$.

b. Given a vector $\mathbf{a} \neq \mathbf{0}$, we can "normalize" it to have unit Euclidean length by creating a vector **b**, defined by $\mathbf{b} = \mathbf{a}/\|\mathbf{a}\| = +\mathbf{a}/\sqrt{\mathbf{a}^T\mathbf{a}}$.

Example 2.8

Consider a set containing m nonzero n-vectors. Let every two of these vectors be pairwise orthogonal. Then the set of m vectors is linearly independent.

Let us prove this for $m = 3$ and $n = 4$. The proof for general m and n follows similarly. Consider three 4-vectors that are pairwise orthogonal. Denote them by $\mathbf{u}_1$, $\mathbf{u}_2$, and $\mathbf{u}_3$. We need to investigate if scalars c_1, c_2, c_3, not all zero, exist, so that

$$c_1\mathbf{u}_1 + c_2\mathbf{u}_2 + c_3\mathbf{u}_3 = \mathbf{0}. \tag{2.8}$$

Example 2.8 (continued)

Denote $\mathbf{u}_i^T\mathbf{u}_i = a_i > 0$, take the transpose of equation (2.8) and multiply on the right by $\mathbf{u}_1$ to get

$$(c_1\mathbf{u}_1^T + c_2\mathbf{u}_2^T + c_3\mathbf{u}_3^T)\mathbf{u}_1 = c_1a_1 = 0, \tag{2.9}$$

yielding $c_1 = 0$. Note that in equation (2.9), $\mathbf{u}_2^T\mathbf{u}_1 = \mathbf{u}_3^T\mathbf{u}_1 = 0$, because of the pairwise orthogonality of the vectors $\mathbf{u}_1$, $\mathbf{u}_2$, and $\mathbf{u}_3$. Again, taking the transpose of (2.8) and multiplying it to the right by $\mathbf{u}_2$ yields $c_2 = 0$. In a similar fashion by multiplying by $\mathbf{u}_3$ we can show that $c_3 = 0$. Thus for equation (2.8) to be valid, $c_1 = c_2 = c_3 = 0$. Hence the three vectors are linearly independent. The proof for an orthogonal set of m nonzero n-vectors is similar. To acquire greater familiarity, the reader may want to think through this.

ORTHOGONAL AND ORTHONORMAL BASIS If $\{\mathbf{v}_1, \mathbf{v}_2, \ldots, \mathbf{v}_m\}$ is a basis (set) for V_n such that every pair of two different vectors of the basis set is orthogonal, then the basis is defined to be an orthogonal basis for V_n. That is, we require $\mathbf{v}_i^T\mathbf{v}_j = 0$, for $i \neq j$.

If, in addition, the vectors are such that $\mathbf{v}_i^T\mathbf{v}_i = 1$, for $i = 1, 2, \ldots, m$, the basis set is defined to be orthonormal.

One can find an orthonormal basis set for every vector space V_n except $\{\mathbf{0}\}$.

Example 2.9

The vectors $\mathbf{e}_j$, $j = 1, 2, \ldots, n$, defined in Example 2.5 form an orthogonal basis for R_n.

Consider n mutually orthogonal n-vectors in R_n. The vectors, being orthogonal, are linearly independent, as seen in Example 2.8. Thus we have n linearly independent vectors in R_n, a space of dimension n. These vectors must therefore form a basis for R_n. In fact, this basis set is orthonormal.

SUM OF TWO VECTOR SPACES Let S_1 and S_2 be two (vector) subspaces of the vector space V_n. The set of vectors S, denoted, $S = S_1 \oplus S_2$, is called the sum of the vector spaces S_1 and S_2, and is defined by

$$S = \{\mathbf{y}: \mathbf{y} = \mathbf{x}_1 + \mathbf{x}_2;\ \mathbf{x}_1 \text{ belongs to } S_1;\ \mathbf{x}_2 \text{ belongs to } S_2\}. \tag{2.10}$$

ORTHOGONAL VECTOR SPACES Let S_1 and S_2 be two subspaces of the vector space V_n. If $\mathbf{x}_1^T\mathbf{x}_2 = 0$ for every vector $\mathbf{x}_1$ in S_1 and every vector $\mathbf{x}_2$ in S_2, then S_1 and S_2 are defined to be orthogonal subspaces in V_n, and are denoted by $S_1 \perp S_2$.

Example 2.10

Consider the subspace S_1 spanned by the vectors $\mathbf{e}_j$, $j = 1, 2, \ldots, r$, and, the subspace S_2 spanned by $\mathbf{e}_j, j = r+1, r+2$. The vectors $\mathbf{e}_j$ are defined as in Example 2.5 and belong to R_n. The spaces S_1 and S_2 are orthogonal subspaces of R_n.

Any vector in S_1 can be expressed as $\mathbf{y} = \sum_{i=1}^{r} a_i\mathbf{e}_i$. Any vector in S_2 can be expressed as $\mathbf{z} = \sum_{j=r+1}^{r+2} a_j\mathbf{e}_j$. Hence,

$$\mathbf{y}^T\mathbf{z} = \left(\sum_{i=r}^{r} a_i\mathbf{e}_i\right)^T \sum_{j=r+1}^{r+2} a_j\mathbf{e}_j = \sum_{j=r+1}^{r+2}\sum_{i=1}^{r} a_ia_j\left(\mathbf{e}_i^T\mathbf{e}_j\right) = 0. \quad (2.11)$$

ORTHOGONAL COMPLEMENT OF A VECTOR SPACE IN V_n Let S_1 and S_2 be two subspaces of the vector space V_n. The vector subspace S_2 is defined as the orthogonal complement of S_1 in V_n if and only if: (1) $V_n = S_1 \oplus S_2$ and (2) $S_1 \perp S_2$. The orthogonal complement of the subspace S_1 is denoted by $S_1^\perp$.

Example 2.11

Let the set of vectors $\{\mathbf{v}_1, \mathbf{v}_2, \ldots, \mathbf{v}_r, \ldots, \mathbf{v}_n\}$ form an orthogonal basis (set) for V_n. Let the subspace spanned by the vectors $\{\mathbf{v}_1, \mathbf{v}_2, \ldots, \mathbf{v}_r\}$ be denoted by S_1, and the subspace spanned by the remaining vectors $\{\mathbf{v}_{r+1}, \mathbf{v}_2, \ldots, \mathbf{v}_n\}$ be denoted by S_2.

Then $S_2 = S_1^\perp$.

If a vector $\mathbf{y}$ belongs to V_n, then it can be uniquely decomposed into two vectors $\mathbf{x}_1$ and $\mathbf{x}_2$, where the vector $\mathbf{x}_1$ belongs to the subspace S_1 and the vector $\mathbf{x}_2$ belongs to $S_1^\perp$. This is because the set $\{\mathbf{v}_1, \mathbf{v}_2, \ldots, \mathbf{v}_m\}$ is an orthogonal basis set, and therefore

Example 2.11 (continued)

$$\mathbf{y} = \sum_{i=1}^{i=r} \mathbf{c}_i \mathbf{v}_i + \sum_{i=r+1}^{i=n} \mathbf{c}_i \mathbf{v}_i = \mathbf{x}_1 + \mathbf{x}_2. \tag{2.12}$$

Note that $\mathbf{x}_1^T \mathbf{x}_2 = 0$, proving that $S_1 \perp S_2$.

Example 2.12

Consider two subspaces S_1 and S_2 of R_n. Suppose that every vector orthogonal to the space S_1 is also orthogonal to the space S_2 and vice versa. Then S_1 and S_2 are one and the same subspace.

Let a set of orthogonal vectors that span the space S_1 be $\{\mathbf{v}_1, \mathbf{v}_2, \ldots, \mathbf{v}_r\}$, and a set that spans S_2 be $\{\mathbf{w}_1, \mathbf{w}_2, \ldots, \mathbf{w}_p\}$. Then the vectors orthogonal to S_1 are $\{\mathbf{v}_{r+1}, \mathbf{v}_2, \ldots, \mathbf{v}_n\}$. Since these vectors are also orthogonal to vectors $\{\mathbf{w}_1, \mathbf{w}_2, \ldots, \mathbf{w}_p\}$, each of the $\mathbf{w}_i$, $i = 1, 2, \ldots, p$, must be a linear combination of the remaining vectors $\mathbf{v}_i$, $i = 1, 2, \ldots, k$, $k \le r$ which span the space R_n. Hence S_2 must be a subspace of S_1. Similarly, since the vectors $\{\mathbf{w}_{p+1}, \mathbf{w}_{p+2}, \ldots, \mathbf{w}_n\}$ are orthogonal to S_2, they are also orthogonal to $\{\mathbf{v}_1, \mathbf{v}_2, \ldots, \mathbf{v}_r\}$. Hence every vector $\mathbf{v}_i$, $i = 1, 2, \ldots, r$, is a linear combination of $\mathbf{w}_i$, $i = 1, 2, \ldots, h$, $h \le p$. Thus S_1 must be a subspace of S_2. Since S_1 is a subspace of S_2, and $S_2$is a subspace of S_1, hence, $S_1 = S_2$.

Types of Matrices

DIAGONAL MATRIX An n by n matrix whose only nonzero elements are those along the diagonal. It is denoted by $Diag\,\{a_{11}, a_{22}, \ldots, a_{nn}\}$ where $a_{11}, a_{22}, \ldots, a_{nn}$ are the elements along the diagonal of the matrix.

IDENTITY MATRIX An n by n diagonal matrix with 1's on the diagonal is defined as an identity matrix and is denoted by I_n. Where there is no source of confusion regarding the dimension of the matrix, we simply write I.

TRANSPOSE OF A MATRIX The transpose, A^T, of an m by n matrix A is an n by m matrix obtained by interchanging the rows and columns of the matrix A. Note that $(AB)^T = B^T A^T$.

SYMMETRIC MATRIX An n by n real matrix is symmetric if $A^T = A$.

SKEW-SYMMETRIC MATRIX An n by n real matrix is skew-symmetric if $A^T = -A$.

ORTHOGONAL MATRIX An n by n orthogonal matrix is defined as a matrix for which $A^TA = AA^T = I$. The columns (rows) of an orthogonal matrix are pairwise orthonormal.

SINGULAR MATRIX An n by n matrix whose determinant is zero is called a singular matrix.

INVERSE OF A SQUARE MATRIX If an n by n matrix A is nonsingular, then there exists a matrix A^{-1} such that $AA^{-1} = A^{-1}A = I$. The matrix A^{-1} is called the inverse of the matrix A.

POSITIVE SEMIDEFINITE MATRIX A symmetric n by n matrix A is defined as positive semidefinite if for any nonzero n-vector $\mathbf{x}$, $\mathbf{x}^TA\mathbf{x} \geq \mathbf{0}$. When strict inequality occurs, the matrix A is called positive definite.

IDEMPOTENT MATRIX If a matrix A is such that $A^2 = A$, it is defined to be idempotent, where $A^2 = AA$.

Example 2.13

a. Sometimes it is convenient to represent a matrix by juxtaposing two or more matrices in a partitioned form. Thus we can represent a partitioned matrix as

$$A = \begin{bmatrix} P & Q \\ R & S \end{bmatrix}, \tag{2.13}$$

where the rows of the matrices P and Q are equal in number, the columns of P and R are equal in number and so on. The transpose of A is then

$$A^T = \begin{bmatrix} P^T & R^T \\ Q^T & S^T \end{bmatrix}. \tag{2.14}$$

The product of two partitioned matrices is obtained by the usual row by column rule, treating the matrices as elements. Thus, we have

Example 2.13 (continued)

$$AB = \begin{bmatrix} P & Q \\ R & S \end{bmatrix} \begin{bmatrix} E \\ G \end{bmatrix} = \begin{bmatrix} PE + QG \\ RE + SG \end{bmatrix} \tag{2.15}$$

and

$$AC = \begin{bmatrix} P & Q \\ R & S \end{bmatrix} \begin{bmatrix} E & F \\ G & H \end{bmatrix} = \begin{bmatrix} PE + QG & PF + QH \\ RE + SG & RF + SH \end{bmatrix} \tag{2.16}$$

provided the products PE, etc., exist.

Example 2.14

Consider the m by n matrix A. Then the m by m matrix AA^T is a symmetric, positive semidefinite matrix.

AA^T is a symmetric because $(AA^T)^T = AA^T$. Now consider any arbitrary m-vector $\mathbf{x}$. Set $\mathbf{y} = A^T\mathbf{x}$. Then we have $\mathbf{y}^T = \mathbf{x}^T A$, and $\mathbf{y}^T\mathbf{y} \geq 0$. Hence

$$\mathbf{y}^T\mathbf{y} = \mathbf{x}^T AA^T \mathbf{x} = \mathbf{x}^T (AA^T)\mathbf{x} \geq 0 \tag{2.17}$$

Hence the matrix AA^T is also positive semidefinite.

Similarly, the matrix $A^T A$ is also a symmetric, positive semidefinite matrix.

Some Properties of Matrices

EIGENVALUE AND EIGENVECTOR OF A SQUARE MATRIX Consider an m by m matrix A. The complex scalar λ, which satisfies the equation

$$A\mathbf{x} = \lambda\mathbf{x} \tag{2.18}$$

for a nonzero m-vector $\mathbf{x}$, is called an eigenvalue of A; the nonzero vector $\mathbf{x}$ for which this equation (2.18) is valid is called the eigenvector corresponding to the eigenvalue λ. It should be noted that, in general, the eigenvalue of a real matrix A can be a complex number, as can the m components of the corresponding eigenvector $\mathbf{x}$. We note that if the vector $\mathbf{x}$ satisfies this equation for a given λ, then so does the vector $\alpha\mathbf{x}$, where α is an arbitrary scalar. Hence the eigenvectors can only be ascertained up to a multiplicative scalar.

The eigenvalues of A are obtained by setting the determinant of the matrix $A-\lambda I$ to zero, i.e., $Det(A - \lambda I) = 0$. This condition requires that a polynomial of order m in λ be set to zero. Since a polynomial of order m in λ has m (complex) roots, an m by m matrix must have m eigenvalues. These m eigenvalues are, in general, complex numbers.

Example 2.15

a. The eigenvalues of the m by m matrix A are identical to the eigenvalues of the matrix $T^{-1}AT$.

This is because

$$Det\left[T^{-1}AT - \lambda I\right] = Det\left[T^{-1}(A - \lambda I)T\right] = Det\left[(A - \lambda I)T^{-1}T\right]$$
$$= Det(A - \lambda I)Det(T^{-1}T)$$
$$= Det(A - \lambda I)Det(I) = Det(A - \lambda I).$$

b. If T is an orthogonal matrix then the eigenvalues of A and T^TAT are identical.

This follows using (a) above, because for an orthogonal matrix $T^T = T^{-1}$.

Example 2.16

If the matrices A and B are orthogonal, then AB is orthogonal. This is because

$$(AB)^T AB = B^T A^T AB = B^T IB = B^T B = I. \tag{2.19}$$

PROPERTIES OF SYMMETRIC MATRICES We list here some of the properties of real symmetric matrices. These results are important and will be used by us later. In what follows we take the real symmetric matrix A to be an m by m matrix.

a. The eigenvalues of a real symmetric matrix are real numbers.

Let the eigenvalue of the symmetric matrix A be $\lambda + i\mu$ and the corresponding eigenvector $\mathbf{q} + i\mathbf{r}$. Here $i = \sqrt{-1}$. Hence by equation (2.18), the eigenvalue problem can be stated as

$$A(\mathbf{q} + i\mathbf{r}) = (\lambda + i\mu)(\mathbf{q} + i\mathbf{r}), \tag{2.20}$$

which implies that $A\mathbf{q} = \lambda\mathbf{q} - \mu\mathbf{r}$, and $A\mathbf{r} = \lambda\mathbf{r} + \mu\mathbf{q}$. Premutiplying the first of these expressions by $\mathbf{r}^T$ and the second by $\mathbf{q}^T$, we get

$$\mathbf{r}^T A\mathbf{q} = \lambda\mathbf{r}^T\mathbf{q} - \mu\mathbf{r}^T\mathbf{r} \tag{2.21}$$

and

$$\mathbf{q}^T A\mathbf{r} = \lambda\mathbf{q}^T\mathbf{r} + \mu\mathbf{q}^T\mathbf{q}. \tag{2.22}$$

But since $\mathbf{r}^T A\mathbf{q}$ is a scalar, $\mathbf{r}^T A\mathbf{q} = (\mathbf{r}^T A\mathbf{q})^T = \mathbf{q}^T A^T\mathbf{r} = \mathbf{q}^T A\mathbf{r}$, where the last equality follows because A is symmetric. Subtracting equation (2.21) from equation (2.22) and noting that $\mathbf{r}^T\mathbf{q}$ is similarly a scalar, we get $\mu(\mathbf{r}^T\mathbf{r} + \mathbf{q}^T\mathbf{q}) = 0$ from which it follows that $\mu = 0$.

b. The eigenvectors may be taken to be real.

Since the eigenvalues are all real, we can take the eigenvectors to be real. Remember that if $\mathbf{x}$ is an eigenvector, then $\alpha\mathbf{x}$ is also an eigenvector for any (in general, complex) α.

c. The eigenvectors corresponding to distinct eigenvalues are orthogonal.

Let $A\mathbf{x}_k = \lambda_k\mathbf{x}_k$ and $A\mathbf{x}_j = \lambda_j\mathbf{x}_j$ where $\lambda_k \neq \lambda_j$. Premultiplication of the first of these relations by $\mathbf{x}_j^T$ and the second by $\mathbf{x}_k^T$ and subtraction yields $(\lambda_k - \lambda_j)\mathbf{x}_k^T\mathbf{x}_j = 0$, so that $\mathbf{x}_k^T\mathbf{x}_j = 0$.

d. If the vector $\mathbf{x}$ is an arbitrary nonzero vector, there exists an eigenvector $\mathbf{y}$ of A belonging to the linear space spanned by the vectors $\{\mathbf{x}, A\mathbf{x}, A^2\mathbf{x}, \ldots\}$.

Since the vectors $\mathbf{x}, A\mathbf{x}, A^2\mathbf{x}, \ldots$ belong to R_m, they cannot all be linearly independent. Let us say that k is the smallest value for which

$$A^k\mathbf{x} + b_1A^{k-1}\mathbf{x} + \cdots + b_k\mathbf{x} = \mathbf{0}. \tag{2.23}$$

Factorizing this, we get

$$(A - \mu_1 I)(A - \mu_2 I)\cdots(A - \mu_k I)\mathbf{x} = (A - \mu_1 I)\mathbf{y} = 0, \tag{2.24}$$

where the vector

$$\mathbf{y} = (A - \mu_2 I)\cdots(A - \mu_k I)\mathbf{x} \neq \mathbf{0}, \tag{2.25}$$

or else k would not be the smallest integer for which (2.23) holds. Since $(A - \mu_1 I)\mathbf{y} = \mathbf{0}$, $\mathbf{y}$ is an eigenvector of A corresponding to the eigenvalue

μ_1. Since A is a real symmetric matrix, μ_1 is real. Similarly all the other μ_i are real. Hence, by equation (2.25), $\mathbf{y}$ belongs to the space spanned by $\{A^{k-1}\mathbf{x}, A^{k-2}\mathbf{x}, \ldots, \mathbf{x}\}$.

e. Let A be an m by m real symmetric matrix. Then there exists an orthogonal matrix W such that $W^TAW = \Lambda$ or $A = W\Lambda W^T$, where the matrix Λ is a diagonal matrix and has the eigenvalues of A along its diagonal.

We will show that if $\lambda_1 \geq \lambda_2 \geq \lambda_3 \geq \ldots \geq \lambda_m$ are the m eigenvalues of A, including the multiplicities, then the ith diagonal element of Λ is λ_i and the ith column of W is $\mathbf{w}_i$, an eigenvector of A corresponding to λ_i.

Suppose that there exist s orthonormal vectors $\mathbf{w}_i$, $i = 1, 2, \ldots, s$ such that

$$A\mathbf{w}_i = \lambda_i\mathbf{w}_i,\ i = 1, 2, \ldots, s. \tag{2.26}$$

Then $A^2\mathbf{w}_i = \lambda_i A\mathbf{w}_i = \lambda_i^2\mathbf{w}_i$, and similarly for any $r > 0$, we find that $A^r\mathbf{w}_i = \lambda_i A^{r-1}\mathbf{w}_i \ldots = \lambda_i^r\mathbf{w}_i$. Choose a vector $\mathbf{x}$ which is orthogonal to the space spanned by $\{\mathbf{w}_1, \mathbf{w}_2, \mathbf{w}_3, \ldots, \mathbf{w}_s\}$. Then, for $i = 1, 2, \ldots, s$,

$$(\mathbf{x}^TA^r)\mathbf{w}_i = \mathbf{x}^T(A^r\mathbf{w}_i) = \mathbf{x}^T\lambda_l^r\mathbf{w}_i = \lambda_l^r\mathbf{x}^T\mathbf{w}_i = 0 \text{ for all values of } r > 0. \tag{2.27}$$

Hence the space spanned by the vectors $\{\mathbf{x}, A\mathbf{x}, A^2\mathbf{x}, \ldots\}$ is orthogonal to the space spanned by the vectors $\{\mathbf{w}_1, \mathbf{w}_2, \mathbf{w}_3, \ldots, \mathbf{w}_s\}$. But from our result in part (d) we find that there must exist an eigenvector $\mathbf{w}_{s+1}$ which belongs to the space spanned by the vectors $\{\mathbf{x}, A\mathbf{x}, A^2\mathbf{x}, \ldots\}$, and this must, by our previous statement, be orthogonal to the set $\{\mathbf{w}_1, \mathbf{w}_2, \mathbf{w}_3, \ldots, \mathbf{w}_s\}$. Now this vector $\mathbf{w}_{s+1}$ can be normalized to have unit length.

Since the vector $\mathbf{w}_1$ can be chosen to be any eigenvector to start with, we have established the existence of m mutually orthogonal eigenvectors such that

$$A\mathbf{w}_i = \lambda_i\mathbf{w}_i,\ i = 1, 2, \ldots, m. \tag{2.28}$$

This may be rewritten as

$$AW = W\Lambda,\ W^TW = I \tag{2.29}$$

where W is an orthogonal matrix with $\mathbf{w}_i$ as its ith column, and Λ is a diagonal matrix with λ_i as its ith diagonal element.

f. An alternative way of representing the result in part (e) is:

$$A = W\Lambda W^T = \lambda_1\mathbf{w}_1\mathbf{w}_1^T + \lambda_2\mathbf{w}_2\mathbf{w}_2^T + \cdots + \lambda_m\mathbf{w}_m\mathbf{w}_m^T, \tag{2.30}$$

$$I_m = WW^T = \mathbf{w}_1\mathbf{w}_1^T + \mathbf{w}_2\mathbf{w}_2^T + \cdots + \mathbf{w}_m\mathbf{w}_m^T. \tag{2.31}$$

This is called the spectral decomposition of the symmetric matrix A.

g. If A is a positive semidefinite matrix, then all its eigenvalues must be nonnegative.

Since A is positive semidefinite, for any vector $\mathbf{y}$, $\mathbf{y}^TA\mathbf{y} \geq 0$. Let $\mathbf{y}$ be the eigenvector $\mathbf{x}_i$ of A, so that $A\mathbf{x}_i = \lambda_i\mathbf{x}_i$. Then $\mathbf{x}_i^TA\mathbf{x}_i = \mathbf{x}_i^T\lambda_i\mathbf{x}_i = \lambda_i\mathbf{x}_i^T\mathbf{x}_i \geq 0$, which requires $\lambda_i \geq 0$, since $\mathbf{x}_i^T\mathbf{x}_i > 0$. Hence a symmetric positive semidefinite matrix A can be expressed as $A=W\Lambda W^T$, where the elements of the diagonal matrix Λ are all nonnegative. In a similar way we can show that each eigenvalue of a positive definite matrix is a real positive number.

h. If A is a positive definite matrix, we can define

$$A^{1/2} = W\Lambda^{1/2}W^T \text{ and } A^{-1/2} = W\Lambda^{-1/2}W^T, \tag{2.32}$$

where the matrices W and Λ are defined in equation (2.29).

Since by part (g) above, the eigenvalues of A are positive, we can define a real diagonal matrix $\Lambda^{1/2}$ whose diagonal elements are the positive square-roots of the diagonal elements of Λ. Thus

$$\begin{aligned} \Lambda^{1/2} &= Diag\{\lambda_1^{1/2}, \lambda_2^{1/2}, \ldots, \lambda_n^{1/2}\}, \text{ and} \\ \Lambda^{-1/2} &= Diag\{\lambda_1^{-1/2}, \lambda_2^{-1/2}, \ldots, \lambda_n^{-1/2}\}. \end{aligned} \tag{2.33}$$

Note that $A^{1/2}A^{1/2} = W\Lambda^{1/2}W^TW\Lambda^{1/2}W^T = W\Lambda W^T = A$, and also $A^{-1/2}A^{-1/2} = A^{-1}$. Similarly, $A^{1/2}A^{-1/2} = W\Lambda^{1/2}W^TW\Lambda^{-1/2}W^T = W\Lambda^{1/2}I\Lambda^{-1/2}W^T = WIW^T = I = A^{-1/2}A^{1/2}$.

Hence if the m by m matrix $A = kI_m$, where k is a positive constant, then $A^{1/2} = k^{1/2}I_m$, and $A^{-1/2} = k^{-1/2}I_m$.

RANK OF A MATRIX Let A be an m by n matrix. The rank of A, often denoted *Rank*(A), is the number of linearly independent columns (or rows) of A. We can denote the n columns of a matrix A as $A = [\mathbf{a}_1\ \mathbf{a}_2\ \mathbf{a}_3 \ldots \mathbf{a}_n]$. The vector space spanned by the n columns of A is defined as the column space of A. The dimension of the column space is the number of linearly independent columns of A, which equals the rank of A. The rank of A is the same as the rank of A^T.

RANK OF THE PRODUCT OF MATRICES If A is an m by n matrix and B is an n by k matrix, then the rank of AB cannot exceed the rank of A. Let C be the matrix product AB. Then we have

$$C = \left[\mathbf{c}_1 \ \mathbf{c}_2 \ \mathbf{c}_3 \ldots \mathbf{c}_n\right] = \left[\sum_i \mathbf{a}_i b_{i1} \ \sum_i \mathbf{a}_i b_{i2} \ldots \sum_i \mathbf{a}_i b_{ik}\right], \tag{2.34}$$

so that every column of C is a linear combination of the columns of A and therefore the columns of C lie in the subspace spanned by the columns of A. Hence the column space of C is a subspace of the column space of A. Hence $Rank(AB) \le Rank(A)$.

Example 2.17

a. Consider the m by n matrix $B_1 = \mathbf{v}_1\mathbf{a}^T$, where $\mathbf{v}_1$ is an m by 1 vector and $\mathbf{a}$ is an n by 1 vector. Then the matrix B_1 can be expressed as $B_1 = [a_1\mathbf{v}_1 \ a_2\mathbf{v}_1 \ldots a_m\mathbf{v}_1]$, where a_i, $i = 1, 2, \ldots, n$ are the n components of the column vector $\mathbf{a}$. Each column of the matrix B_1 is made up of a scalar times the vector $\mathbf{v}_1$. Hence there is only one linearly independent column in the matrix B_1, and so the matrix has rank 1.

b. Let $\mathbf{v}_1$ and $\mathbf{v}_2$ be two m by 1 vectors which are linearly independent (or mutually orthogonal). Consider the matrix $B_2 = \lambda_1\mathbf{v}_1\mathbf{a}^T + \lambda_2\mathbf{v}_2\mathbf{b}^T$ where the vectors $\mathbf{a}$ and $\mathbf{b}$ are each n by 1, and λ_1 and λ_2 are scalars. This matrix can be expressed in column form as

$$B_2 = [(\lambda_1 a_1\mathbf{v}_1 + \lambda_2 b_1\mathbf{v}_2)\ (\lambda_1 a_2\mathbf{v}_1 + \lambda_2 b_2\mathbf{v}_2)\cdots(\lambda_1 a_n\mathbf{v}_1 + \lambda_2 b_n\mathbf{v}_2)] \tag{2.35}$$

where a_i, $i = 1, 2, \ldots, n$, and b_i, $i = 1, 2, \ldots, n$ are the n components of the column vectors $\mathbf{a}$ and $\mathbf{b}$ respectively. Hence each column of the matrix B_2 is a linear combination of the two vectors $\mathbf{v}_1$ and $\mathbf{v}_2$. Hence the column space of the matrix B_2 is spanned by the vectors $\mathbf{v}_1$ and $\mathbf{v}_2$, and the matrix therefore has rank 2.

c. We can generalize the result in part (b) above to the matrix $B_r = \sum_{i=1}^{r} \lambda_i\mathbf{v}_i\mathbf{w}_i^T$, where the independent vectors $\mathbf{v}_i$ are each m by 1, and the vectors $\mathbf{w}_i$ are each n by 1. The λ_i are scalars, r in number. In similar fashion, we can show that the column space of the matrix B_r is spanned by the vectors $\mathbf{v}_1, \mathbf{v}_2, \ldots, \mathbf{v}_r$.

d. From part (c) above, we see that the spectral decomposition (2.30) of a real m by m symmetric matrix implies that the column space of the

Example 2.17 (continued)

matrix is spanned by its m eigenvectors. Note how the decomposition (2.30) represents the matrix A as a linear sum of m matrices, each of which has unit rank.

Example 2.18

a. Show that if $C = AB$, $Rank(C)$ cannot exceed either $Rank(A)$ or $Rank(B)$.

We have already proved that $Rank(C) \leq Rank(A)$. Since $(AB)^T = B^T A^T$, we get $Rank(C) = Rank(AB) = Rank((AB)^T) = Rank(B^T A^T) \leq Rank(B^T) = Rank(B)$. Hence the result.

b. Show that $Rank(AB) = Rank(A)$, if B is a square and nonsingular matrix.

We know that $Rank(AB) \leq Rank(A)$, but $A = (AB)B^{-1}$, and so $Rank(A) \leq Rank(AB)$. Hence the result.

Similarly, $Rank(CA) = Rank(A)$, if C is a square nonsingular matrix.

Example 2.19

Let A be an m by n matrix of rank r.

a. If the vector $\mathbf{b}$ is such that $\mathbf{b}^T A = \mathbf{0}$, then the vector $\mathbf{b}$ is orthogonal to every column of the matrix A. Thus $\mathbf{b}$ is orthogonal to the column space of A.

We have

$$\mathbf{b}^T A = \mathbf{b}^T \left[\mathbf{a}_1 \ \ \mathbf{a}_2 \cdots \mathbf{a}_n\right] = \left[\mathbf{b}^T \mathbf{a}_1 \ \ \mathbf{b}^T \mathbf{a}_2 \cdots \ \mathbf{b}^T \mathbf{a}_n\right] = \mathbf{0}. \qquad (2.36)$$

This implies that $\mathbf{b}^T \mathbf{a}_1 = \mathbf{b}^T \mathbf{a}_2 \cdots = \mathbf{b}^T \mathbf{a}_n = 0$, hence $\mathbf{b}$ is orthogonal to any vector $\mathbf{y}$ of the form $\mathbf{y} = c_1 \mathbf{a}_1 + c_2 \mathbf{a}_2 + \cdots + c_n \mathbf{a}_n$, where c_1, $c_2, \ldots, c_n$ are arbitrary numbers.

b. If a vector $\mathbf{b}$ is such that it is orthogonal to the column space of A, then it is also orthogonal to the column space of AA^T.

Consider any vector $\mathbf{b}$ such that $\mathbf{b}^T A = \mathbf{0}$. This implies, by part (a) above, that $\mathbf{b}$ is orthogonal to the column space of A. But $\mathbf{b}^T A = \mathbf{0}$ implies that $\mathbf{b}^T AA^T = \mathbf{b}^T (AA^T) = \mathbf{0}$. Hence every vector $\mathbf{b}$ which is orthogonal to the column space of A is also orthogonal to the column space of AA^T.

c. If a vector **d** is such that it is orthogonal to the column space of AA^T, then it is also orthogonal to the column space of A.

Consider any vector **d** such that it is orthogonal to the column space of the matrix AA^T. Hence we have $\mathbf{d}^T AA^T = \mathbf{0}$. But this implies that $\mathbf{d}^T AA^T \mathbf{d} = 0$, which in turn implies that $\mathbf{d}^T AA^T\mathbf{d} = (\mathbf{d}^T A)(\mathbf{d}^T A)^T = 0$ requiring $\mathbf{d}^T A = \mathbf{0}$. This means that **d** is orthogonal to the column space of A.

d. $Rank(AA^T) = Rank(A) = Rank(A^T) = r.$

By part (b) and (c) above, every vector which is orthogonal to the column space of A is also orthogonal to the column space of AA^T, and vice versa. By Example 2.12, the column spaces of A and AA^T must then be the same, as also their dimensions. Hence the result.

e. $Rank(A^T A) = r.$

Interchanging A^T and A in part (d) above, we get the result.

PROPERTIES OF THE MATRIX AA^T Let A be an m by n matrix of rank r. Then the m by m matrix AA^T has the following properties:

a. The matrix is symmetric and positive semidefinite.

This is already proved in Example 2.14.

b. The matrix has rank r.

This is proved in Example 2.19, part (d).

c. The matrix has r positive eigenvalues. The remaining $(m - r)$ eigenvalues are all zero.

Since the matrix is symmetric, it can be represented as $AA^T = W\Lambda W^T$, where Λ is a diagonal matrix containing the eigenvalues of the matrix AA^T and W is an orthogonal matrix. Since W is an orthogonal matrix, its inverse exists and it is therefore nonsingular. Using the result in Example 2.18, part (b), we get

$$\begin{aligned} Rank(AA^T) &= Rank(W\Lambda W^T) = Rank(W(\Lambda W^T)) \\ &= Rank(\Lambda W^T) = Rank(\Lambda). \end{aligned} \tag{2.37}$$

But by part (b) above, $Rank(AA^T) = r$, so that $Rank(\Lambda) = r$. But Λ is a diagonal matrix, and therefore there must be r nonzero values along its diagonal. Indeed, these r eigenvalues must be positive because the matrix is positive semidefinite. Since the rank of Λ is r, the remaining $(m - r)$ eigenvalues must be zero.

Singular Value Decomposition of an m by n Matrix Let A be an m by n matrix of rank r. Then A can be represented as

$$A = W\Lambda V^T = \lambda_1 \mathbf{w}_1 \mathbf{v}_1^T + \lambda_2 \mathbf{w}_2 \mathbf{v}_2^T + \cdots + \lambda_r \mathbf{w}_r \mathbf{v}_r^T \tag{2.38}$$

where $\lambda_1 \geq \lambda_2 \geq \cdots \geq \lambda_r > 0$, $\{\mathbf{w}_1, \mathbf{w}_2, \mathbf{w}_3, \ldots, \mathbf{w}_r\}$ is an orthonormal set of vectors in R_m, and $\{\mathbf{v}_1, \mathbf{v}_2, \mathbf{v}_3, \ldots, \mathbf{v}_r\}$ is an orthonormal set of vectors in R_n. The λ_i are called the singular values of A.

The m *by* m matrix AA^T is symmetric, positive semidefinite and has rank r. Hence it has m orthonormal eigenvectors (see equation (2.29)). Denote the r positive eigenvalues of AA^T by λ_i^2, $i = 1, 2, \ldots, r$. Thus there are r orthonormal eigenvectors in R_m corresponding to these r positive eigenvalues such that

$$AA^T \mathbf{w}_i = \lambda_i^2 \mathbf{w}_i, \; i = 1, 2, \ldots, r. \tag{2.39}$$

Furthermore, the remaining $m - r$ eigenvalues of AA^T which are zero are related to the remaining orthonormal eigenvectors of AA^T by the relation

$$AA^T \mathbf{w}_i = \mathbf{0}, \; i = r + 1, \; r + 2, \ldots, m. \tag{2.40}$$

Equation (2.40) implies that $\mathbf{w}_i^T AA^T = \mathbf{0}$, $i = r + 1$, $r + 2, \ldots, m$. Hence, by Example 2.19, part (a), the vectors $\mathbf{w}_i$, $r + 1, r + 2, \ldots, m$ are orthogonal to the column space of AA^T. But this implies (see Example 2.19, part (c)) that they must also be orthogonal to the column space of A. Hence

$$\mathbf{w}_i^T A = \mathbf{0}, \; i = r + 1, \; r + 2, \ldots, m. \tag{2.41}$$

Using (2.31), we can write A as

$$A = IA = \left[\mathbf{w}_1 \mathbf{w}_1^T + \mathbf{w}_2 \mathbf{w}_2^T + \cdots + \mathbf{w}_m \mathbf{w}_m^T\right] A, \tag{2.42}$$

which by equation (2.41) becomes

$$A = \left[\mathbf{w}_1 \mathbf{w}_1^T + \mathbf{w}_2 \mathbf{w}_2^T + \cdots + \mathbf{w}_r \mathbf{w}_r^T\right] A. \tag{2.43}$$

Now consider the vectors $\mathbf{v}_i$ belonging to R_n, defined by

$$v_i = \frac{1}{\lambda_i} A^T \mathbf{w}_i, \; i = 1, 2, \ldots, r \tag{2.44}$$

so that

$$\lambda_i \mathbf{v}_i^T = \mathbf{w}_i^T A, \; i = 1, 2, \ldots, r \tag{2.45}$$

In equation (2.44) we take, $\lambda_i = +\sqrt{\lambda_i^2}$, $i = 1, 2, \ldots, r$. Multiplying both sides of equation (2.45) by $\mathbf{v}_j$ we get

$$\mathbf{v}_i^T \mathbf{v}_j = \frac{1}{\lambda_i \lambda_j} \mathbf{w}_i^T A A^T \mathbf{w}_j = \delta_{ij} \tag{2.46}$$

where $\delta_{ij} = 0$ for $i \neq j$, and $\delta_{ij} = 1$ for $i = j$. The last equality above follows from equation (2.39) and because the vectors $\mathbf{w}_i$, $i = 1, 2, \ldots, r$ are orthonormal; for, $\mathbf{w}_i^T A A^T \mathbf{w}_j = \lambda_j^2 \mathbf{w}_i^T \mathbf{w}_j = \lambda_j^2 \delta_{ij}$. Hence, by equation (2.46), the vectors $\mathbf{v}_i$, $i = 1, 2, \ldots, r$ are orthonormal.

Using equation (2.45) in equation (2.43), the expression for A now becomes

$$A = \lambda_1 \mathbf{w}_1 \mathbf{v}_1^T + \lambda_2 \mathbf{w}_2 \mathbf{v}_2^T + \ldots + \lambda_r \mathbf{w}_r \mathbf{v}_r^T. \tag{2.47}$$

If we denote the m by r matrix $W = [\mathbf{w}_1 \ \ \mathbf{w}_2 \ldots \mathbf{w}_r]$, and the n by r matrix $V = [\mathbf{v}_1 \ \ \mathbf{v}_2 \ldots \mathbf{v}_r]$, then equation (2.47) can be written for short as

$$A = W \Lambda V^T, \tag{2.48}$$

where Λ is a diagonal r by r matrix along whose diagonal are the positive numbers $\lambda_1, \lambda_2, \ldots, \lambda_r$. The squares of these numbers are the positive eigenvalues of AA^T. Because of the orthonormality conditions we note that

$$W^T W = I_r \text{ and } V^T V = I_r. \tag{2.49}$$

We can also represent A alternatively as (we assume that $r < n$ and $r < m$),

$$A = \begin{bmatrix} W & \tilde{W} \end{bmatrix} \begin{bmatrix} \Lambda & 0 \\ 0 & 0 \end{bmatrix} \begin{bmatrix} V^T \\ \tilde{V}^T \end{bmatrix} = P \begin{bmatrix} \Lambda & 0 \\ 0 & 0 \end{bmatrix} Q^T \tag{2.50}$$

Here P is an m by m orthogonal matrix, and Q is an n by n orthogonal matrix. The first r columns of P are made up of the matrix W defined above; the remaining $(m - r)$ columns, denoted $\tilde{W}$, are found so as to make P an orthogonal matrix. Similarly, the first r columns of Q are made up of the matrix V defined above; the remaining $(n - r)$ columns, denoted $\tilde{V}$, are chosen to make Q an orthogonal n by n matrix.

The reader would do well to be conversant with the three alternative forms of the decomposition of the m by n matrix A given by equations (2.47), (2.48) and (2.50).This decomposition of A is called the singular value decomposition (SVD), and the positive numbers $\lambda_1, \lambda_2, \ldots, \lambda_r$ are called the singular values of A. It would be of interest for the reader to go through the steps outlined in Problem 2.1 at the end of this chapter where an alternative proof of equation (2.50) is provided. This proof, in addition, yields some insights into the properties of the matrices $\tilde{W}$ and $\tilde{V}$ – insights which we will have occasion to use later on in this chapter and beyond.

Generalized Inverse of a Matrix

The methods developed in this book critically depend on our understanding of the solution of the algebraic set of equations

$$A\mathbf{x} = \mathbf{b} \tag{2.51}$$

where A is an m by n matrix, $\mathbf{x}$ is an n by 1 vector and $\mathbf{b}$ is an m by 1 vector. If $m = n$ and the matrix A is a nonsingular matrix, then the unique solution to this set of equations is obtained as $\mathbf{x} = A^{-1}\mathbf{b}$. However, we shall find that in the study of constrained motion, the matrix A will usually be non-square. We therefore need to generalize our notion of the inverse of a matrix, to develop a "generalized" inverse which will handle non-square matrices as well as square matrices that are singular. In what follows we will be dealing with, in general, a non-square matrix A. We will learn that there are different types of generalized inverses of matrices. Specifically, we shall define three types of generalized inverses: the G-inverse, the L-inverse and the MP-inverse of a non-square matrix. The G-inverse arises quite naturally when we want to solve a consistent set of linear equations of the form $A\mathbf{x} = \mathbf{b}$, where A is no longer a square matrix; the L-inverse arises when we want to solve the least squares problem of finding $\mathbf{x}$ so that $\|A\mathbf{x} - \mathbf{b}\|^2$ is minimized, again for a non-square matrix A. For our purposes, the MP-inverse will perhaps be the most useful, since it is both a G-inverse and an L-inverse, and a good bit of our discussion will center on it. Later on towards the end of this section, after the reader is familiarized with the general notions of these inverses, we deal with some of the less common generalized inverses. Knowledge of some of these inverses, we will see, will be required to deepen our understanding of analytical dynamics.

The reader who has skipped the previous section dealing with the preliminaries of vector and matrix algebra is advised to, at first reading, skim the last two subsections (dealing with the singular value decomposition of a non-square matrix) and be aware of the main results. The proofs provided there can be digested slowly as (s)he gets more familiar with the various concepts involved.

Moore–Penrose Generalized Inverse

Consider the m by n matrix A which has rank r. We will denote the n by m matrix A^+, called the Moore-Penrose (MP) inverse of the matrix A, if the matrix A^+ satisfies all the following four conditions:

1. $AA^+A = A.$ (2.52)

2. $A^+AA^+ = A^+$ (2.53)

3. $AA^+ = (AA^+)^T$, i.e., the matrix AA^+ is symmetric. (2.54)

4. $A^+A = (A^+A)^T$, i.e., the matrix A^+A is symmetric. (2.55)

We shall call these conditions, taken in the order specified above, the MP conditions.

G-Inverse of a Matrix Consider the m by n matrix A which has rank r. Any n by m matrix A^G for which the first MP condition (i.e., equation (2.52)) is satisfied is called a G-inverse of the matrix A. Thus the G-inverse of A, denoted A^G, satisfies the condition

$$AA^GA = A. \tag{2.52a}$$

L-Inverse of a Matrix Consider the m by n matrix A which has rank r. Any n by m matrix A^L for which the first and third MP conditions are satisfied is called an L-inverse of the matrix A. Thus the L-inverse of A, denoted A^L, satisfies the two conditions

$$AA^LA = A, \tag{2.52b}$$

and

$$AA^L = (AA^L)^T. \tag{2.54b}$$

We observe that the L-inverse must satisfy one additional condition which the G-inverse does not have to. Every L-inverse of the matrix A is therefore automatically a G-inverse of A.

Note that there can be more than one G-inverse and more than one L-inverse of a given matrix A. Hence for any given matrix A, we have then a set of matrices which all satisfy equation (2.52a), and therefore any one of these matrices is a G-inverse of A; similarly we have a set of matrices which are L-inverses of A. However, we shall show that the MP-inverse of any given matrix A is unique. Because of the way we have defined these inverses, any matrix which is an MP-inverse of A is also an L-inverse of A, and any matrix which is an L-inverse of A is also a G-inverse of A.

Example 2.20

a. If A is a square nonsingular matrix, then the usual inverse of A, the matrix A^{-1}, satisfies all four of the MP conditions above, and hence A^{-1} can be thought of as an MP-inverse of A when A is square and nonsingular.

b. The MP-inverse of the m by n matrix $A = \mathbf{0}$ is the n by m matrix $A^+ = \mathbf{0}$.

This follows because the matrix A^+ satisfies the four MP conditions.

EXISTENCE OF THE MP-INVERSE For every m by n matrix A, there exists an n by m matrix A^+.

Let the rank of the matrix A be r. If the singular value decomposition of A is given by (see equation (2.47)),

$$A = W\Lambda V^T = \lambda_1 \mathbf{w}_1 \mathbf{v}_1^T + \lambda_2 \mathbf{w}_2 \mathbf{v}_2^T + \ldots + \lambda_r \mathbf{w}_r \mathbf{v}_r^T \tag{2.56}$$

then we define

$$A^+ = V\Lambda^{-1}W^T = \frac{1}{\lambda_1}\mathbf{v}_1\mathbf{w}_1^T + \frac{1}{\lambda_2}\mathbf{v}_2\mathbf{w}_2^T + \ldots + \frac{1}{\lambda_r}\mathbf{v}_r\mathbf{w}_r^T \tag{2.57}$$

That A^+, so defined, is the MP-inverse, can be verified by checking that A^+ satisfies the four MP conditions (2.52)–(2.55). For example, we observe that (see equation (2.49))

$$AA^+ = W\Lambda V^T V\Lambda^{-1}W^T = W\Lambda I_r \Lambda^{-1} W^T = WI_r W^T = WW^T \tag{2.58}$$

and hence

$$AA^+A = WW^TA = WW^TW\Lambda V^T = WI_r\Lambda V^T = W\Lambda V^T = A. \tag{2.59}$$

Thus A^+ as defined in equation (2.57) satisfies the first MP condition (2.52). That AA^+ is symmetric follows from the right hand side of equation (2.58). Hence the third MP condition given by equation (2.54) is satisfied. The other two MP conditions can similarly be shown to be satisfied.

UNIQUENESS OF THE MP-INVERSE Assume that these are two different MP-inverses, say A_1^+ and A_2^+, of an m by n matrix A. We will show that if this is true then $A_1^+ = A_2^+$. We first show that $AA_1^+ = AA_2^+$ and $A_2^+A = A_1^+A$.

Since A_1^+ is an MP-inverse, $A = AA_1^+A$. Hence, multiplying on the right by A_2^+ we get

$$AA_2^+ = AA_1^+AA_2^+. \tag{2.60}$$

But by the MP conditions, AA_2^+ is symmetric. Hence the right hand side of equation (2.60) must also be symmetric, and so $AA_1^+AA_2^+ = (AA_1^+AA_2^+)^T$. Thus we get

$$\begin{aligned} AA_2^+ = AA_1^+A\,A_2^+ &= \left((AA_1^+)(A\,A_2^+)\right)^T \\ &= (A\,A_2^+)^T(AA_1^+)^T \\ &= (A\,A_2^+)(AA_1^+) = (A\,A_2^+A)A_1^+ = AA_1^+. \end{aligned} \tag{2.61}$$

Note that $(AA_2^+)^T = (AA_2^+)$ because A_2^+ is an MP-inverse and the third MP-condition of equation (2.54) must therefore be satisfied.

Again since $A = AA_1^+A$, multiplying on the left by A_2^+ we get

$$A_2^+A = A_2^+AA_1^+A. \tag{2.62}$$

But by the MP conditions, the left hand side of (2.62) is symmetric and therefore the right hand side of (2.62) must also be symmetric. Hence, $A_2^+AA_1^+A = (A_2^+AA_1^+A)^T$. Then

$$\begin{aligned} A_2^+A = A_2^+AA_1^+A = \left((A_2^+A)(A_1^+A\,)\right)^T &= (A_1^+A)^T(A_2^+A\,)^T \\ &= (A_1^+A)(A_2^+A) = A_1^+A. \end{aligned} \tag{2.63}$$

Noting the results in equations (2.61) and (2.63) we get

$$A_2^+ = A_2^+(AA_2^+) = A_2^+(AA_1^+) = (A_2^+A)A_1^+ = (A_1^+A)A_1^+ = A_1^+AA_1^+ = A_1^+ \tag{2.64}$$

It should be noted that though the MP-inverse of any matrix A is unique, the G-inverse is not. For a given matrix A, there are, in general, several matrices A^G which will satisfy the first MP condition given by equation (2.52). We shall see this in greater detail a little later.

Some Properties of the MP-Inverse Let A be an m by n matrix. We present here some properties of the MP-inverse of A which will be useful to us in future when we deal with describing the motion of constrained mechanical systems. All these properties can be proved using the singular value decomposition of A, and the definition of A^+ given in equation (2.57).

a. $\left(A^T\right)^+ = \left(A^+\right)^T$ (2.65)

Using the singular value decomposition of A we have $A = W\Lambda V^T$, so that

$$\left(A^T\right)^+ = \left(V\Lambda W^T\right)^+ = W\Lambda^{-1}V^T = \left(V\Lambda^{-1}W^T\right)^T = \left(A^+\right)^T. \qquad (2.66)$$

b. $\left(A^+\right)^+ = A$ (2.67)

Again using the singular value decomposition we get $(A^+)^+ = (V\Lambda^{-1}W^T)^+ = W\Lambda V^T = A$.

c. $Rank(A^+) = Rank(A)$ (2.68)

Using the result of Example 2.18, part (a), $Rank(A) = Rank(AA^+A) \leq Rank(A^+)$. Also, $Rank(A^+) = Rank(A^+AA^+) \leq Rank(A^+A) \leq Rank(A)$. Hence the result.

d. $Rank(A) = Rank(AA^+) = Rank(A^+A) = Rank(A^+AA^+) = Rank(AA^+A)$

(2.69)

We prove here the first relation. The others follow in a similar manner:

$Rank(AA^+) \leq Rank(A)$ and $Rank(A) = Rank(A^+AA^+) \leq Rank(AA^+)$. Therefore $Rank(A) = Rank(AA^+)$.

e. When dealing with MP-inverses, it is *not* always true that $(AB)^+ = B^+A^+$. But for any matrix A, $(A^TA)^+ = A^+(A^T)^+$, $(AA^+)^+ = AA^+$ and $(A^+A)^+ = A^+A$. These relations can be easily proved using the singular value decomposition of A.

f. Given two non-square matrices A and B, then

if $AB = \mathbf{0}$, then $B^+A^+ = \mathbf{0}$. (2.70)

We observe that $AB = \mathbf{0}$ implies that $A^+ABB^+ = \mathbf{0}$. Transposing both sides of this equation, we get

$$\mathbf{0} = ((A^+A)(BB^+))^T = (BB^+)^T(A^+A)^T = (BB^+)(A^+A).$$

Note that we have used equations (2.54) and (2.55) – the last two MP conditions – above. Premultiplying the above equation by B^+ and postmultiplying it by A^+, we then obtain

$$\mathbf{0} = B^+BB^+A^+AA^+ = B^+A^+,$$

where we have made use of the second MP condition given by equation (2.53). We shall use this observation later on in Chapter 3.

g. The matrices A^+A, $(I - A^+A)$, AA^+, $(I - AA^+)$ are each idempotent. (2.71)

Recall that a matrix X is idempotent if $X^2 = X$. These properties follow directly from equations (2.52)–(2.55). We prove this result for the matrices A^+A and $(I - A^+A)$. By equation (2.55) we see that A^+A is a symmetric matrix. Also, $(A^+A)(A^+A) = (A^+AA^+)(A) = A^+A$, proving that A^+A is idempotent.

Again using equation (2.55), we have $(I - A^+A)^T = I^T - (A^+A)^T = I - A^+A$, so that the matrix $(I - A^+A)$ is symmetric. Using equation (2.53), we prove idempotence because we have $(I - A^+A)(I - A^+A) = I - 2(A^+A) + A^+AA^+A = I - 2A^+A + A^+A = I - A^+A$. Proofs for the matrices AA^+ and $(I - AA^+)$ are similar, using equations (2.52) and (2.54), and are left as exercises for the reader.

h. $A^+ = A^T$, if and only if A^TA is idempotent.

First let us assume that A^TA is idempotent. We know from Example 2.14 that A^TA is a symmetric positive semidefinite matrix. Let the rank of A be r. Using the singular value decomposition of A, we get

$$A^TA = V\Lambda W^TW\Lambda V^T = V\Lambda I_r\Lambda V^T = V\Lambda^2V^T, \tag{2.72}$$

where we have used the relation $W^TW = I_r$ (equation (2.49)). Similarly,

$$A^TAA^TA = V\Lambda^2V^TV\Lambda^2V^T = V\Lambda^4V^T. \tag{2.73}$$

Since A^TA is assumed idempotent, the left hand sides of equations (2.72) and (2.73) are equal; hence their right hand sides must also be equal. But this requires that $\lambda_i^4 = \lambda_i^2$, $i = 1, 2, \ldots, r$. The roots of this equation yield $\lambda_i = 0, \pm 1$, $i = 1, 2, \ldots, r$. But the singular values of A must all be positive. Hence all the singular values of A must be unity. This implies that $\Lambda = \Lambda^{-1} = I_r$, so that $A^+ = V\Lambda^{-1}W^T = V\Lambda W^T = A^T$. Therefore, if A^TA is idempotent, then $A^+ = A^T$.

Next assume that $A^+ = A^T$. Multiplying both sides of this equation on the right by A, we get $A^+A = A^TA$. But by part (g) above, we know that A^+A is idempotent. Hence A^TA must also be idempotent.

i. If A is symmetric and idempotent, then $A^+ = A$. The MP-inverse of A is then A itself.

If A is idempotent then $AA = A$. Thus $(A^TA)(A^TA) = A^TA(AA) = A^TA(A) = A^TA$. Hence A^TA is idempotent, and by part (h) above, $A^+ = A^T = A$, the last equality following because A is symmetric.

Example 2.21

a. The MP-inverse of each of the following matrices is the matrix itself: $A^+A, I - A^+A, AA^+$, and $I - AA^+$.

Since each of these matrices is symmetric and idempotent (see part (g) above), by part (i), we get

$$(A^+A)^+ = A^+A, \tag{2.74}$$

$$(I - A^+A)^+ = I - A^+A, \tag{2.75}$$

$$(AA^+)^+ = AA^+, \tag{2.76}$$

$$(I - AA^+)^+ = I - AA^+. \tag{2.77}$$

b. The quantity

$$I - (I - A^+A)^+(I - A^+A) = A^+A. \tag{2.78}$$

This follows because

$$\begin{aligned} I - (I - A^+A)^+(I - A^+A) &= I - (I - A^+A)(I - A^+A) \\ &= I - (I - A^+A) = A^+A. \end{aligned} \tag{2.79}$$

We have used equation (2.75) and relation (2.71) in the above simplifications.

c. The following identity is true:

$$(I - A^+A)A^+A = \mathbf{0}. \tag{2.80}$$

Again using equation (2.53), $(I - A^+A)A^+A = A^+A - A^+AA^+A = A^+A - A^+A = \mathbf{0}$.

DETERMINATION OF THE MP-INVERSE We provide here some MP-inverses for commonly occurring matrices. Except for result (f) below, all the following results can be verified directly by the reader.

a. $(cA)^+ = \frac{1}{c}A^+$ where c is a nonzero scalar. (2.81)

b. If $\mathbf{a}$ is a nonzero 1 by n row vector, and $\mathbf{b}$ is a nonzero n-by 1 column vector then,

$$\mathbf{a}^+ = \frac{1}{(\mathbf{a}\mathbf{a}^T)}\mathbf{a}^T, \tag{2.82}$$

and

$$\mathbf{b}^+ = \frac{1}{(\mathbf{b}^T\mathbf{b})}\mathbf{b}^T. \tag{2.83}$$

c. If $\mathbf{a}$ is a nonzero 1 by n vector, then

$$(\mathbf{a}^T\mathbf{a})^+ = \frac{1}{(\mathbf{a}\mathbf{a}^T)^2}\mathbf{a}^T\mathbf{a}. \tag{2.84}$$

d. If

$$A = \begin{bmatrix} B & 0 \\ 0 & C \end{bmatrix}, \text{ then } A^+ = \begin{bmatrix} B^+ & 0 \\ 0 & C^+ \end{bmatrix}. \tag{2.85}$$

e. If A has rows $i_1, i_2, \ldots, i_q$ which are proportional, then these same columns in A^+ will also be proportional. In particular if q rows of A are zero, then the corresponding q columns of A^+ are also zero.

f. Recursive determination of A^+

We provide here results by Greville (without proof) for finding the MP-inverse of large matrices using a recursive scheme. We will see later that this recursive scheme gives considerable insight into the effect each individual constraint has in affecting the motion of a dynamical system.

Let A be an i by n matrix, and let A_{i-1} be the $(i-1)$ by n matrix that consists of the first $(i-1)$ rows of A. Let the ith row of A be $\mathbf{a}_i$ so that $A = \begin{bmatrix} A_{i-1} \\ \mathbf{a}_i \end{bmatrix}$. Then the MP-inverse of A is given by

$$A^+ = \left[(A_{i-1}^+ - \mathbf{b}_i^+ \mathbf{a}_i A_{i-1}^+) \qquad \mathbf{b}_i^+\right], \tag{2.86}$$

where the n by 1 vector $\mathbf{b}_i^+$ is the MP-inverse of the vector $\mathbf{b}_i$ which is defined as

$$\mathbf{b}_i = \begin{cases} \mathbf{a}_i(I - A_{i-1}^+ A_{i-1}) & \text{if } \mathbf{a}_i \neq \mathbf{a}_i A_{i-1}^+ A_{i-1} \\ \mathbf{a}_i (A_{i-1}^T A_{i-1})^+ \dfrac{[1 + \mathbf{a}_i (A_{i-1}^T A_{i-1})^+ \mathbf{a}_i^T]}{[\mathbf{a}_i (A_{i-1}^T A_{i-1})^+ (A_{i-1}^T A_{i-1})^+ \mathbf{a}_i^T]} & \text{if } \mathbf{a}_i = \mathbf{a}_i A_{i-1}^+ A_{i-1}. \end{cases} \tag{2.87}$$

The proof of this result may be found in reference G2.

g. For any m by n matrix A, whatever its rank, the following relations hold:

$$A^+ = A^T (AA^T)^+ \tag{2.88}$$

$$A^+ = (A^T A)^+ A^T \tag{2.89}$$

h. If the m by n matrix A has rank n then

$$A^+ = (A^T A)^{-1} A^T \text{ and } A^+ A = I \tag{2.90}$$

By Example 2.19 (e), the n by n matrix $A^T A$ has rank n, and so $(A^T A)^+ = (A^T A)^{-1}$, the ordinary inverse. Using equation (2.89), the result follows. Also, $A^+ A = (A^T A)^{-1} A^T A = I$.

Example 2.22

Let A be an m by n matrix and $\mathbf{u}$ be a nonzero n-vector.

$$\text{If } A\mathbf{u} = \mathbf{0}, \text{ then } \mathbf{u}^T A^+ = \mathbf{0} \tag{2.91}$$

By relation (2.70), we see that $A\mathbf{u} = \mathbf{0}$ implies $\mathbf{u}^+ A^+ = \mathbf{0}$. Using equation (2.83) we get $\mathbf{u}^+ = (\mathbf{u}^T \mathbf{u})^{-1} \mathbf{u}^T$, so that $\mathbf{u}^T A^+ = \mathbf{0}$ because $(\mathbf{u}^T \mathbf{u}) > \mathbf{0}$. We shall have occasion to use this result in the next chapter.

Example 2.23

a. Consider the 1 by n matrix A given by

$$A = [a_1 \; a_2 \; \ldots \; a_n]. \tag{2.92}$$

By equation (2.82) the MP-inverse of A is

$$A^+ = \frac{1}{\sum_{i=1}^{n} a_i^2} \begin{bmatrix} a_1 \\ a_2 \\ \cdot \\ \cdot \\ a_n \end{bmatrix}. \tag{2.93}$$

b. Let A be an m by n matrix. Let the matrix $B = kI_m$ and the matrix $C = kI_n$ where k is a nonzero scalar. Then

$$(BA)^+ = \frac{1}{k} A^+ \tag{2.94}$$

and

$$(AC)^+ = \frac{1}{k} A^+. \tag{2.95}$$

This follows using equation (2.81) because $(BA)^+ = (kI_mA)^+ = (kA)^+ = \frac{1}{k}A^+$. Also, $(AC)^+ = (kAI_n)^+ = (kA)^+ = \frac{1}{k}A^+$.

Example 2.24

Consider next the 2 by n matrix composed of the two rows $\mathbf{a}_1 \neq 0$, and $\mathbf{a}_2$, so that

$$A = \begin{bmatrix} \mathbf{a}_1 \\ \mathbf{a}_2 \end{bmatrix}. \tag{2.96}$$

Then the MP-inverse of A is given by

$$A^+ = [(\mathbf{a}_1^+ - \beta \mathbf{b}_2^+) \;\; \mathbf{b}_2^+], \tag{2.97}$$

where

$$\mathbf{a}_1^+ = \frac{1}{(\mathbf{a}_1\mathbf{a}_1^T)} \mathbf{a}_1^T, \; \beta = \frac{(\mathbf{a}_1\mathbf{a}_2^T)}{(\mathbf{a}_1\mathbf{a}_1^T)}, \tag{2.98}$$

and

Example 2.24 (continued)

$$\mathbf{b}_2^+ = \begin{cases} (\mathbf{a}_2 - \beta\mathbf{a}_1)^+ = \dfrac{1}{(\mathbf{a}_2\mathbf{a}_2^T - \beta^2\mathbf{a}_1\mathbf{a}_1^T)}(\mathbf{a}_2^T - \beta\mathbf{a}_1^T) & \text{if } \mathbf{a}_2 \neq \beta\mathbf{a}_1 \\ \dfrac{\beta}{1+\beta^2}\mathbf{a}_1^+ & \text{if } \mathbf{a}_2 = \beta\mathbf{a}_1. \end{cases} \tag{2.99}$$

When $\mathbf{a}_2 = \beta\mathbf{a}_1$, A^+ simplifies to

$$A^+ = \begin{bmatrix} \dfrac{1}{1+\beta^2}\mathbf{a}_1^+ & \dfrac{\beta}{1+\beta^2}\mathbf{a}_1^+ \end{bmatrix}. \tag{2.100}$$

To prove this, we use equations (2.86) and (2.87) with $i = 2$ and $A_1 = \mathbf{a}_1$. We use the following results:

1. $\mathbf{a}_2 A_1^+ A_1 = \dfrac{(\mathbf{a}_2\mathbf{a}_1^T)}{(\mathbf{a}_1\mathbf{a}_1^T)}\mathbf{a}_1 = \beta\mathbf{a}_1,$

2. $(A_1^T A_1)^+ = (\mathbf{a}_1^T\mathbf{a}_1)^+ = \dfrac{(\mathbf{a}_1^T\mathbf{a}_1)}{(\mathbf{a}_1\mathbf{a}_1^T)^2},$

3. $1 + \mathbf{a}_2(A_1^T A_1)^+\mathbf{a}_2^T = 1 + \dfrac{(\mathbf{a}_2\mathbf{a}_1^T)(\mathbf{a}_1\mathbf{a}_2^T)}{(\mathbf{a}_1\mathbf{a}_1^T)^2} = 1 + \beta^2,$

4. $\mathbf{a}_2(A_1^T A_1)^+ = \dfrac{(\mathbf{a}_2\mathbf{a}_1^T)\mathbf{a}_1}{(\mathbf{a}_1\mathbf{a}_1^T)^2} = \dfrac{\beta\mathbf{a}_1}{\mathbf{a}_1\mathbf{a}_1^T},$

5. $$\mathbf{a}_2(A_1^T A_1)^+(A_1^T A_1)^+\mathbf{a}_2^T = \frac{(\mathbf{a}_2\mathbf{a}_1^T)(\mathbf{a}_1\mathbf{a}_1^T)(\mathbf{a}_1\mathbf{a}_2^T)}{(\mathbf{a}_1\mathbf{a}_1^T)^4} = \frac{\beta^2(\mathbf{a}_1\mathbf{a}_1^T)^3}{(\mathbf{a}_1\mathbf{a}_1^T)^4} = \frac{\beta^2}{(\mathbf{a}_1\mathbf{a}_1^T)}.$$

Substituting these relations in (2.86) and (2.87) we obtain the result.

Consistency of the System of Equations $A\mathbf{x} = \mathbf{b}$ The system of equations given by equation (2.51) is consistent if and only if

$$AA^G\mathbf{b} = \mathbf{b}\,, \tag{2.101}$$

where A^G is *any* G-inverse of A (see equation (2.52a) for the definition of G-inverse).

1. Assume that the system is consistent and let the vector $\mathbf{x}$ satisfy the system of equations. Then $A\mathbf{x} = \mathbf{b}$. Multiplying this last equation on the left by AA^G, we then get

$$AA^GA\mathbf{x} = AA^G\mathbf{b} \tag{2.102}$$

Using equation (2.52a), the left hand side of equation (2.102) is simply $AA^GA\mathbf{x} = A\mathbf{x} = \mathbf{b}$. By setting the left hand side of (2.102) to the right hand side, we get

$$\mathbf{b} = AA^G\mathbf{b} \tag{2.103}$$

if $\mathbf{x}$ is a solution to the consistent equation $A\mathbf{x} = \mathbf{b}$.

2. Now assume that $\mathbf{b} = AA^G\mathbf{b}$. Let $\mathbf{x} = A^G\mathbf{b}$. Substituting for this value of $\mathbf{x}$ in the equation $A\mathbf{x} = \mathbf{b}$, we get $A(A^G\mathbf{b}) = \mathbf{b}$. Hence $\mathbf{x} = A^G\mathbf{b}$ is a solution of the equation $A\mathbf{x} = \mathbf{b}$, and the result follows.

Example 2.25

The equation $A\mathbf{x} = \mathbf{b}$ is consistent if and only if

$$AA^+\mathbf{b} = \mathbf{b}. \tag{2.104}$$

Note that the MP-inverse is a G-inverse, and hence the result in equation (2.101) is true for $A^G = A^+$.

General Solution of the Consistent Equation $A\mathbf{x} = \mathbf{b}$ Let A be an m by n matrix. The general solution $\mathbf{x}$ of the equation $A\mathbf{x} = \mathbf{b}$, assuming it exists, is

$$\mathbf{x} = A^G\mathbf{b} + (I - A^GA)\mathbf{h} \tag{2.105}$$

where $\mathbf{h}$ is any n by 1 vector. A^G is *any* G-inverse of A. Furthermore, every solution of $A\mathbf{x} = \mathbf{b}$ can be expressed in the form of equation (2.105) for some n by 1 vector $\mathbf{h}$.

Since the equation is consistent, we have, by equation (2.101), $AA^G\mathbf{b} = \mathbf{b}$. To prove that $\mathbf{x}$ is a solution of the equation, multiply (2.105) by A. This gives

$$A\mathbf{x} = AA^G\mathbf{b} + A(I - A^GA)\mathbf{h} = \mathbf{b} + (A - A)\mathbf{h} = \mathbf{b} \tag{2.106}$$

where we have made use of the fact that $AA^GA = A$.

Next suppose that we have a solution $\mathbf{x}$ to the equation $A\mathbf{x} = \mathbf{b}$. We must show that it can always be put in the form shown in equation (2.105). Since $\mathbf{x}$ is a solution, $A\mathbf{x} = \mathbf{b}$, and hence $A^GA\mathbf{x} = A^G\mathbf{b}$, so that $\mathbf{0} = A^G\mathbf{b} - A^GA\mathbf{x}$. If we add $\mathbf{x}$ to both sides we get $\mathbf{x} = A^G\mathbf{b} + (I - A^GA)\mathbf{x}$, which is of the form of equation (2.105) with $\mathbf{h} = \mathbf{x}$. Hence the result.

Example 2.26

The general solution to the consistent equation $A\mathbf{x} = \mathbf{b}$, where A is an m by n matrix, is

$$\mathbf{x} = A^+\mathbf{b} + (I - A^+A)\mathbf{h} \tag{2.107}$$

where $\mathbf{h}$ is an arbitrary n by 1 vector.

As before, the MP-inverse is a G-inverse and hence the result follows.

Example 2.27

If the m by n matrix A has rank n then the solution to the consistent equation $A\mathbf{x} = \mathbf{b}$, is unique and is given by

$$\mathbf{x} = A^+\mathbf{b} = (A^TA)^{-1}A^T\mathbf{b}. \tag{2.108}$$

We note from relation (2.90) that when the rank of A is n, $A^+ = (A^TA)^{-1}A^T$ and $A^+A = I$. Because $A^+A = I$, we see that the second term on the right hand side of equation (2.107) vanishes, and the result follows. Uniqueness of the solution follows from the observation that A^+ is unique.

Example 2.28

Consider equations (1.51) and (1.52) in Chapter 1, which represent the constraint that the pendulum bob is at a fixed distance from its point of suspension at all times. Thus the general solution $\ddot{\mathbf{x}}$ to the equation

$$A\ddot{\mathbf{x}} = [x(t) \quad y(t)] \begin{bmatrix} \ddot{x}(t) \\ \ddot{y}(t) \end{bmatrix} = [-\dot{x}^2 - \dot{y}^2] = \mathbf{b} \tag{2.109}$$

is given by the equation (2.107) as

$$\ddot{\mathbf{x}} = -A^+[\dot{x}^2 + \dot{y}^2] + (I - A^+A)\mathbf{h} \tag{2.110}$$

where $\mathbf{h}$ is an arbitrary vector, and A^+ is given by equation (2.83) to be

$$A^+ = \frac{1}{[x^2(t) + y^2(t)]} \begin{bmatrix} x(t) \\ y(t) \end{bmatrix}. \tag{2.111}$$

Hence the two components of acceleration $\ddot{x}(t)$ and $\ddot{y}(t)$ are,

$$\begin{bmatrix} \ddot{x}(t) \\ \ddot{y}(t) \end{bmatrix} = \frac{-(\dot{x}^2 + \dot{y}^2)}{(x^2 + y^2)} \begin{bmatrix} x(t) \\ y(t) \end{bmatrix} + \begin{bmatrix} \frac{y^2}{(x^2 + y^2)} & -\frac{xy}{(x^2 + y^2)} \\ -\frac{xy}{(x^2 + y^2)} & \frac{x^2}{(x^2 + y^2)} \end{bmatrix} \begin{bmatrix} h_1(t) \\ h_2(t) \end{bmatrix} \tag{2.112}$$

Note that the constraint equation (1.51), when solved, leaves the quantities $h_1(t)$ and $h_2(t)$ undetermined. We will show how to determine these quantities in the following chapters. These quantities must be found so that the resulting motion of the mechanical system, in the presence of the given forces, is consistent with the agreed upon principles of mechanics.

Example 2.29

Consider the constraint described in Example 1.10 (of Chapter 1) by equation (1.55). The general solution to this constraint equation (see equation (1.57))

$$[-z \quad 1 \quad 0] \begin{bmatrix} \ddot{x} \\ \ddot{y} \\ \ddot{z} \end{bmatrix} = \dot{z}\dot{x} \tag{2.113}$$

is

$$\begin{bmatrix} \ddot{x} \\ \ddot{y} \\ \ddot{z} \end{bmatrix} = A^+\dot{z}\dot{x} + (I - A^+A)\mathbf{h} \tag{2.114}$$

Example 2.29 (continued)

where,

$$A^{+} = \frac{1}{[1+z^{2}(t)]}\begin{bmatrix}-z\\1\\0\end{bmatrix} \tag{2.115}$$

and $\mathbf{h}$ is an arbitrary 3-vector. Hence,

$$\begin{bmatrix}\ddot{x}\\ \ddot{y}\\ \ddot{z}\end{bmatrix} = \frac{\dot{z}\dot{x}}{[1+z^{2}(t)]}\begin{bmatrix}-z\\1\\0\end{bmatrix} + \begin{bmatrix}\frac{1}{1+z^{2}} & \frac{z}{1+z^{2}} & 0\\ \frac{z}{1+z^{2}} & \frac{z^{2}}{1+z^{2}} & 0\\ 0 & 0 & 1\end{bmatrix}\begin{bmatrix}h_1(t)\\h_2(t)\\h_3(t)\end{bmatrix}. \tag{2.116}$$

Again the quantities $h_1(t)$, $h_2(t)$ and $h_3(t)$ are arbitrary and we need additional information to determine them. We shall see that this information comes from the principles of mechanics.

Least Squares Solution and the L-inverse Given the m by n matrix A, and the m-vector $\mathbf{b}$, we want to find the n-vector $\mathbf{x}$ such that

$$\mathcal{Z}(x) = \|A\mathbf{x}-\mathbf{b}\|^{2} = (A\mathbf{x}-\mathbf{b})^{T}(A\mathbf{x}-\mathbf{b}) \tag{2.117}$$

is a minimum. We assert that the vector $\mathbf{x}$ which minimizes $\mathcal{Z}(x)$ is given by the equation

$$\mathbf{x} = A^{L}\mathbf{b} + (I - A^{L}A)\mathbf{h}, \tag{2.118}$$

where A^{L} is *any* L-inverse of the matrix A, and $\mathbf{h}$ is an arbitrary n-vector.

We observe that

$$A\mathbf{x}-\mathbf{b} = (A\mathbf{x}-AA^{L}\mathbf{b}) + (AA^{L}\mathbf{b}-\mathbf{b}) = A\mathbf{y} + (AA^{L}\mathbf{b}-\mathbf{b}) \tag{2.119}$$

where $\mathbf{y}$ is the n-vector $(\mathbf{x}-A^{L}\mathbf{b})$. Now noting that A^{L} satisfies equations (2.52b) and (2.54b), we get,

$$\begin{aligned}(AA^{L}\mathbf{b}-\mathbf{b})^{T}(A\mathbf{y}) &= \mathbf{b}^{T}(AA^{L})^{T}A\mathbf{y} - \mathbf{b}^{T}A\mathbf{y}\\ &= \mathbf{b}^{T}(AA^{L})A\mathbf{y} - \mathbf{b}^{T}A\mathbf{y}\\ &= \mathbf{b}^{T}AA^{L}A\mathbf{y} - \mathbf{b}^{T}A\mathbf{y} = \mathbf{b}^{T}A\mathbf{y} - \mathbf{b}^{T}A\mathbf{y} = 0.\end{aligned} \tag{2.120}$$

Using equation (2.119) and noting that the vector $A\mathbf{y}$ is orthogonal to $(AA^L\mathbf{b} - \mathbf{b})$, we have,

$$\|A\mathbf{x} - \mathbf{b}\|^2 = \left\|A\mathbf{x} - AA^L\mathbf{b}\right\|^2 + \left\|AA^L\mathbf{b} - \mathbf{b}\right\|^2. \tag{2.121}$$

The minimum value of $\|A\mathbf{x} - \mathbf{b}\|^2$ occurs when $\mathbf{x} = A^L\mathbf{b} + (I - A^LA)\mathbf{h}$, where $\mathbf{h}$ is an arbitrary n-vector. We note that the minimum of $\|A\mathbf{x} - \mathbf{b}\|^2$ for this $\mathbf{x}$ is $\left\|AA^L\mathbf{b} - \mathbf{b}\right\|^2$. We note that this minimum value does not depend on the arbitrary vector $\mathbf{h}$.

Example 2.30

Given the m by n matrix A and the m-vector $\mathbf{b}$, an n-vector $\mathbf{x}$ which minimizes

$$\mathcal{Z}(x) = \|A\mathbf{x} - \mathbf{b}\|^2 = (A\mathbf{x} - \mathbf{b})^T(A\mathbf{x} - \mathbf{b}) \tag{2.122}$$

is given by

$$\mathbf{x} = A^+\mathbf{b} + (I - A^+A)\mathbf{h} \tag{2.123}$$

where A^+ is the MP-inverse of the matrix A and $\mathbf{h}$ is an arbitrary n-vector.

We note that the MP-inverse of A is also an L-inverse of A. So setting $A^L = A^+$ in equation (2.118), the result follows.

Example 2.31

Consider the determination of the vector x which minimizes the quantity

$$\mathcal{Z}_{M^{-1}}(\mathbf{x}) = (A\mathbf{x} - \mathbf{b})^T M^{-1}(A\mathbf{x} - \mathbf{b}) \tag{2.124}$$

where A is an m by n matrix, and M is an m by m symmetric positive definite matrix. Since M is symmetric and positive definite, we can define the matrix $M^{-1/2}$ as in equation (2.32). Hence we have

$$\begin{aligned}\mathcal{Z}_{M^{-1}}(\mathbf{x}) &= (A\mathbf{x} - \mathbf{b})^T M^{-1/2}M^{-1/2}(A\mathbf{x} - \mathbf{b})\\ &= [M^{-1/2}(A\mathbf{x} - \mathbf{b})]^T[M^{-1/2}(A\mathbf{x} - \mathbf{b})]\\ &= [(M^{-1/2}A)\mathbf{x} - (M^{-1/2}\mathbf{b})]^T[(M^{-1/2}A)\mathbf{x} - (M^{-1/2}\mathbf{b})].\end{aligned} \tag{2.125}$$

The vector $\mathbf{x}$ that minimizes $\mathcal{Z}_{M^{-1}}(\mathbf{x})$ is obtained by using equation (2.123) and becomes

Example 2.31 (continued)

$$\mathbf{x} = (M^{-1/2}A)^{+}(M^{-1/2}\mathbf{b}) + [I - (M^{-1/2}A)^{+}(M^{-1/2}A)]\mathbf{h} \qquad (2.126)$$

where $\mathbf{h}$ is an arbitrary vector.

Column Spaces of Matrices Involving the MP-Inverse

We list here a set of results, all of which can be proved by resorting to the singular value decomposition of the m by n matrix A which has rank r.

a. The column spaces of A and AA^{+} are the same.

b. The column spaces of A^{+} and $A^{+}A$ are the same.

c. The column space of $(I - A^{+}A)$ is the orthogonal complement of the column space of A^{T}.

d. The column space of $(I - A^{+}A)$ is the same as the null space of A.

e. The column space of A^{T} is the same as that of A^{+}.

The null space of a matrix A is the space spanned by all those vectors $\mathbf{x}$ which satisfy the equation $A\mathbf{x} = \mathbf{0}$.

We illustrate here the proof of part (c) and part (d). Since $A = W\Lambda V^{T}$ and $A^{+} = V\Lambda^{-1}W^{T}$, we have

$$A^{+}A = \sum_{i=1}^{r} \mathbf{v}_i\mathbf{v}_i^{T}. \qquad (2.127)$$

But the n-vectors $\mathbf{v}_i$, $i = 1, 2, \ldots, n$ can be chosen to form an orthonormal set so that $I = \sum_{i=1}^{n} \mathbf{v}_i\mathbf{v}_i^{T}$. Hence,

$$I - A^{+}A = \sum_{i=r+1}^{n} \mathbf{v}_i\mathbf{v}_i^{T} \qquad (2.128)$$

and the column space of $I - A^{+}A$ is then the space spanned by the vectors $\mathbf{v}_i$, $i = r+1, r+2, \ldots, n$. Now $A^{T} = \sum_{i=1}^{r} \lambda_i\mathbf{v}_i\mathbf{w}_i^{T}$. So, by Example 2.17, part (c), the column space of A^{T} is spanned by the vectors $\mathbf{v}_i$, $i = 1, 2, \ldots, r$. Hence the column space of A^{T} is the orthogonal complement of the column space of $I - A^{+}A$. Part (c) is now proved.

We note that the matrix $I - A^+A$ is symmetric; hence its column space is the same as its row space. Furthermore, the column space of A^T is the row space of A. Thus the row space of A is the orthogonal complement of the column space of $I - A^+A$.

We next prove part (d). Note that any vector $\mathbf{z}$ belonging to the column space of $I - A^+A$ can be written as $\mathbf{z} = (I - A^+A)\mathbf{y}$, and since $A\mathbf{z} = A(I - A^+A)\mathbf{y} = (A - AA^+A)\mathbf{y} = (A - A)\mathbf{y} = \mathbf{0}$, the vector $\mathbf{z}$ belongs to the null space of A. Hence the column space of $I - A^+A$ is a subset of the null space of A. On the other hand, if the vector $\mathbf{z}$ belongs to the null space of A, $\mathbf{z}^TA^T = \mathbf{0}$, and hence $\mathbf{z}$ is orthogonal to the column space of A^T. But by part (c) the column space of A^T is the orthogonal complement of the column space of $I - A^+A$. Thus $\mathbf{z}$ must belong to the space spanned by $I - A^+A$. The null space of A is then a subset of the column (row) space of $I - A^+A$, and part (d) is proved.

Example 2.32

Let B be an m by n matrix. Consider any n–vector $\mathbf{u}$ which belongs to the row space of the matrix B. Then $(I - B^+B)\mathbf{u} = \mathbf{0}$.

Since the vector $\mathbf{u}$ belongs to the row space of B, it can be written as a linear combination of the rows of the matrix B. Thus we can express $\mathbf{u}$ as $\mathbf{u} = B^T\mathbf{v}$, for some m-vector $\mathbf{v}$. Note that the rows of B are the columns of B^T. Now using equation (2.52) we have

$$[(I - B^+B)\mathbf{u}]^T = \mathbf{u}^T(I - B^+B)^T = \mathbf{v}^TB(I - B^+B) = \mathbf{v}^T(B - BB^+B). \tag{2.129}$$

Hence $(I - B^+B)\mathbf{u} = \mathbf{0}$. We will have occasion to use this result in a later chapter.

Example 2.33

Let B be an m by n matrix. Consider the n-vector $\mathbf{z}$ formed by multiplying the n by m matrix B^+ with the m-vector $\mathbf{e}$, so that $\mathbf{z} = B^+\mathbf{e}$. Show that there always exists an m-vector $\boldsymbol{\lambda}$ such that $B^T\boldsymbol{\lambda} = \mathbf{z}$.

We have shown in part (e) of the last sub-section that the column space spanned by the matrix B^+ is identical to that spanned by B^T.

The vector $\mathbf{z}$ belongs to the column space of B^+, because $\mathbf{z} = \sum_{i=1}^{m} \mathbf{b}_i^+ e_i$

where we have denoted the ith columns of the matrix B^+ by $\mathbf{b}_i^+$. Hence this vector $\mathbf{z}$ also belongs to the column space of B^T. Hence there must

Example 2.33 (continued)

exist a vector $\boldsymbol{\lambda}$ so that we can express the vector $\mathbf{z}$ as $\mathbf{z} = B^T\boldsymbol{\lambda} = \sum_{i=1}^{m} \mathbf{b}_i^T \lambda_i$ where we have denoted here the i-th column of B^T by $\mathbf{b}_i^T$. We shall use this result in Chapter 4.

Some Other Generalized Inverses

Though we have so far concentrated primarily on the MP generalized inverse of the m by n matrix A, we have indicated to the reader the existence of other types of generalized inverses. In fact, we have already talked about the G-inverse of an m by n matrix A which is any n by m matrix A^G which satisfies the first of the four MP conditions listed (i.e., equation (2.52) on page 45), namely $AA^GA = A$. We often denote A^G also as $A^{\{1\}}$, signifying that such a matrix satisfies just the first of the four MP conditions. Similarly any n by m matrix A^L which satisfies only the first and third MP conditions (equations (2.52) and (2.54)) is often denoted as $A^{\{1,3\}}$. More generally, we can denote a generalized inverse which satisfies a subset of the four MP conditions, say conditions i, j and k, by writing it as $A^{\{i,j,k\}}$. For example, the matrix $A^{\{1,2,3,4\}}$ satisfies all four of the MP conditions and is the MP inverse of A, which we have denoted as A^+.

The reader will note that the MP inverse of the matrix A was defined in equation (2.57), with no indication how one arrived at such a relation! We now need to go a little deeper to understand how this relation came about. In the process, we shall see much more. We shall be uncovering the underlying structures of a whole host of generalized inverses of the matrix A. In particular, in this book we will need to understand two of these inverses – the $A^{\{1,2,4\}}$ and the $A^{\{1,4\}}$ – in some detail. Before going further, we suggest that the reader look over our discussion of singular value decomposition, and walk through Problem 2.1, whose notation we use below.

On page 44 we showed that any m by n matrix A of rank r (we assume that $r < m$ and $r < n$) can be expressed as (see equation (2.50))

$$A = P\begin{bmatrix} \Lambda & 0 \\ 0 & 0 \end{bmatrix} Q^T, \tag{2.130}$$

where both P and Q are orthogonal matrices, m by m and n by n respectively. The diagonal matrix Λ is of dimension r by r and contains the (positive) singular values of A. Using these matrices P and Q which are obtained in the singular value decomposition of A, we can express any n by m matrix B in the form

$$B = Q\begin{bmatrix} N & K \\ L & M \end{bmatrix}P^T \tag{2.131}$$

for suitable matrices *K, L, M* and *N*. Now for *B* to be $A^{\{1\}}$, we must require it to satisfy the first MP condition, namely $ABA = A$. Using the expressions for A and B from equations (2.130) and (2.131) respectively on both sides of this relation, we get

$$P\begin{bmatrix} \Lambda & 0 \\ 0 & 0 \end{bmatrix}Q^T Q\begin{bmatrix} N & K \\ L & M \end{bmatrix}P^T P\begin{bmatrix} \Lambda & 0 \\ 0 & 0 \end{bmatrix}Q^T = P\begin{bmatrix} \Lambda & 0 \\ 0 & 0 \end{bmatrix}Q^T. \tag{2.132}$$

Noting that P and Q are both orthogonal matrices, relation (2.132) reduces to

$$\begin{bmatrix} \Lambda & 0 \\ 0 & 0 \end{bmatrix}\begin{bmatrix} N & K \\ L & M \end{bmatrix}\begin{bmatrix} \Lambda & 0 \\ 0 & 0 \end{bmatrix} = \begin{bmatrix} \Lambda & 0 \\ 0 & 0 \end{bmatrix}, \tag{2.133}$$

from which it follows that $\Lambda N \Lambda = \Lambda$, or that $N = \Lambda^{-1}$! Hence, given the m by n matrix A whose singular value decomposition is given by equation (2.130), $A^{\{1\}}$ is given by

$$A^{\{1\}} = Q\begin{bmatrix} \Lambda^{-1} & K \\ L & M \end{bmatrix}P^T, \tag{2.134}$$

where the matrices *K, L,* and *M* are *arbitrary.* For different matrices *K, L,* and *M*, we will have different $A^{\{1\}}$ generalized inverses of the matrix A! Thus the $\{1\}$-inverse of the matrix A is not unique. In fact, in general, an infinite number of $\{1\}$-inverses exist.

If we seek the form of $A^{\{1,2\}}$, we can now write it as

$$A^{\{1,2\}} = Q\begin{bmatrix} \Lambda^{-1} & K \\ L & M \end{bmatrix}P^T, \tag{2.135}$$

and further require that it satisfy the second MP condition, i.e., that $A^{\{1,2\}}AA^{\{1,2\}} = A^{\{1,2\}}$. Using equations (2.135) and (2.130) for $A^{\{1,2\}}$ and A respectively in this relation yields

$$\begin{bmatrix} \Lambda^{-1} & K \\ L & L\Lambda K \end{bmatrix} = \begin{bmatrix} \Lambda^{-1} & K \\ L & M \end{bmatrix}, \tag{2.136}$$

indicating that $M = L\Lambda K$. Using this relation in equation (2.135), the structure of $A^{\{1,2\}}$ is given by

$$A^{\{1,2\}} = Q\begin{bmatrix} \Lambda^{-1} & K \\ L & L\Lambda K \end{bmatrix} P^T, \tag{2.137}$$

where the matrices K and L are again arbitrary. Again, for different choices of the matrices K and L, different $\{1,2\}$-inverses of A will be obtained.

Again starting with equation (2.134), and applying the fourth MP condition, we can show that every $\{1,4\}$-inverse of A has the structure

$$A^{\{1,4\}} = Q\begin{bmatrix} \Lambda^{-1} & K \\ 0 & M \end{bmatrix} P^T \tag{2.138}$$

where the matrices K and M are arbitrary. Each specific choice of the matrices K and M generates a specific $\{1,4\}$-inverse of A. Thus the $\{1,4\}$ inverse of the matrix A is also not unique.

By starting with equation (2.138) and then imposing the second MP condition, the reader can show that the $\{1,2,4\}$-inverse of A has the structure

$$A^{\{1,2,4\}} = Q\begin{bmatrix} \Lambda^{-1} & K \\ 0 & 0 \end{bmatrix} P^T, \tag{2.139}$$

where K is an arbitrary matrix. Again, different choices of K yield different $\{1,2,4\}$-inverses. Thus there are, in general, many matrices which would be $\{1,2,4\}$-inverses of a given matrix A.

Lastly, the $A^{\{1,2,3,4\}}$-inverse (or what we call the MP-inverse) of the matrix A may be obtained by starting with the form given by equation (2.139) and then imposing the third MP condition. This yields (see Problem 2.1 for notation)

$$A^{\{1,2,3,4\}} = Q\begin{bmatrix} \Lambda^{-1} & 0 \\ 0 & 0 \end{bmatrix} P^T = Q_1 \Lambda^{-1} P_1^T. \tag{2.140}$$

Notice that we have obtained equation (2.57), which we had introduced earlier, but now with some justification! Moreover, there are no arbitrary quantities left in the expression on the right hand side of equation (2.140). The $\{1,2,3,4\}$-inverse of A is uniquely determined by A!

Example 2.34

Consider any $\{1,2,4\}$-inverse of the matrix A. Show that the following three conditions are always satisfied by this inverse.

$$\begin{aligned}&(1)\ AA^{\{1,2,4\}}A = A.\\&(2)\ \text{The equation } A\mathbf{x} = \mathbf{b} \text{ is consistent}\\&\qquad \text{if and only if } AA^{\{1,2,4\}}\mathbf{b} = \mathbf{b}.\\&(3)\ \text{If } A\mathbf{v} = \mathbf{0}, \text{ then } \mathbf{v}^T A^{\{1,2,4\}} = \mathbf{0}.\end{aligned} \tag{2.141}$$

The first condition follows because the {1,2,4}-inverse of A is also a {1}-inverse of A. The second condition also follows because of the same reason (see equation (2.101)). We therefore need to prove just the third relation. Now $A\mathbf{v} = \mathbf{0}$ implies $A^{\{1,2,4\}}A\mathbf{v} = \mathbf{0}$. Taking the transpose on both sides of this equation yields

$$\mathbf{0} = \mathbf{v}^T(A^{\{1,2,4\}}A)^T = \mathbf{v}^T(A^{\{1,2,4\}}A) = \mathbf{v}^T A^{\{1,2,4\}}AA^{\{1,2,4\}} = \mathbf{v}^T A^{\{1,2,4\}}. \tag{2.142}$$

Note that in the above we have successively used the fact that the {1,2,4}-inverse of A satisfies the fourth and the second MP conditions.

Example 2.35

One choice of the {1,4}-inverse of the m by n matrix A is

$$A^{\{1,4\}} = A^T(AA^T)^{\{1\}}. \tag{2.143}$$

First we shall show that if $A^{\{1\}}$ is any {1}-inverse of A, then $(A^{\{1\}})^T$ is a {1}-inverse of A^T. This follows by taking the transpose of the relation $AA^{\{1\}}A = A$, which gives $A^T(A^{\{1\}})^T A^T = A^T$. From this relation, we therefore see that $(A^{\{1\}})^T = (A^T)^{\{1\}}$.

Now we need to show that the right hand side of equation (2.143) satisfies the first and the fourth MP conditions.

For the first MP condition, we consider the matrix $X = AA^{\{1,4\}}A = AA^T(AA^T)^{\{1\}}A$ and show that it equals A. Alternately we must prove that $(X - A)(X - A)^T = \mathbf{0}$. Using the expression for X, we have

$$\begin{aligned}XX^T &= AA^T(AA^T)^{\{1\}}AA^T\{(AA^T)^{\{1\}}\}^T AA^T = AA^T\{(AA^T)^T\}^{\{1\}}AA^T\\&= AA^T(AA^T)^{\{1\}}AA^T = AA^T\end{aligned} \tag{2.144}$$

and

$$XA^T = AA^T(AA^T)^{\{1\}}AA^T = AA^T, \tag{2.145}$$

Example 2.35 (continued)

so that $AX^T = (XA^T)^T = AA^T$. Using these expressions we get

$$(X - A)(X - A)^T = XX^T - XA^T - AX^T + AA^T = \mathbf{0}, \tag{2.146}$$

and therefore $X = A$. We next show that the fourth MP condition is also satisfied. Since

$$\begin{aligned}\left[A^{\{1,4\}}A\right]^T &= \left[A^T(AA^T)^{\{1\}}A\right]^T = A^T\left\{(AA^T)^{\{1\}}\right\}^T A = A^T\left\{(AA^T)^T\right\}^{\{1\}} A\\ &= A^T\left\{AA^T\right\}^{\{1\}} A = A^{\{1,4\}}A,\end{aligned} \tag{2.147}$$

the result follows.

Example 2.36

Consider any vector $\mathbf{b}$ which belongs to the column space of the matrix A. Then the vector $\mathbf{c} = A^{\{1,4\}}\mathbf{b}$ is unique no matter which particular $\{1,4\}$-inverse of A is used.

Since $\mathbf{b}$ belongs to the column space of A, we can express it as $\mathbf{b} = A\mathbf{z}$. Noting equation (2.138), which yields the general structure of any $\{1,4\}$-inverse of A, we get

$$\begin{aligned}A^{\{1,4\}}\mathbf{b} &= A^{\{1,4\}}A\mathbf{z} = [Q_1 \ \ Q_2]\begin{bmatrix}\Lambda^{-1} & K\\ 0 & M\end{bmatrix}\begin{bmatrix}P_1^T A\\ P_2^T A\end{bmatrix}\mathbf{z}\\ &= Q_1\Lambda^{-1}P_1^T A\mathbf{z} = Q_1Q_1^T\mathbf{z}\end{aligned} \tag{2.148}$$

where we have made use of the fact that $P_2^T A = \mathbf{0}$ (see Problem 2.1, part (a)) and $Q_1^T = \Lambda^{-1}P_1^T A$ (see Problem 2.1, part (c)). Since Q_1 is well determined from the singular value decomposition of A, the result follows.

Incidentally, we have also hereby shown, using (2.143), that $A^T(AA^T)^{\{1\}}A$ is invariant with respect to the particular $\{1\}$-inverse that is used.

THE MINIMUM NORM GENERALIZED INVERSE SOLUTION OF THE CONSISTENT EQUATION $A\mathbf{x} = \mathbf{b}$ Let $A\mathbf{x} = \mathbf{b}$ be a consistent set of equations and let Y be a generalized inverse of A such that $Y\mathbf{b}$ is a solution of the equation $A\mathbf{x} = \mathbf{b}$ and $Y\mathbf{b}$ has minimum length. Then Y satisfies the first and fourth MP conditions and is some $\{1,4\}$-inverse of A.

The general solution of the equation $A\mathbf{x} = \mathbf{b}$ is given by equation (2.105) as

$$\mathbf{x} = Y\mathbf{b} + (I - YA)\mathbf{h} = Y\mathbf{b} + \mathbf{z} \tag{2.149}$$

where Y is any {1}-inverse (or alternately stated, any G-inverse) of A. The vector $\mathbf{h}$ is arbitrary. From all the possible {1}-inverses we could choose, we now want to pick that {1}-inverse which renders the length $Y\mathbf{b}$ a minimum. This requires that

$$\|Y\mathbf{b}\| \le \|Y\mathbf{b} + (I - YA)\mathbf{h}\| \tag{2.150}$$

for all $\mathbf{b}$ which lie in the column space of A, and arbitrary $\mathbf{h}$. But this will hold if and only if for all $\mathbf{b}$ belonging to the column space of A, $Y\mathbf{b}$ is orthogonal to $(I - YA)\mathbf{h}$ for all $\mathbf{h}$ (see Problem 2.12). This requires then that the columns of YA be orthogonal to $(I - YA)$, i.e., that

$$(YA)^T(I - YA) = \mathbf{0}, \text{ or, } (YA)^T = (YA)^T(YA). \tag{2.151}$$

Taking the transpose of the last relation, it follows that $(YA)^T = YA$.

Hence the minimum length solution of the consistent equation $A\mathbf{x} = \mathbf{b}$ is simply $\mathbf{x} = A^{\{1,4\}}\mathbf{b}$. Furthermore, since the equation $A\mathbf{x} = \mathbf{b}$ is consistent, $\mathbf{b}$ belongs to the column space of A and hence, by Example 2.36, $\mathbf{x}$ is unique no matter which particular {1,4}-inverse of A is chosen. Hence, the vector $\mathbf{x}$ which satisfies the consistent equation set $A\mathbf{x} = \mathbf{b}$ and is of minimum length is uniquely given by $A^{\{1,4\}}\mathbf{b}$.

The Minimum Length Least Squares Generalized Inverse

Let $A\mathbf{x} = \mathbf{b}$ be a possibly inconsistent equation and let G be a matrix such that $G\mathbf{b}$ has minimum length in the set of those vectors $\mathbf{x}$ for which $\|A\mathbf{x} - \mathbf{b}\|$ is a minimum. Then G is the MP inverse of the matrix A.

We have already shown (see equation (2.118)) that the vector $\mathbf{x}$ which minimizes $\|A\mathbf{x} - \mathbf{b}\|$ can be expressed as $\mathbf{x} = G\mathbf{b} + (I - GA)\mathbf{h}$ where the matrix G is the {1,3}-inverse of A and the vector $\mathbf{h}$ is arbitrary. We now want that

$$\|G\mathbf{b}\| \le \|G\mathbf{b} + (I - GA)\mathbf{h}\| \text{ for all } m\text{-vectors } \mathbf{b}, \text{ and all } \mathbf{h}. \tag{2.152}$$

But this will occur if and only if $G\mathbf{b}$ is orthogonal to $(I - GA)\mathbf{h}$ for all vectors $\mathbf{b}$ and $\mathbf{h}$. This is equivalent to requiring that $G^T(I - GA) = \mathbf{0}$. This is equivalent to the satisfaction of the second and fourth MP conditions. The last equivalence we leave as an exercise for the reader (see Problem 2. 13). Hence the matrix G is the {1,2,3,4}-inverse of A, i.e., the MP inverse.

Note that the MP inverse is unique and hence the vector $\mathbf{x}$ which is of minimum length and which minimizes $\|A\mathbf{x} - \mathbf{b}\|$ is therefore unique.

PROBLEMS

2.1 We present here an alternative proof for the singular value decomposition (SVD) of an m by n matrix A. Consider the matrix AA^T, which is a positive semidefinite matrix. Then we know that there exists an orthogonal matrix $P = [P_1 \quad P_2]$ such that

$$P^T AA^T P = \begin{bmatrix} \Lambda^2 & 0 \\ 0 & 0 \end{bmatrix}. \tag{2.153}$$

Here Λ^2 is, in general, an r by r diagonal matrix with positive elements.

- **a.** Show that this relation yields the two relations: $P_1^T AA^T P_1 = \Lambda^2$, and $P_2^T A = \mathbf{0}$. Note that any matrix X for which $XX^T = \mathbf{0}$ is a zero matrix.
- **b.** Since $PP^T = I$, show that $P_1 P_1^T = I - P_2 P_2^T$.
- **c.** Define $Q_1^T = \Lambda^{-1} P_1^T A$. Note that $Q_1^T Q_1 = \Lambda^{-1} P_1^T AA^T P\Lambda^{-1} = I_r$.
- **d.** By parts (a) and (b) above: $A = A - P_2 P_2^T A = (I - P_2 P_2^T)A = P_1 P_1^T A$.
- **e.** Using the definition in part (c) then $A = P_1 \Lambda Q_1^T$, which is the SVD of A.
- **f.** Another useful form of the SVD is

$$A = \begin{bmatrix} P_1 & P_2 \end{bmatrix} \begin{bmatrix} \Lambda & 0 \\ 0 & 0 \end{bmatrix} \begin{bmatrix} Q_1^T \\ Q_2^T \end{bmatrix} = P \begin{bmatrix} \Lambda & 0 \\ 0 & 0 \end{bmatrix} Q^T \tag{2.154}$$

  where $Q = [Q_1 \quad Q_2]$ is chosen to be an orthogonal matrix, so that $Q_1^T Q_2 = \mathbf{0}$.
- **g.** Note that $Q_1^T Q_2 = \mathbf{0}$ implies by (c), $P_1^T A Q_2 = \mathbf{0}$ and hence that $P_1 P_1^T A Q_2 = \mathbf{0}$. Using part (d) this implies that $AQ_2 = \mathbf{0}$.

2.2 Verify that the matrix $V\Lambda^{-1}W^T$ of equation (2.57) is the MP inverse of A by showing that it satisfies the four MP conditions.

2.3 Prove that for any m by n matrix A, $(A^T A)^+ = A^+(A^T)^+$, $(AA^+)^+ = AA^+$ and $(A^+A)^+ = A^+A$.

2.4 Show that $(I - A^+A)^+ = (I - A^+A)$ by directly verifying that the four MP conditions are satisfied.

2.5 Let A be an m by n matrix of rank r, and let M be an n by n matrix which is symmetric and positive definite. To find the MP-inverse of $AM^{-1/2}$, consider the symmetric matrix $B = AM^{-1/2}(AM^{-1/2})^T = AM^{-1}A^T$. Show that:

- **a.** the rank of B is r, and that B is symmetric and positive semidefinite.
- **b.** if $\mathbf{w}_i$, $i = 1, 2, \ldots, r$ are the r orthonormal eigenvectors corresponding to the r positive eigenvalues of B, then $\mathbf{w}_i^T(AM^{-1/2}) = 0$, for $i > r$.
- **c.** if $\mathbf{v}_i = \frac{1}{\lambda_i}(AM^{-1/2})^T \mathbf{w}_i$, $i = 1, 2, \ldots, r$, and λ_i, $i = 1, 2, \ldots, r$ are the positive square-roots of the positive eigenvalues of B, then we can write

$$AM^{-1/2} = W\Lambda V^T = \sum_{i=1}^{r} \lambda_i \mathbf{w}_i \mathbf{v}_i^T, \tag{2.155}$$

where the vectors $\mathbf{v}_i^T \mathbf{v}_j = \delta_{ij}$.

d. Show that the MP-inverse of the matrix $AM^{-1/2}$ is given by

$$(AM^{-1/2})^+ = V\Lambda^{-1}W^T = \sum_{i=1}^{r} \frac{1}{\lambda_i} \mathbf{v}_i \mathbf{w}_i^T. \tag{2.156}$$

This problem is a generalization of the singular value decomposition presented in this chapter and will be useful to us later on.

e. Show further that if the rank r of the matrix A equals m, the number of its rows, then by equation (2.88),

$$(AM^{-1/2})^+ = M^{-1/2}A^T(AM^{-1}A^T)^{-1}. \tag{2.157}$$

Note that this result is not valid, in general, when $r < m$.

2.6 Let A be an m by i matrix, and let A_{i-1} be the m by $(i-1)$ matrix that consists of the first $(i-1)$ columns of A. Let the ith column of A be $\mathbf{a}_i$ so that $A = [A_{i-1}\ \mathbf{a}_i]$. Then the MP-inverse of A is given by

$$A^+ = \begin{bmatrix} A_{i-1}^+ - A_{i-1}^+ \mathbf{a}_i \mathbf{b}_i^+ \\ \mathbf{b}_i^+ \end{bmatrix}, \tag{2.158}$$

where the 1 by m vector $\mathbf{b}_i^+$ is the MP-inverse of the vector $\mathbf{b}_i$ defined as

$$\mathbf{b}_i = \begin{cases} (I - A_{i-1}A_{i-1}^+)\mathbf{a}_i & \text{if } \mathbf{a}_i \neq A_{i-1}A_{i-1}^+\mathbf{a}_i \\ \dfrac{[1 + \mathbf{a}_i^T (A_{i-1}A_{i-1}^T)^+ \mathbf{a}_i]\,(A_{i-1}A_{i-1}^T)^+ \mathbf{a}_i}{\mathbf{a}_i^T (A_{i-1}A_{i-1}^T)^+ (A_{i-1}A_{i-1}^T)^+ \mathbf{a}_i} & \text{if } \mathbf{a}_i = A_{i-1}A_{i-1}^+\mathbf{a}_i. \end{cases} \tag{2.159}$$

2.7 Use the singular value decomposition to prove relations (2.88) and (2.89).

2.8 Prove that the matrices AA^+ and $(I - AA^+)$ are symmetric and idempotent.

2.9 Let A be an m by n matrix. Prove that the column space spanned by A^+ is the same as that spanned by A^T.

2.10 Let A be an m by n matrix. Show that for all vectors $\mathbf{v}$ such that $A\mathbf{v} = \mathbf{0}$, we must have $\mathbf{v}^T A^{\{1,4\}}\mathbf{z} = 0$ for all vectors $\mathbf{z}$ which belong to the column space of A.

2.11 Particularize all the results related to the "structure" of the different generalized inverses which we have discussed when the rank r of the matrix A equals m, the number of rows of A. Find analogs for equations (2.134) to (2.140).

2.12 Prove that $\|Y\mathbf{b}\| \le \|Y\mathbf{b} + (I - YA)\mathbf{h}\|$ for all $\mathbf{b}$ which lie in the column space of A, and arbitrary $\mathbf{h}$, if and only if for all $\mathbf{b}$ belonging to the column space of A, $Y\mathbf{b}$ is orthogonal to $(I - YA)\mathbf{h}$ for all $\mathbf{h}$.

2.13 Prove that any matrix X which satisfies the condition $X^T(I - XA) = \mathbf{0}$ is a $\{2,4\}$-inverse of the matrix A.

2.14 Show that $(A^T)^{\{1,4\}} = (A^{\{1,3\}})^T$.

2.15 Show that if B and C are non null matrices, then $BA^{\{1\}}C$ is invariant with respect to the specific choice of $A^{\{1\}}$ if and only if: (1) the column space of B^T is a subset of the column space of A^T, and (2) the column space of C is a subset of the column space of A.

For Further Reading

There are many good books on matrix algebra. We list here just a few of them to give the reader a deeper grasp of the type of material which we will be using in this book, if (s)he decides to dig deeper.

1. G1 This book is what we have used as a primary reference in our brief treatment of matrix analysis in this chapter. The first few chapters of this book will be useful to the reader who is unfamiliar with the concepts introduced in this chapter. The discussion on generalized inverses can be found in Chapter 6.

2. G2 The recursive determination of the MP-inverse of a matrix may be found in the paper by Greville.

3. B1 This is an excellent book on matrix analysis. Several of the exercises in it are, however, at a level which the authors would not consider introductory.

4. HJ1 This book has an excellent discussion on positive definite and semidefinite matrices as well as on the singular value decomposition (Chapter 6).

5. LH1 This book is excellent in its discussion on how to numerically determine the G-, L- and MP-inverses of matrices.

6. R2 The first chapter of this book, though very condensed, is well worth reading. It constitutes a rapid and intense treatment of the fundamentals of matrix analysis.

3 The Fundamental Equation

Having taken the small detour into matrix algebra in the last chapter, we suggest to the reader at this point to go back to the first chapter and skim over it again, especially so that this time the meaning of equations (1.65) to (1.70) is better understood. In this chapter, the equation of motion for constrained mechanical systems will be formulated. The reader will find that it is not necessary to digest all the intricacies that are broached in the previous chapter. Only a few key results from it will be needed. However, for a deeper understanding of the material in this chapter, we recommend that Chapter 2 be carefully read; for, in doing so, a better grounding of the basics, which we will be using in later chapters, will be obtained.

We begin by starting our discussion on the nature of some elementary constraints. We will introduce the concepts of holonomic and nonholonomic Pfaffian constraints and, in this chapter, we will limit our discussion only to these two types of constraints for they form the backbone of Lagrangian mechanics. However, we will show that the fundamental equation is valid for a much larger class of constraints. Next we will introduce Gauss's principle, which may be taken as our starting point. This principle is somewhat less popular that the principles of Lagrange, Hamilton, Gibbs and Appell; we assure the reader that what Gauss's principle lacks in popularity, it more than makes up for in its tremendous sweep of applicability. The principle gives a clear description of the general nature of constrained motion in terms of the minimization of a function of the accelerations of the particles of a system. It is from this basic principle that we will obtain the fundamental equation of motion that describes the dynamics of constrained systems. The chapter ends with several examples, some of which were introduced in the first chapter followed by a reassessment of the fundamental equation.

Configuration Space

Consider a system of n point-particles of masses $m_1, m_2, \ldots, m_n$. Let the position of the ith mass, in an inertial rectangular frame of reference, be x_i, y_i, z_i. The position of each particle requires thus three coordinates to be specified, and so the configuration of the entire system of n particles is specified at any time t by $3n$ coordinates. One can stack these coordinates into a $3n$-vector $\mathbf{x} = [x_1, y_1, z_1, x_2, y_2, z_2, \ldots, x_n, y_n, z_n]^T$. Each $3n$-vector $\mathbf{x}$ (or set of $3n$ coordinates) describes a unique configuration of the system, and hence a unique "point" in this $3n$-dimensional space. This $3n$-dimensional space is called the configuration space. Each "point" in this space refers to some specific configuration of the system. As the system evolves in time, a curve is traced by this "point" in this configuration space. This curve is often called a C-trajectory.

Let the three components of the forces acting (impressed) on the ith particle be F_{i_x}, F_{i_y} and F_{i_z}. Then the equations of motion for the system are given as in equation (1.28) by

$$M\ddot{\mathbf{x}}(t) = \mathbf{F}(t), \tag{3.1}$$

where, as before, $M = Diag\{m_1, m_1, m_1, m_2, \ldots, m_n, m_n, m_n\}$, and the $3n$-vector $\mathbf{F} = [F_{1_x}, F_{1_y}, F_{1_z}, F_{2_x}, F_{2_y}, F_{2_z}, \ldots, F_{n_x}, F_{n_y}, F_{n_z}]^T$ contains the components of the impressed forces on the particles. The components of the vector $\mathbf{F}$ will always be considered as known functions of $\mathbf{x}$, $\dot{\mathbf{x}}$ and t.

It will often be convenient for notational purposes to represent the $3n$-vector $\mathbf{x}$ as $\mathbf{x} = [x_1, x_2, x_3, x_4, \ldots, x_{3n-2}, y_{3n-1}, x_{3n}]^T$. We shall do this from time to time in the following sections to make the equations simpler. We will clearly state what the vector $\mathbf{x}$ denotes in these cases.

Constraints

The reason we begin our study of dynamical systems with the study of constraints is because to understand constrained motion we must understand the nature and types of constraints that we will be dealing with in this book. Also, we want to tie in our formulation of the equations of motion with those already known. Furthermore, it is essential that the reader be introduced to the common language in the field. We will find that our approach to obtaining the equations of motion of a constrained system emphasizes different aspects of the nature of constraints from those usually emphasized.

Holonomic Constraints

Consider now a single particle of mass m moving along a straight line which is aligned along, say, the X-axis of an inertial rectangular coordinate system. One can imagine a bead moving on a straight wire. We usually write the equation of motion of the particle as

$$m\ddot{x} = F_x(x, \dot{x}, t), \tag{3.2}$$

where $x(t)$ is the position of the particle along the line, measured from a fixed point on it, and $F_x(x, \dot{x}, t)$ is the impressed force acting on the particle. However, to describe the problem more fully, we should write the three equations of motion as

$$\begin{aligned} m\ddot{x} &= F_x(x, \dot{x}, y, \dot{y}, z, \dot{z}, t), \\ m\ddot{y} &= F_y(x, \dot{x}, y, \dot{y}, z, \dot{z}, t), \\ m\ddot{z} &= F_z(x, \dot{x}, y, \dot{y}, z, \dot{z}, t), \end{aligned} \tag{3.3}$$

and add the constraints

$$y(t) = 0 \text{ and } z(t) = 0. \tag{3.4}$$

If we differentiate these constraint equations we obtain, $\dot{y}(t) = \dot{z}(t) = 0$; one more differentiation yields $\ddot{y}(t) = \ddot{z}(t) = 0$. Using the last two equations of the set (3.3), these equations imply that $F_y(t) = F_z(t) = 0$, and the three equations in the set (3.3) reduce to the single equation

$$m\ddot{x} = F_x(x, \dot{x}, 0, 0, 0, 0; t), \tag{3.5}$$

which is identical to equation (3.2), the equation we started with.

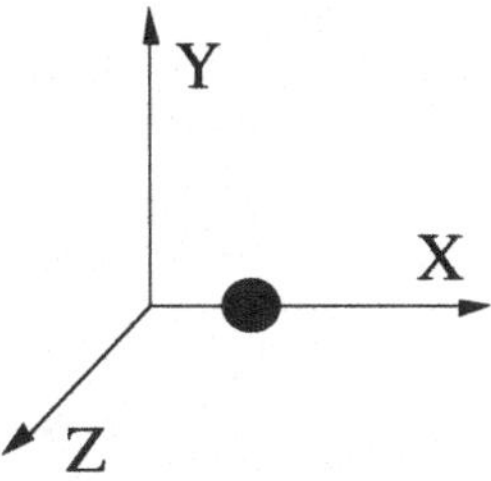

Figure 3.1. Rectilinear motion of a particle

A constraint like $z(t) = 0$ is a special case of a constraint whose general form is

$$f(x, y, z, t) = 0. \tag{3.6}$$

For example, a particle moving on a plane that moves parallel to itself would be expressed by the relation $x(t) + y(t) = b(t)$ where $b(t)$ is a given function.

More generally, if we have a system of n particles described by the $3n$ coordinates x_i, y_i, z_i, $i = 1, 2, \ldots, n$, then a constraint of the form

$$f(x_1, y_1, z_1, x_2, \ldots, x_n, y_n, z_n, t) = 0, \tag{3.7}$$

or a constraint which can be reduced to this form, is called a holonomic constraint. If the holonomic constraint does not have time explicitly in it, it is called a scleronomic constraint; else, it is called rheonomic. (These are Greek names; the word "sclero" means rigid, the word "rheo" means flowing.) Clearly, rheonomic constraints constitute the more general variety.

To get a feel for holonomic constraints, consider first a single particle and a simple scleronomic constraint written as

$$f(x_1, y_1, z_1) = 0. \tag{3.8}$$

The configuration space of this particle is 3-dimensional. The constraint represents a 2-dimensional surface in this 3-dimensional configuration space. When we impose a constraint like (3.8) on the particle, we mean that the particle whose position at time t is $x_1(t)$, $y_1(t)$, $z_1(t)$ must satisfy the equation (3.8), and therefore must lie on the 2-dimensional surface $f = 0$. Thus the dimensionality of the space of accessible configurations is reduced from 3 to 2.

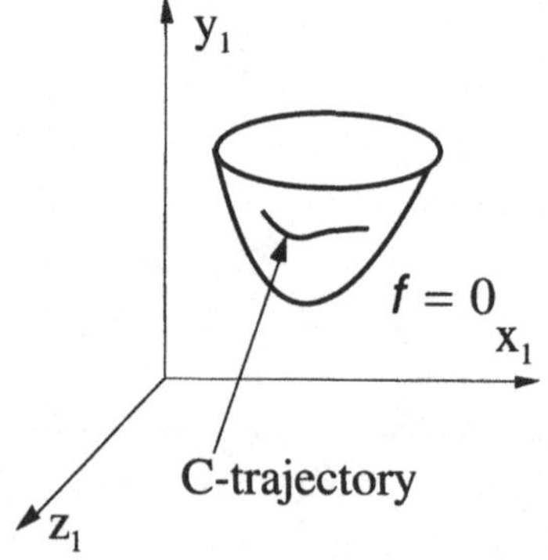

Figure 3.2. The 2-dimensional surface $f = 0$ of equation (3.8)

We can generalize this idea to the more general scleronomic constraint

$$f(x_1, y_1, z_1, x_2, \ldots, x_n, y_n, z_n) = 0 \tag{3.9}$$

as meaning that the configuration point which lies in the $3n$-dimensional space must lie in the $(3n-1)$-dimensional surface lying in this configuration space. The surface represented by equation (3.9) is rigid, and does not change its shape in time. We can similarly interpret the general rheonomic constraint (3.7) as again representing a surface in configuration space on which the configuration point is constrained to lie, but now the surface changes and deforms in time.

If the function f in equation (3.7) has first partial derivatives, then taking the total differential and using the chain rule we get,

$$\sum_{i=1}^{n} \frac{\partial f}{\partial x_i} dx_i + \sum_{i=1}^{n} \frac{\partial f}{\partial y_i} dy_i + \sum_{i=1}^{n} \frac{\partial f}{\partial z_i} dz_i + \frac{\partial f}{\partial t} dt = 0. \tag{3.10}$$

Hence the infinitesimal displacements dx_i, dy_i, dz_i of a system subjected to rheonomic constraints are obliged to satisfy this relation. Note that while equation (3.7) places restrictions on the *finite* displacements (i.e., the x_i's, y_i's, and z_i's) of the particle, equation (3.10) involves restrictions on the *infinitesimal* displacements.

Equation (3.10) is said to be in Pfaffian form. However, it is integrable, its integral obviously being the equation (3.7) from which we obtained (3.10) by differentiation! One can also differentiate equation (3.7) with respect to time to give the equivalent form involving the velocities as

$$\sum_{i=1}^{n} \frac{\partial f}{\partial x_i} \dot{x}_i + \sum_{i=1}^{n} \frac{\partial f}{\partial y_i} \dot{y}_i + \sum_{i=1}^{n} \frac{\partial f}{\partial z_i} \dot{z}_i + \frac{\partial f}{\partial t} = 0. \tag{3.11}$$

Any set of $3n$ quantities $\{\dot{x}_i, \dot{y}_i, \dot{z}_i\}$, $i = 1, 2, \ldots, n$, which satisfy the equation (3.11) is called a *possible* set of velocities, since the system can have these velocities without violating the constraint (3.7).

Note that though in a holonomic constraint like (3.7) velocities do not directly appear, constraints *are* placed on the velocities of the system of particles. This is because the holonomic constraint must be satisfied at *all* times.

Example 3.1

Consider a pendulum of constant length L. Let the coordinates of the bob of the pendulum be (x, y, z). Then the constraint on the motion of the pendulum bob is that

Example 3.1 (continued)

$$x^2 + y^2 + z^2 = L^2 \tag{3.12}$$

The configuration space is 3-dimensional. This is a scleronomic constraint and implies that infinitesimal configuration changes must satisfy the relation

$$xdx + ydy + zdz = 0. \tag{3.13}$$

Differentiating equation (3.12) with respect to time yields a constraint on the velocity of the pendulum given by

$$x\dot{x} + y\dot{y} + z\dot{z} = 0, \tag{3.14}$$

and a further differentiation gives

$$x\ddot{x} + y\ddot{y} + z\ddot{z} = -(\dot{x}^2 + \dot{y}^2 + \dot{z}^2), \tag{3.15}$$

which can be put in matrix form as

$$[x \quad y \quad z]\begin{bmatrix}\ddot{x}\\ \ddot{y}\\ \ddot{z}\end{bmatrix} = -(\dot{x}^2 + \dot{y}^2 + \dot{z}^2). \tag{3.16}$$

Though we start with a 3-dimensional configuration space, we find that the bob cannot access any point in this 3-dimensional space while still satisfying the constraint (3.12). This constraint has actually limited the accessible space of configurations simply to the 2-dimensional surface of the sphere $x^2 + y^2 + z^2 = L^2$, thereby reducing the dimensionality of the configuration space from 3 to 2.

Example 3.2

The plane of support of the pendulum in the previous example is moved so that $f(t)$ is the distance moved by the point of support in the Y-direction, and $g(t)$ is the distance moved in the Z-direction (see Figure 3.3).

Then

$$x^2 + [y - f(t)]^2 + [z - g(t)]^2 = L^2. \tag{3.17}$$

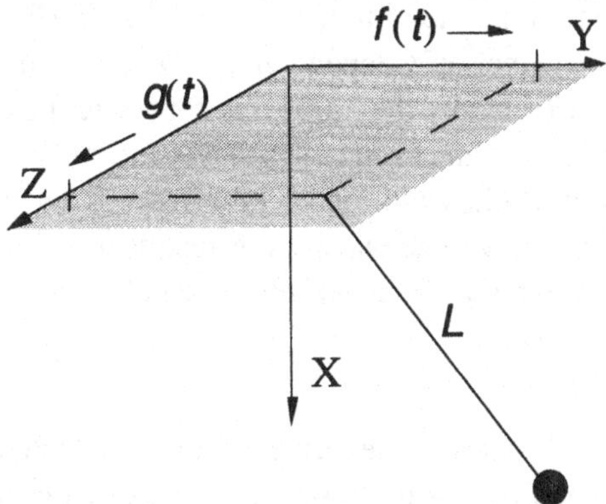

Figure 3.3. Pendulum with moving support

The configuration space is again 3-dimensional. The constraint is now rheonomic, and implies that the infinitesimal changes in the configuration of the bob must satisfy the relation

$$\begin{aligned} &x dx + [y - f(t)]dy + [z - g(t)]dz - [y - f(t)]\frac{df}{dt}\,dt \\ &- [z - g(t)]\frac{dg}{dt}\,dt = 0. \end{aligned} \tag{3.18}$$

The constraint on the velocities is obtained, by dividing throughout by dt, as

$$x\dot{x} + [y - f(t)]\dot{y} + [z - g(t)]\dot{z} - [y - f(t)]\dot{f} - [z - g(t)]\dot{g} = 0. \tag{3.19}$$

At each instant in time, the dimensionality of the space of accessible configurations has been reduced from 3 to 2; at each instant of time the bob must lie on the surface of a sphere, as before. But now this surface keeps changing with time.

Were we to have a particle subjected to two independent constraints,

$$f(x_1, y_1, z_1) = 0 \text{ and } g(x_1, y_1, z_1) = 0, \tag{3.20}$$

instead of just one constraint, then the particle is constrained to lie on the curve of intersection of the two surfaces. With three linearly independent constraints, the particle would have to remain at the point of intersection of the three surfaces. If the constraints were rheonomic, their intersection would have been a point that would move in time in a manner only dependent on the constraints. The particle's motion would not be affected by the impressed forces acting on it. Thus the number of independent constraints must be less that the dimension of the configuration space for the particle's motion to be governed by the impressed forces acting on it. We are ready to generalize our results now.

In general, h holonomic constraints can be described by the set of equations

$$f_i(\mathbf{x},t) = 0,\ i = 1,2,\ldots,h, \tag{3.21}$$

where $\mathbf{x} = [x_1, x_2, x_3, \ldots, x_{3n-1}, x_{3n}]^T$ denotes the $3n$-vector of particle positions whose jth component is x_j. Then taking the total derivatives of the set (3.21), and using the chain rule, we obtain the Pfaffian representation of these equations

$$\sum_{j=1}^{3n} d_{ij}(\mathbf{x},t)dx_j + g_i(\mathbf{x},t)dt = 0,\ i = 1,2,\ldots,h, \tag{3.22}$$

or equivalently in terms of the *possible* velocities as

$$\sum_{j=1}^{3n} d_{ij}(\mathbf{x},t)\dot{x}_j + g_i(\mathbf{x},t) = 0, i = 1,\ 2,\ldots,h. \tag{3.23}$$

where,

$$d_{ij}(\mathbf{x},t) = \frac{\partial f_i(\mathbf{x},t)}{\partial x_j} \text{ and } g_i(\mathbf{x},t) = \frac{\partial f_i(\mathbf{x},t)}{\partial t} \tag{3.24}$$

The equations (3.23) and (3.24) are clearly integrable, their integrals being the h equations (3.21). As we saw before, when h independent holonomic constraints are imposed, the dimensionality of the space of accessible configurations is reduced to ($3n$–h). These constraints can be construed as placing restrictions on the finite displacements of the system, as prescribed by equation (3.21); yet, they also *always* place restrictions on the infinitesimal displacements and the velocities, as prescribed by equations (3.22) and (3.23).

We shall see in our development of the equations of motion that the key concept related to the equations of constraint is not whether they are linearly independent or not, but whether they are consistent or not. We demand that the constraints be consistent, i.e., that fulfillment of any one constraint does not preclude the fulfillment of another. A trivial, though useful, observation is that holonomic constraints, defined by equation (3.21), can be put in the form (3.23) by differentiating them once with respect to time.

Nonholonomic Constraints

Any constraint that cannot be put in the form of equation (3.7) is a nonholonomic constraint. For example, if a particle rests on a horizontal surface, then with the Z-direction pointing upwards and normal to the surface, we must have

$$z(t) \geq 0, \tag{3.25}$$

for the particle cannot penetrate the surface! Such an inequality constraint is nonholonomic. Though inequality constraints can be important in many practical situations, in this book we shall not be dealing with such constraints.

Nonholonomic constraints are often categorized as rheonomic and scleronomic (just like holonomic constraints) according to whether they do, or they do not, explicitly depend on the time t.

An important class of nonholonomic equality constraints is that represented by the equations

$$\sum_{j=1}^{3n} d_{ij}(\mathbf{x},t)dx_j + g_i(\mathbf{x},t)dt = 0, \; i = 1, 2, \ldots, r, \tag{3.26}$$

where the system of equations (3.26) does *not* possess any integrals. Here again, as in equation (3.21), we have denoted, for convenience, the components of the $3n$-vector $\mathbf{x}$ of position as $\mathbf{x} = [x_1, x_2, x_3, \ldots, x_{3n-1}, x_{3n}]^T$.

As mentioned earlier in the context of equation (3.10), equations in this form are defined as Pfaffian forms. Some Pfaffian forms, like the set of equations (3.22), can be integrated, others cannot. It is the nonintegrable Pfaffian forms that lead to nonholonomic constraints. These constraints prescribe restrictions on the infinitesimal displacements of the system; our inability to integrate them indicates that we cannot find the corresponding restrictions on the finite displacements of the system.

Whether the Pfaffian forms are integrable, as in the case of the h equations (3.22), or nonintegrable, as in the case of r equations (3.26), they can of course always be differentiated, provided the functions $d_{ij}(\mathbf{x},t)$ and $g_i(\mathbf{x},t)$ are sufficiently smooth, to yield the set of $m = h + r$ equations

$$\begin{aligned} &\sum_{j=1}^{3n} d_{ij}(\mathbf{x},t)\ddot{x}_j + \sum_{j=1}^{3n}\sum_{k=1}^{3n} \frac{\partial d_{ij}(\mathbf{x},t)}{\partial x_k}\dot{x}_k\dot{x}_j + \sum_{j=1}^{3n} \frac{\partial d_{ij}(\mathbf{x},t)}{\partial t}\dot{x}_j \\ &+\sum_{k=1}^{3n} \frac{\partial g_i(\mathbf{x},t)}{\partial x_k}\dot{x}_k + \frac{\partial g_i(\mathbf{x},t)}{\partial t} = 0, \; i = 1, 2, \ldots, m. \end{aligned} \tag{3.27}$$

We can express these m equations, involving the accelerations, in matrix form as

$$A(\mathbf{x},t)\ddot{\mathbf{x}} = \mathbf{b}(\mathbf{x},\dot{\mathbf{x}},t) \tag{3.28}$$

where the i–jth element of the m by $3n$ matrix A is $d_{ij}(\mathbf{x},t)$, and $b_i(\mathbf{x},\dot{\mathbf{x}},t)$, the ith row element of the m-vector **b,** is given by

$$\begin{aligned} b_i(\mathbf{x},\dot{\mathbf{x}},t) = &-\sum_{j=1}^{3n}\sum_{k=1}^{3n}\frac{\partial d_{ij}(\mathbf{x},t)}{\partial x_k}\dot{x}_k\dot{x}_j - \sum_{j=1}^{3n}\frac{\partial d_{ij}(\mathbf{x},t)}{\partial t}\dot{x}_j \\ &-\sum_{k=1}^{3n}\frac{\partial g_i(\mathbf{x},t)}{\partial x_k}\dot{x}_k - \frac{\partial g_i(\mathbf{x},t)}{\partial t}. \end{aligned} \tag{3.29}$$

The $3n$-vector of acceleration in (3.28) is $\ddot{\mathbf{x}} = [\ddot{x}_1, \ddot{x}_2, \ddot{x}_3, \ldots, \ddot{x}_{3n-1}, \ddot{x}_{3n}]^T$.

Notice that for holonomic and nonholonomic Pfaffian equality constraints, the matrix A depends, in general, on $\mathbf{x}$ and t. The vector $\mathbf{b}$ depends on $\mathbf{x}$, $\dot{\mathbf{x}}$ and t. Most of analytical mechanics deals with the two types of equality constraints described by equations (3.21) and (3.26). For our purposes, it suffices to observe that both types of constraints can be put in the form (3.28), assuming that the functions involved are sufficiently smooth.

The number of coordinates needed to describe the configuration of a system less the number of independent constraints is called the number of degrees of freedom of the system. As seen before, we can express both holonomic and nonholonomic constraints in the form of equation (3.28). Let us say that k out of these m equations are linearly independent. This means that the rank of the matrix A is k. We can thus arbitrarily prescribe $3n - k$ of the components $\ddot{x}_i$ of the $3n$-vector $\ddot{\mathbf{x}}$ in equation (3.28), and we can solve for the remaining k of the $\ddot{x}_i$'s from that equation. The number of degrees of freedom $d = 3n - k$ can then be also thought of as the number of components of the acceleration $3n$-vector $\ddot{\mathbf{x}}$ which can be given arbitrary values.

To get a better feel for nonholonomic Pfaffian constraints of the form (3.26), let us take up a simple situation where we have only one constraint. Consider a particle moving in the 3-dimensional configuration space under the constraint

$$dy = z dx. \tag{3.30}$$

This constraint places a restriction on the infinitesimal displacements of the particle. Equation (3.30) cannot be integrated to yield a constraint on the finite displacements of the particle because the equation does not admit an integrating factor.

It is interesting to find how such a nonintegrable constraint limits the configuration space of the particle. Is the particle capable of accessing any point in the 3-dimensional configuration space while still satisfying at all times this constraint? In other words, can we find a path which leads from, say, the origin to any *arbitrary* point (x_e, y_e, z_e) while still satisfying equation (3.30)? The answer to this question is, yes!

To see why this is so, consider the path described by the equation $y = f(x)$, and $z = f'(x)$ where $f(x)$ is sufficiently smooth. Since $dy = f'(x)dx$, equation (3.30) is always satisfied along this path. Now we have only to choose the function $f(x)$ to be such that $f(0) = 0$, $f'(0) = 0$, $f(x_e) = y_e$ and $f'(x_e) = z_e$. In fact there are an infinite number of functions that will do the job! Thus the dimensionality of the space of accessible configurations is not affected by the presence of a nonintegrable constraint such as equation (3.30). It can be shown that the general nonintegrable Pfaffian equality constraint can always be reduced to (3.30) and hence our result regarding the dimensionality of the accessible space of configurations becomes a general result.

Example 3.3

Consider a constraint on a particle of the form

$$\dot{x} = z^2\dot{y}. \tag{3.31}$$

It is nonintegrable. This nonholonomic constraint can be expressed in the form of equation (3.28) by differentiating with respect to time, to give,

$$\ddot{x} = z^2\ddot{y} + 2z\dot{z}\dot{y}, \tag{3.32}$$

or in the matrix form $A(\mathbf{x}, t)\ddot{\mathbf{x}} = \mathbf{b}(\mathbf{x}, \dot{\mathbf{x}}, t)$ as

$$\begin{bmatrix} 1 & -z^2 & 0 \end{bmatrix} \begin{bmatrix} \ddot{x} \\ \ddot{y} \\ \ddot{z} \end{bmatrix} = 2z\dot{z}\dot{y}. \tag{3.33}$$

The matrix A is 1 by 3, and the vector $\mathbf{b}$ is a scalar.

For nonholonomic constraints of the Pfaffian type, all the configurations accessible in the absence of these nonholonomic constraints are also accessible in their presence. This is a major difference between holonomic constraints and Pfaffian nonholonomic equality constraints. Thus while the space of accessible configurations is reduced by holonomic constraints, it is left unaltered by Pfaffian nonholonomic constraints.

For our purposes, there are three important ideas we need to carry with us about both holonomic and nonholonomic constraints:

1. In the description of any given set of constraints, the equations must be consistent; as such, we do not really care whether the constraints are linearly independent or not.

2. By differentiating nonholonomic constraints (equation (3.26)) once with respect to time, and holonomic constraints (equation (3.21)) twice, we obtain a set of constraint equations which are linear in the accelerations, given by the matrix equation $A\ddot{\mathbf{x}} = \mathbf{b}$, where both A and $\mathbf{b}$ are known functions of time t, if $\mathbf{x}$ and $\dot{\mathbf{x}}$ are known at that time. We will assume throughout this book that the functions $d_{ij}(\mathbf{x}, t)$ and $g_i(\mathbf{x}, t)$ are sufficiently smooth to allow such differentiation.

3. From our perspective there is not much difference between holonomic and nonholonomic Pfaffian equality constraints in the way they enter into the fundamental equation of motion of constrained systems. The approach presented in this book handles these constraints with equal ease, making their distinction unnecessary, and perhaps somewhat obsolete. Similar remarks apply to scleronomic and rheonomic constraints – there appears no need to distinguish between them in our approach to constrained motion.

One might well wonder why we have spent so much time explaining the differences between the two types of constraint, if indeed they are treated with an even hand by the methods we expound in this book. We answer this query in the following way: our discussion has been aimed at exposing the student to the qualitative nature of these constraints, and more importantly, at introducing the necessary terminology needed to understand any book on constrained motion. Lagrangian dynamics usually deals with constraints of the type described by equations (3.21) and (3.26). We shall show in what follows that our results are applicable to a much wider arena of constraint equations than these, equations which have the general form $f(\mathbf{x}, \dot{\mathbf{x}}, t) = 0$, and which may or may not be integrable.

Gauss's Principle

With these preliminary statements on the nature of constraints, we are now ready to take up Gauss's principle, one of the most aesthetic principles in the field of mechanics. We may consider this principle as the starting point for our study of constrained motion.

Consider a system of n particles of masses $m_1, m_2, \ldots, m_n$. Let the 3-vector $\mathbf{x}_i = [x_i, y_i, z_i]^T$ represent the position of the ith particle in a rectangular inertial frame of reference. We assume that the ith particle is subjected to the given impressed force $\mathbf{F}_i(t)$, so that its acceleration, were no constraints present, would be given by the 3-vector $\mathbf{a}_i = \frac{1}{m_i}\mathbf{F}_i(t)$. The three components of the vector $\mathbf{a}_i$ would correspond to the accelerations of the ith particle in the three mutually perpendicular coordinate directions, and the three components of $\mathbf{F}_i$ would be the forces on that particle in those corresponding three directions. We assume, though, that the particles are constrained through certain interconnections, possibly the requirement that some lie on certain surfaces in configuration space, or that some satisfy some nonholonomic Pfaffian constraints. Our task is to determine the actual accelerations of the particles at any time t, as a result of the given impressed forces *and* the constraints, given that we know the position and velocity of each particle at time t.

Thus the equation of motion, were there no constraints on the particles of the system, can be simply written as

$$M\ddot{\mathbf{a}} = \mathbf{F}(\mathbf{x}(t), \dot{\mathbf{x}}(t), t) \tag{3.34}$$

where the $3n$-vector $\mathbf{F}$ of the given impressed forces is obtained by stacking the known impressed forces $\mathbf{F}_i(t)$ acting on each of the particles, so that $\mathbf{F}(t) = [\mathbf{F}_1^T, \mathbf{F}_2^T, \ldots, \mathbf{F}_n^T]^T$; the $3n$-vector $\mathbf{a}$ is obtained by stacking the corresponding acceleration vectors $\mathbf{a}_i(t)$ of each of the particles so that $\mathbf{a}(t) = [\mathbf{a}_1^T, \mathbf{a}_2^T, \ldots, \mathbf{a}_n^T]^T$; and, the $3n$ by $3n$ matrix M is, as usual, diagonal with the masses occurring down the main diagonal in sets of three so that $M = Diag\{m_1, m_1, m_1, m_2, \ldots, m_n, m_n, m_n\}$. Similarly, as usual, we have the $3n$-vector of position given by $\mathbf{x}(t) = [\mathbf{x}_1^T, \mathbf{x}_2^T, \ldots, \mathbf{x}_n^T]^T$. To say that the forces $\mathbf{F}_i$ are given means that they are known functions of $\mathbf{x}$, $\dot{\mathbf{x}}$ and t.

But in the presence of the constraints, the acceleration of the particles at time t will differ from $\mathbf{a}(t)$ and we denote this acceleration by the $3n$-vector $\ddot{\mathbf{x}}(t)$, which is obtained, as usual, by stacking the corresponding accelerations of each particle, so that $\ddot{\mathbf{x}}(t) = [\ddot{\mathbf{x}}_1^T, \ddot{\mathbf{x}}_2^T, \ldots, \ddot{\mathbf{x}}_n^T]^T$. We assume that $\mathbf{x}$ and $\dot{\mathbf{x}}$ are known at time t and therefore so also the entire impressed force vector $\mathbf{F}(\mathbf{x}, \dot{\mathbf{x}}, t)$ at that time. Furthermore we assume that both vectors $\mathbf{x}$ and $\dot{\mathbf{x}}$ are compatible with the given constraints. We note that the matrix M consists of positive entries down the diagonal and is therefore positive definite.

Gauss's principle now asserts that among all the accelerations that the system can have at time t which *are compatible with the constraints*, the ones that actually materialize are those that minimize the following quantity:

$$G(\ddot{\mathbf{x}}) = (\ddot{\mathbf{x}} - \mathbf{a})^T M(\ddot{\mathbf{x}} - \mathbf{a}) = (M^{1/2}\ddot{\mathbf{x}} - M^{1/2}\mathbf{a})^T(M^{1/2}\ddot{\mathbf{x}} - M^{1/2}\mathbf{a}). \quad (3.35)$$

We shall refer to this scalar, for short, as the Gaussian, G. Gauss stated this general fundamental principle of mechanics in a short paper in 1829. The principle is applicable to *any* type of kinematical constraint that the system may be subjected to.

We note that the quantity $\Delta\ddot{\mathbf{x}} = \ddot{\mathbf{x}} - \mathbf{a}$ is simply the deviation of the acceleration of the constrained system from what it would be, had there been no constraints on it. Thus the quantity G can be thought of as the square of the normalized length of the vector $\Delta\ddot{\mathbf{x}}$, normalized with respect to the matrix M. Clearly, when there are no constraints, the minimum of $G(\ddot{\mathbf{x}})$ is achieved when $\ddot{\mathbf{x}} = \mathbf{a}$, the acceleration of the system when it is unconstrained, and $\Delta\ddot{\mathbf{x}} = \mathbf{0}$.

We will consider only those constraints which can be expressed as linear equality relations between the accelerations of the particles of the system. Thus the constraints that we will deal with in this book will always be of the standard form

$$A(\mathbf{x}, \dot{\mathbf{x}}, t)\ddot{\mathbf{x}} = \mathbf{b}(\mathbf{x}, \dot{\mathbf{x}}, t) \quad (3.36)$$

where the matrix A is m by $3n$ and the vector $\mathbf{b}$ is an m-vector. (Recall that inequality constraints like those expressed by equation (3.25) cannot be put into this form.)

We allow the elements of the matrix A to be functions of $\mathbf{x}$, t, as well as of $\dot{\mathbf{x}}$. These constraint equations are therefore slightly more general (see equation (3.28)) than (1) those obtained by differentiating (with respect to time) Pfaffian nonholonomic constraints, i.e., those described by equations (3.26), and (2) those obtained by twice differentiating holonomic constraints, i.e., those described by the equation set (3.21). They may be thought of as being obtained, as usual, by differentiating the m constraint equations $\varphi_i(\mathbf{x}, \dot{\mathbf{x}}, t) = 0$, $i = 1, 2, \ldots, m$.

In what follows we shall find that the matrix $AM^{-1/2}$ will play a central role. We shall call this matrix the Constraint Matrix, B.

Thus the constraints that we will be dealing with in this book comfortably accommodate all of what goes under the usual rubric of Lagrangian mechanics (more on that later). We assume that the equation set (3.36) is consistent, *though the equations contained in it need not be linearly independent.*

The Fundamental Equation

We assert that at each instant of time t, the actual acceleration $3n$-vector $\ddot{\mathbf{x}}(t)$ of the system of n particles, in the presence of the constraints which we have expressed by equation (3.36), is given by

$$\boxed{\ddot{\mathbf{x}} = \mathbf{a} + M^{-1/2}(AM^{-1/2})^{+}(\mathbf{b} - A\mathbf{a})}, \tag{3.37}$$

where $(AM^{-1/2})^{+}$ is the unique MP-inverse of the Constraint Matrix $AM^{-1/2}$.

The verification is straightforward. We first show that the acceleration given by equation (3.37) satisfies the constraint equation (3.36); then we show that of all the acceleration vectors which satisfy equation (3.36), it is the unique vector that minimizes the Gaussian G which has been defined in equation (3.35). If the acceleration vector $\ddot{\mathbf{x}}(t)$ satisfies these two conditions, Gauss's principle then assures us that this acceleration is then the correct acceleration of the particles constituting the constrained system.

1. To begin with, we note that equation (3.36) can be expressed as

$$AM^{-1/2}(M^{1/2}\ddot{\mathbf{x}}) = AM^{-1/2}(\mathbf{y}) = \mathbf{b}. \tag{3.38}$$

For this equation to be consistent, i.e., for $\ddot{\mathbf{x}}$ to be a solution of this equation, we require, by equation (2.104), that

$$AM^{-1/2}(AM^{-1/2})^{+}\mathbf{b} = \mathbf{b}. \tag{3.39}$$

Now substituting the expression for $\ddot{\mathbf{x}}$ given in equation (3.37) into the left hand side of equation (3.36) we get

$$\begin{aligned} A\ddot{\mathbf{x}} &= A\mathbf{a} + AM^{-1/2}(AM^{-1/2})^{+}(\mathbf{b} - A\mathbf{a}) \\ &= [I - AM^{-1/2}(AM^{-1/2})^{+}]A\mathbf{a} + AM^{-1/2}(AM^{-1/2})^{+}\mathbf{b} \\ &= [I - AM^{-1/2}(AM^{-1/2})^{+}](AM^{-1/2})M^{1/2}\mathbf{a} + AM^{-1/2}(AM^{-1/2})^{+}\mathbf{b}. \end{aligned} \tag{3.40}$$

By the MP conditions (see equation (2.52)), the first term on the right hand side (of the last expression above) vanishes, so we get

$$A\ddot{\mathbf{x}} = AM^{-1/2}(AM^{-1/2})^{+}\mathbf{b} = \mathbf{b}, \tag{3.41}$$

where the last equality follows because of equation (3.39). Hence the acceleration $\ddot{\mathbf{x}}$ as defined by equation (3.37) satisfies the constraint equation (3.36).

2. Let us consider an acceleration vector different from $\ddot{\mathbf{x}}$, where $\ddot{\mathbf{x}}$ is given by equation (3.37). Thus consider the vector $\ddot{\mathbf{u}} = \ddot{\mathbf{x}} + \mathbf{v}$, where $\mathbf{v}$ is any vector such that $\ddot{\mathbf{u}}$ satisfies the constraint equation (3.36) at time t. We will show that $G(\ddot{\mathbf{u}}) > G(\ddot{\mathbf{x}})$ for any such vector $\mathbf{v} \neq \mathbf{0}$.

Since $A\ddot{\mathbf{u}} = A(\ddot{\mathbf{x}} + \mathbf{v}) = A\ddot{\mathbf{x}} + A\mathbf{v} = \mathbf{b}$, and $A\ddot{\mathbf{x}} = \mathbf{b}$, it follows that $A\mathbf{v} = \mathbf{0}$. Also, $A\mathbf{v} = AM^{-1/2}(M^{1/2}\mathbf{v}) = \mathbf{0}$ implies from relation (2.91) that

$$(M^{1/2}\mathbf{v})^T(AM^{-1/2})^+ = \mathbf{0}. \tag{3.42}$$

Now noting that $\ddot{\mathbf{u}} = \ddot{\mathbf{x}} + \mathbf{v}$, and that $\ddot{\mathbf{x}}$ is given by equation (3.37), we get

$$\begin{aligned}\mathrm{G}(\ddot{\mathbf{u}}) &= [(AM^{-1/2})^+(\mathbf{b} - A\mathbf{a}) + M^{1/2}\mathbf{v}]^T[(AM^{-1/2})^+(\mathbf{b} - A\mathbf{a}) + M^{1/2}\mathbf{v}]\\ &= [(AM^{-1/2})^+(\mathbf{b} - A\mathbf{a})]^T[(AM^{-1/2})^+(\mathbf{b} - A\mathbf{a})]\\ &\quad + [(AM^{-1/2})^+(\mathbf{b} - A\mathbf{a})]^T M^{1/2}\mathbf{v} + (M^{1/2}\mathbf{v})^T[(AM^{-1/2})^+(\mathbf{b} - A\mathbf{a})]\\ &\quad + (M^{1/2}\mathbf{v})^T(M^{1/2}\mathbf{v}).\end{aligned} \tag{3.43}$$

In the last equation, the first term on the right hand side is simply $\mathrm{G}(\ddot{\mathbf{x}})$; the third term is zero because of equation (3.42) and the second term is the transpose of the third term. From this we obtain

$$\mathrm{G}(\ddot{\mathbf{u}}) = \mathrm{G}(\ddot{\mathbf{x}}) + (M^{1/2}\mathbf{v})^T(M^{1/2}\mathbf{v}). \tag{3.44}$$

Since the matrix M is positive definite, the second term on the right hand side of equation (3.44) is positive for all $\mathbf{v} \neq \mathbf{0}$, and so that $\mathrm{G}(\ddot{\mathbf{u}}) > \mathrm{G}(\ddot{\mathbf{x}})$. We have thus proved that the minimum of G occurs when $\ddot{\mathbf{u}} = \ddot{\mathbf{x}}$.

Since no assumption is made regarding the magnitude of the vector $\mathbf{v}$, we have proved that the acceleration $\ddot{\mathbf{x}}$ given by equation (3.37) yields a global minimum of the scalar G. Notice that we have shown that the unique solution for the acceleration $\ddot{\mathbf{x}}$ to Gauss's minimization problem is given by equation (3.37).

Example 3.4

Consider the particle described in Example 1.2 moving in a horizontal plane subjected to the externally impressed forces $F_x(t)$ and $F_y(t)$. It is constrained to lie on a line inclined to the X-axis at a fixed angle α. We want to describe the motion of this constrained system. The unconstrained motion of the particle can be written as

$$\begin{bmatrix} m & 0 \\ 0 & m \end{bmatrix} \mathbf{a} = \begin{bmatrix} F_x(t) \\ F_y(t) \end{bmatrix}. \tag{3.45}$$

The constraint equation $y = x \tan \alpha$ implies $\ddot{y} = \ddot{x} \tan \alpha$. Hence $A = [-\tan \alpha \quad 1]$, and $\mathbf{b} = 0$, a scalar. From equation (3.45), the vector of acceleration of the unconstrained system is $\mathbf{a} = \begin{bmatrix} \frac{F_x(t)}{m} & \frac{F_y(t)}{m} \end{bmatrix}^T$. By equation (3.37) the equation of motion of the constrained system is then

$$\begin{bmatrix} \ddot{x} \\ \ddot{y} \end{bmatrix} = \begin{bmatrix} \frac{F_x(t)}{m} \\ \frac{F_y(t)}{m} \end{bmatrix} - M^{-1/2}(AM^{-1/2})^+[-\tan \alpha \quad 1] \begin{bmatrix} \frac{F_x(t)}{m} \\ \frac{F_y(t)}{m} \end{bmatrix} \tag{3.46}$$

where $AM^{-1/2} = [-m^{-1/2} \tan \alpha \quad m^{-1/2}]$ and by equation (2.82), we obtain its MP-inverse as $(AM^{-1/2})^+ = \frac{1}{m^{-1} \tan^2 \alpha + m^{-1}} \begin{bmatrix} -m^{-1/2} \tan \alpha \\ m^{-1/2} \end{bmatrix}$. Equation (3.46) then becomes

$$\begin{bmatrix} \ddot{x} \\ \ddot{y} \end{bmatrix} = \begin{bmatrix} \frac{F_x(t)}{m} \\ \frac{F_y(t)}{m} \end{bmatrix} - \begin{bmatrix} m^{-1/2} & 0 \\ 0 & m^{-1/2} \end{bmatrix} \left(\frac{m}{\tan^2 \alpha + 1} \right) \begin{bmatrix} -m^{-1/2} \tan \alpha \\ m^{-1/2} \end{bmatrix} \left[-\frac{F_x(t)}{m} \tan\alpha + \frac{F_y(t)}{m} \right] \tag{3.47}$$

or

$$\begin{bmatrix} m\ddot{x} \\ m\ddot{y} \end{bmatrix} = \begin{bmatrix} F_x(t) \\ F_y(t) \end{bmatrix} + \left(\frac{F_x(t) \tan \alpha - F_y(t)}{\tan^2 \alpha + 1} \right) \begin{bmatrix} -\tan \alpha \\ 1 \end{bmatrix}. \tag{3.48}$$

Observe that the constraint $y = x \tan \alpha$ can be expressed in Pfaffian form as $-(\tan \alpha)dx + (+1)dy = 0$, and that equation (3.48) can be written as

$$\begin{bmatrix} m\ddot{x} \\ m\ddot{y} \end{bmatrix} = \begin{bmatrix} F_x(t) \\ F_y(t) \end{bmatrix} + \lambda(t) \begin{bmatrix} -\tan \alpha \\ 1 \end{bmatrix} \tag{3.49}$$

where the multiplying factor $\lambda(t)$ is given by

$$\lambda(t) = \left(\frac{F_x(t) \tan \alpha - F_y(t)}{\tan^2 \alpha + 1} \right). \tag{3.50}$$

Example 3.4 (continued)

The components of the vector which multiplies $\lambda(t)$ on the right hand side of equation (3.49) are the components of dx and dy in the Pfaffian form of the constraint equation. We will learn the significance of this multiplier $\lambda(t)$ a little later.

As a special case, consider a particle on a wire in a vertical plane under the force of gravity. Hence $F_x(t) = 0$ and $F_y(t) = -mg$. Equation (3.48) becomes

$$\begin{bmatrix} m\ddot{x} \\ m\ddot{y} \end{bmatrix} = \begin{bmatrix} 0 \\ -mg \end{bmatrix} + mg\begin{bmatrix} -\sin\alpha\cos\alpha \\ \cos^2\alpha \end{bmatrix}, \tag{3.51}$$

or,

$$\begin{cases} \ddot{x} = -g\sin\alpha\cos\alpha \\ \ddot{y} = -g\sin\alpha\sin\alpha. \end{cases} \tag{3.52}$$

The reader may verify these equations through the use of vector mechanics. While our procedure of obtaining equation (3.48) is somewhat tedious in algebra, once the matrices A and M and the vectors **a** and **b** are obtained, it becomes perfectly routine. While vector mechanics may be easy to use for this simple example, even slight variations from this easy example can be hard to handle in this way (see Problem 3.1).

Example 3.5

As a slight generalization of the previous example, consider a particle moving on a wire which has the shape $f(x,y) = 0$ in the field of gravity. Gravity is acting in the negative Y-direction. We want to write the equations of motion of the particle.

Differentiating the constraint equation $f(x,y) = 0$ twice we get

$$f_x\ddot{x} + f_y\ddot{y} = -\dot{x}\{f_{xx}\dot{x} + f_{xy}\dot{y}\} - \dot{y}\{f_{yx}\dot{x} + f_{yy}\dot{y}\} \tag{3.53}$$

so that

$$\begin{aligned} A = [f_x \ \ f_y],\ \mathbf{b} &= -\{f_{xx}\dot{x}^2 + 2f_{xy}\dot{x}\dot{y} + f_{yy}\dot{y}^2\} \\ &= \text{scalar, and } \mathbf{a} = [0 \qquad g]^T. \end{aligned} \tag{3.54}$$

Also, we have

$$AM^{-1/2} = [m^{-1/2} f_x \quad m^{-1/2} f_y],$$

$$(AM^{-1/2})^+ = \frac{1}{m^{-1} f_x^2 + m^{-1} f_y^2} \begin{bmatrix} m^{-1/2} f_x \\ m^{-1/2} f_y \end{bmatrix} \text{ and}$$

$$\mathbf{b} - A\mathbf{a} = -\{f_{xx}\dot{x}^2 + 2f_{xy}\dot{x}\dot{y} + f_{yy}\dot{y}^2\} - [f_x \quad f_y] \begin{bmatrix} 0 \\ -g \end{bmatrix}$$

$$= -\{f_{xx}\dot{x}^2 + 2f_{xy}\dot{x}\dot{y} + f_{yy}\dot{y}^2\} + gf_y. \tag{3.55}$$

Therefore the equation of motion (3.37) becomes

$$\begin{bmatrix} \ddot{x} \\ \ddot{y} \end{bmatrix} = \begin{bmatrix} 0 \\ -g \end{bmatrix} + \begin{bmatrix} m^{-1/2} & 0 \\ 0 & m^{-1/2} \end{bmatrix} \frac{1}{m^{-1} f_x^2 + m^{-1} f_y^2} \begin{bmatrix} m^{-1/2} f_x \\ m^{-1/2} f_y \end{bmatrix} [-\{f_{xx}\dot{x}^2 + 2f_{xy}\dot{x}\dot{y} + f_{yy}\dot{y}^2\} + gf_y] \tag{3.56}$$

or

$$\begin{bmatrix} \ddot{x} \\ \ddot{y} \end{bmatrix} = \begin{bmatrix} 0 \\ -g \end{bmatrix} + \frac{[-\{f_{xx}\dot{x}^2 + 2f_{xy}\dot{x}\dot{y} + f_{yy}\dot{y}^2\} + gf_y]}{f_x^2 + f_y^2} \begin{bmatrix} f_x \\ f_y \end{bmatrix}. \tag{3.57}$$

Once again we note that the constraint equation can be expressed in Pfaffian form as $(f_x)dx + (f_y)dy = 0$, and equation (3.57) can be rewritten as

$$\begin{bmatrix} m\ddot{x} \\ m\ddot{y} \end{bmatrix} = \begin{bmatrix} 0 \\ -mg \end{bmatrix} + \lambda(t) \begin{bmatrix} f_x \\ f_y \end{bmatrix} \tag{3.58}$$

where the multiplying factor $\lambda(t)$ is given by

$$\lambda(t) = m \frac{[-\{f_{xx}\dot{x}^2 + 2f_{xy}\dot{x}\dot{y} + f_{yy}\dot{y}^2\} + gf_y]}{f_x^2 + f_y^2}. \tag{3.59}$$

The components of the vector multiplying $\lambda(t)$ in equation (3.58) are the coefficients of dx and dy in the equation of constraint when written in Pfaffian form.

Example 3.6

Consider the pendulum described in Example 1.9. We want to write down the equations of motion for the system and also determine the forces of constraint on the bob so that it maintains a fixed distance L from its point of suspension. Here the unconstrained motion can be described by

$$\begin{bmatrix} m & 0 \\ 0 & m \end{bmatrix}\mathbf{a} = \begin{bmatrix} 0 \\ mg \end{bmatrix}. \tag{3.60}$$

Hence, $\mathbf{a} = [0 \;\; g]^T$. The constraint equation $x^2 + y^2 = L^2$ yields on twice differentiation with respect to time,

$$[x(t) \;\; y(t)]\begin{bmatrix} \ddot{x} \\ \ddot{y} \end{bmatrix} = -(\dot{x}^2 + \dot{y}^2), \tag{3.61}$$

so that $A = [x(t) \; y(t)]$ and $\mathbf{b} = -(\dot{x}^2 + \dot{y}^2)$. From equation (3.37) the equation of motion is

$$\begin{bmatrix} \ddot{x} \\ \ddot{y} \end{bmatrix} = \begin{bmatrix} 0 \\ g \end{bmatrix} + M^{-1/2}(AM^{-1/2})^{+}\left\{\mathbf{b} - A\begin{bmatrix} 0 \\ g \end{bmatrix}\right\} \tag{3.62}$$

where

$$AM^{-1/2} = [m^{-1/2}x \;\; m^{-1/2}y] \tag{3.63}$$

and by equation (2.82), the MP-inverse is given by

$$(AM^{-1/2})^{+} = \frac{1}{m^{-1}x^2 + m^{-1}y^2}\begin{bmatrix} m^{-1/2}x \\ m^{-1/2}y \end{bmatrix}. \tag{3.64}$$

Thus

$$\begin{bmatrix} \ddot{x} \\ \ddot{y} \end{bmatrix} = \begin{bmatrix} 0 \\ g \end{bmatrix} + \begin{bmatrix} m^{-1/2} & 0 \\ 0 & m^{-1/2} \end{bmatrix}\frac{m}{x^2 + y^2}\begin{bmatrix} m^{-1/2}x \\ m^{-1/2}y \end{bmatrix}\{-(\dot{x}^2 + \dot{y}^2) - gy\} \tag{3.65}$$

and the equation of motion becomes

$$\begin{bmatrix} \ddot{x} \\ \ddot{y} \end{bmatrix} = \begin{bmatrix} 0 \\ g \end{bmatrix} - \frac{\dot{x}^2 + \dot{y}^2 + gy}{x^2 + y^2} \begin{bmatrix} x \\ y \end{bmatrix}. \quad (3.66)$$

Noting that $x^2 + y^2 = L^2$, we can rewrite equation (3.66) after multiplying by m as

$$\begin{bmatrix} m & 0 \\ 0 & m \end{bmatrix} \begin{bmatrix} \ddot{x} \\ \ddot{y} \end{bmatrix} = \begin{bmatrix} 0 \\ mg \end{bmatrix} - m \frac{\dot{x}^2 + \dot{y}^2 + gy}{L^2} \begin{bmatrix} x \\ y \end{bmatrix}. \quad (3.67)$$

The Pfaffian form of the constraint equation is $(x)dx + (y)dy = 0$ and the multiplier $\lambda(t)$ is in this case

$$\lambda(t) = -m \frac{\dot{x}^2 + \dot{y}^2 + gy}{L^2}. \quad (3.68)$$

Let us compare equation (3.67) with equation (3.60). Equation (3.67) corresponds to the constrained motion of the system; equation (3.60) to the unconstrained motion. We find that the acceleration has changed from $\mathbf{a}$ to $\ddot{\mathbf{x}}$, and that this change has been effected by an additional term on the right hand side of the equation (3.67). This additional term is the force $\mathbf{F}^{(c)}$ generated by the constraint; this force, in conjunction with the force of gravity, causes the pendulum bob to remain at a fixed distance from its point of suspension. Thus the "force of constraint" is given by

$$\begin{bmatrix} F_x^c \\ F_y^c \end{bmatrix} = -m \frac{\dot{x}^2 + \dot{y}^2 + gy}{L^2} \begin{bmatrix} x \\ y \end{bmatrix} = \lambda(t) \begin{bmatrix} x \\ y \end{bmatrix}. \quad (3.69)$$

We note that both the X- and Y-components of the constraint force depend on g.

The reader may have realized by now that the algebra can be greatly simplified if the matrix $M = mI$, and by using equation (2.95) we get

$$M^{-1/2}(AM^{-1/2})^+ = m^{-1/2}I(m^{-1/2}AI)^+ = A^+. \quad (3.70)$$

When the matrix $M = mI$, the fundamental equation (3.37) thus simplifies to

$$\boxed{\ddot{\mathbf{x}} = \mathbf{a} + A^+(\mathbf{b} - A\mathbf{a})}. \quad (3.71)$$

We note the remarkable simplicity of the description of constrained motion when the matrix $M = mI$, as in the case when we have the constrained three-dimensional motion of a single particle of mass m. One may ask if such a simple form of the fundamental equation is obtainable for a general discrete mechanical system with n particles whose masses may, in general, be different from each other. It turns out that by a suitable "scaling" of the acceleration vectors, the equations of motion can still be put into a simple form like that of (3.71) (see Problem 3.14).

Example 3.7

Consider a particle of mass m moving in a circle of constant radius r with no impressed forces acting on it. We want to write the equations of motion for this constrained system.

The matrix M is given by $Diag\{m, m\}$, and the vector $\mathbf{a} = \mathbf{0}$. The single constraint equation $x^2 + y^2 = r^2$ on two differentiations gives $x\ddot{x} + y\ddot{y} = -(\dot{x}^2 + \dot{y}^2) = -\upsilon^2$, where υ is the speed of the particle. This means that $A = [x \quad y]$ and $\mathbf{b}$ is the scalar $-(\dot{x}^2 + \dot{y}^2)$. Since the matrix M is a constant diagonal matrix, the equation of motion (3.37) then simplifies to (3.71) and yields

$$\begin{bmatrix} \ddot{x} \\ \ddot{y} \end{bmatrix} = \mathbf{0} + A^{+}(\mathbf{b} - A\mathbf{a}) = [x \quad y]^{+}\{-(\dot{x}^2 + \dot{y}^2)\} = -\begin{bmatrix} x \\ y \end{bmatrix}\frac{(\dot{x}^2 + \dot{y}^2)}{(x^2 + y^2)} = -\frac{\upsilon^2}{r}\begin{bmatrix} \frac{x}{r} \\ \frac{y}{r} \end{bmatrix}. \tag{3.72}$$

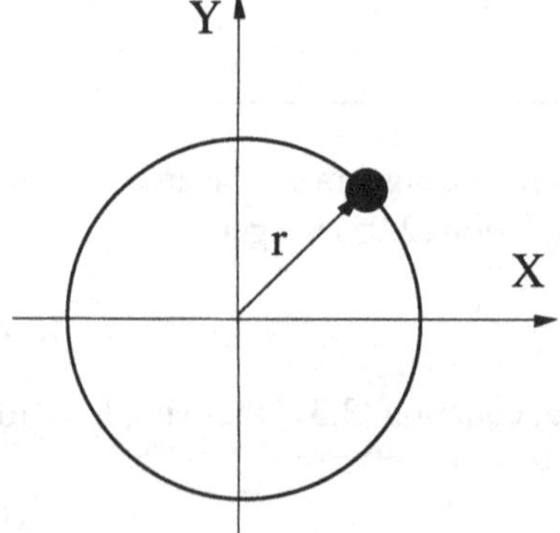

Figure 3.4. Circular motion with no impressed forces

Multiplying both sides by the matrix M, we get

$$\begin{bmatrix} m\ddot{x} \\ m\ddot{y} \end{bmatrix} = \frac{-mv^2}{r} \begin{bmatrix} \frac{x}{r} \\ \frac{y}{r} \end{bmatrix}. \tag{3.73}$$

Comparing this with the equation of the unconstrained system, i.e.,

$$\begin{bmatrix} ma_x \\ ma_y \end{bmatrix} = \begin{bmatrix} 0 \\ 0 \end{bmatrix}, \tag{3.74}$$

we find that the constraint has altered the acceleration from **a** to $\ddot{\mathbf{x}}$ by providing a force given by

$$\begin{bmatrix} F_x^c \\ F_y^c \end{bmatrix} = -\frac{mv^2}{r} \begin{bmatrix} \frac{x}{r} \\ \frac{y}{r} \end{bmatrix}. \tag{3.75}$$

The right hand side of equation (3.75) is easily recognized as the centripetal force acting on the particle and the components of the column vector are simply the components of the unit vector along the radial direction!

Note that because the position and the velocity of the particle must satisfy the equations $x^2 + y^2 = r^2$ and $x\dot{x} + y\dot{y} = 0$, all four quantities cannot be independently specified at any "initial time" t_0. Such constraints on the specification of the initial conditions must be carefully considered in the complete specification of the motion of constrained systems.

The Force of Constraint

We have seen in Examples 3.6 and 3.7 that the presence of the constraint causes the acceleration of the system at every instant of time to deviate from that which it would have had, had there been no constraints. This deviation in the acceleration of the constrained system is brought about by a force that is exerted on the system by virtue of the fact that the unconstrained system must now further satisfy the constraints.

To see how this comes about, consider an instant of time t. We can compare the equation of motion of the unconstrained system, given by

$$M\mathbf{a} = \mathbf{F}(t), \tag{3.76}$$

where the vector **F** consists of the known impressed forces on the system, with the equation of the constrained system, given by

$$M\ddot{\mathbf{x}} = M\mathbf{a} + M^{1/2}(AM^{-1/2})^{+}(\mathbf{b} - A\mathbf{a}). \tag{3.77}$$

Using equation (3.76) in (3.77) we can describe the motion of the constrained system alternately as

$$M\ddot{\mathbf{x}} = \mathbf{F}(t) + M^{1/2}(AM^{-1/2})^{+}(\mathbf{b} - A\mathbf{a}) = \mathbf{F}(t) + \mathbf{F}^{c}(t). \tag{3.78}$$

Thus, at each instant of time t, the constrained system is subjected to an additional "constraint force" $\mathbf{F}^{c}(t)$ given by

$$\mathbf{F}^{c}(t) = M^{1/2}(AM^{-1/2})^{+}(\mathbf{b} - A\mathbf{a}). \tag{3.79}$$

It is this additional force that causes the acceleration of the system at time t to change from its unconstrained value of $\mathbf{a}(t)$ to its constrained value of $\ddot{\mathbf{x}}(t)$.

Notice that nowhere in our discussion do we require that the set of constraint equations be linearly independent. Hence equations (3.37) and (3.79) are valid even when the constraint equations are linearly dependent. Often in complex systems it becomes difficult to ascertain which of the constraints are linearly dependent; this difficulty poses no problem to the approach proposed in this book.

As before, when the matrix M is a constant diagonal matrix so that $M = mI$, then equation (3.79) simplifies to

$$\mathbf{F}^{c}(t) = mA^{+}(\mathbf{b} - A\mathbf{a}). \tag{3.80}$$

Example 3.8

a. Consider a particle of mass m that has no impressed forces acting on it, which is constrained to move along an ellipse. Find the forces of constraint that must be exerted on the particle so that it describes the elliptical trajectory.

We shall describe the elliptical trajectory by placing the origin of the coordinate system at one of the foci of the ellipse. In Figure 3.5 we show the definition of an ellipse, as a curve such that the distance of any point D on it from the focus F is ε times the distance from D to the directrix AB. The directrix is located a distance l from the focus F. The parameter ε is called the eccentricity of the ellipse and it is less than unity. The two parameters l and ε uniquely define any given ellipse.

The equation of the ellipse then becomes

$$\frac{FD}{DN} = \frac{\sqrt{x^2 + y^2}}{x + l} = \varepsilon \tag{3.81}$$

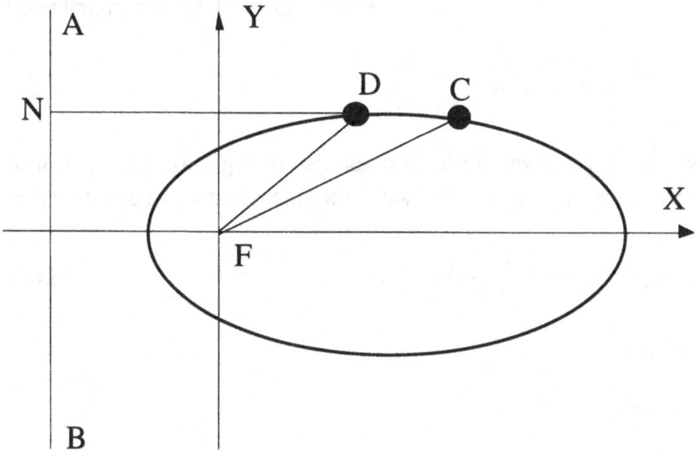

Figure 3.5. Elliptic motion of a particle

or

$$\sqrt{x^2 + y^2} = \varepsilon x + p, \tag{3.82}$$

where the constant $p = \varepsilon l \neq 0$. The constraint on the motion is that it must satisfy this equation. Since the particle has no impressed forces acting on it, the unconstrained acceleration **a** of the particle is zero. Since the matrix M is a constant diagonal matrix and equals $Diag\{m, m\}$, the force of constraint $\mathbf{F}^c$ is therefore simply given by equation (3.80) as

$$\mathbf{F}^c = mA^+\mathbf{b}. \tag{3.83}$$

Squaring both sides of equation (3.82) and differentiating with respect to time we get

$$x\dot{x} + y\dot{y} = (\varepsilon x + p)\varepsilon\dot{x}. \tag{3.84}$$

One more differentiation yields

$$x\ddot{x} + y\ddot{y} = (\varepsilon x + p)\varepsilon\ddot{x} + \varepsilon^2\dot{x}^2 - \dot{x}^2 - \dot{y}^2, \tag{3.85}$$

which through the use of equation (3.82) simplifies to

$$x\ddot{x} + y\ddot{y} - r\varepsilon\ddot{x} = \varepsilon^2\dot{x}^2 - \dot{x}^2 - \dot{y}^2, \tag{3.86}$$

where we have used the notation $r = \sqrt{x^2 + y^2}$, where r is the radial distance of the particle from the focus F of the ellipse. But from equation (3.84)

Example 3.8 (continued)

$$\varepsilon^2\dot{x}^2 = \left(\frac{x\dot{x} + y\dot{y}}{\varepsilon x + p}\right)^2 = \frac{(x\dot{x} + y\dot{y})^2}{r^2}. \tag{3.87}$$

In the last equality above, we have made use of equation (3.82). Using this relationship in equation (3.86) we obtain the equation of constraint as

$$(x - r\varepsilon)\ddot{x} + y\ddot{y} = -\frac{(y\dot{x} - x\dot{y})^2}{r^2}. \tag{3.88}$$

Thus the matrix

$$A = [(x - r\varepsilon) \quad y], \tag{3.89}$$

and **b** is the scalar $-\frac{(y\dot{x} - x\dot{y})^2}{r^2}$. We then obtain

$$A^+ = \frac{1}{[y^2 + (x - r\varepsilon)^2]}\begin{bmatrix} x - r\varepsilon \\ y \end{bmatrix} \tag{3.90}$$

and the constraint force is given by equation (3.83) as

$$\mathbf{F}^c = mA^+\mathbf{b} = -\frac{m(y\dot{x} - x\dot{y})^2}{r^2[y^2 + (x - r\varepsilon)^2]}\begin{bmatrix} x - r\varepsilon \\ y \end{bmatrix}. \tag{3.91}$$

Notice that the force of constraint is pointed in the direction of the normal to the ellipse and not toward its focus. Furthermore, the force is a function of the components of the velocity $\dot{x}$ and $\dot{y}$ of the particle.

b. Now consider a particle of mass m which has no impressed forces acting on it, and which

(1) traces an elliptical trajectory and

(2) moves along the ellipse so that the sector area FCD (see Figure 3.5) which it traces in every unit of time is a constant.

What would the constraint forces need to be so that they are compatible with these two constraints?

We have already seen that the constraint equation (3.88) implies that the trajectory is elliptical. Hence the first constraint on the motion requires equation (3.88) to be satisfied. In addition, the constraint that equal areas be swept out in equal times requires that

$$x\dot{y} - y\dot{x} = c, \tag{3.92}$$

where c is a constant. Differentiating equation (3.92), we get

$$x\ddot{y} - y\ddot{x} = 0. \tag{3.93}$$

The matrix

$$A = \begin{bmatrix} (x - r\varepsilon) & y \\ y & -x \end{bmatrix}, \text{ and } \mathbf{b} = -\begin{bmatrix} c^2 / r^2 \\ 0 \end{bmatrix}. \tag{3.94}$$

Since A is of full rank,

$$\begin{aligned} A^+ = A^{-1} &= \begin{bmatrix} x & y \\ y & -(x - r\varepsilon) \end{bmatrix} \frac{1}{(x^2 + y^2) - rx\varepsilon} \\ &= \frac{1}{rp}\begin{bmatrix} x & y \\ y & -(x - r\varepsilon) \end{bmatrix}, \end{aligned} \tag{3.95}$$

where the denominator in the last equality is simplified using equation (3.82). We therefore obtain, using equation (3.80),

$$\mathbf{F}^c = mA^+\mathbf{b} = -\frac{m}{p}\begin{bmatrix} \frac{x}{r} \\ \frac{y}{r} \end{bmatrix} \frac{c^2}{r^2}. \tag{3.96}$$

The constraint force on the particle is now directed along the focus of the ellipse and it varies inversely with the square of the distance r from the focus.

In this example, comparing our results of part (a) and part (b), we find that the addition of the second constraint not only requires that the force now be central (that is, directed towards the focus of the ellipse), but that it vary inversely as the square of the distance from the focus. We thus find that the central force field where the force on the mass m is inversely proportional to the square of its distance from the focus is the unique constraint force field which will cause the particle to satisfy the two constraints imposed on its motion!

It is exactly this problem that Newton solved, obtaining the celebrated inverse square law of gravitation, when he used Kepler's observations which were nothing other than the two constraints we have provided in part (b) of this example!

Example 3.9

Consider a particle moving in 3-dimensional Euclidean space, free of any impressed forces. Let its position at any time be denoted by (x, y, z). The particle is subjected to the nonholonomic constraint $\dot{y} = z\dot{x}$. The initial conditions are specified so that the particle satisfies this constraint at time $t = 0$. Determine the accelerations of the particle so that it always satisfies the constraint; or, alternatively, find the constraint forces so that the particle satisfies this nonholonomic constraint.

Differentiating the constraint once, we get $\ddot{y} - z\ddot{x} = \dot{z}\dot{x}$, so that $A = [-z \quad 1 \quad 0]$ and $\mathbf{b}$ is the scalar $\dot{z}\dot{x}$. The acceleration of the unconstrained system $\mathbf{a}$ equals zero. Since the matrix $M = Diag\{m, m, m\}$, the equation of motion of the constrained system (see equation (3.71)) is then

$$\ddot{\mathbf{x}} = \mathbf{0} + A^{+}(\dot{z}\dot{x}). \tag{3.97}$$

Since

$$A = [-z \quad 1 \quad 0], \tag{3.98}$$

we obtain

$$A^{+} = \frac{1}{z^{2} + 1}\begin{bmatrix} -z \\ 1 \\ 0 \end{bmatrix}. \tag{3.99}$$

The equation of motion of the constrained system is therefore

$$\begin{bmatrix} \ddot{x} \\ \ddot{y} \\ \ddot{z} \end{bmatrix} = \frac{\dot{z}\dot{x}}{z^{2} + 1}\begin{bmatrix} -z \\ 1 \\ 0 \end{bmatrix}. \tag{3.100}$$

The force of constraint is now given by (3.80) as

$$\mathbf{F}^{c}(t) = mA^{+}(\mathbf{b} - A\mathbf{a}) = mA^{+}(\dot{z}\dot{x}), \tag{3.101}$$

which by equation (3.99) yields

$$\begin{bmatrix} F_{x}^{c} \\ F_{y}^{c} \\ F_{z}^{c} \end{bmatrix} = \frac{m\dot{z}\dot{x}}{(1 + z^{2})}\begin{bmatrix} -z \\ 1 \\ 0 \end{bmatrix} = \lambda(t)\begin{bmatrix} -z \\ 1 \\ 0 \end{bmatrix}. \tag{3.102}$$

Note that the constraint equation can be expressed in Pfaffian form as $(-z)dx + (1)dy + (0)dz = 0$, and that the components of the vector multiplying $\lambda(t)$ are the coefficients of dx, dy and dz corresponding to this Pfaffian form. The multiplier is given by

$$\lambda(t) = \frac{m\dot{z}\dot{x}}{(1+z^2)}. \tag{3.103}$$

We notice that the fundamental equation can be used with equal ease for both nonholonomic and holonomic constraints. If anything, nonholonomic constraints are easier to handle because they require the constraint equations to be differentiated only once (instead of twice in the case of holonomic constraints) to bring them in the standard form of equation (3.36).

Example 3.10

Consider a particle of constant mass m, moving in 2-dimensional Euclidean space. The components of the impressed forces acting on the particle are F_x and F_y. The particle is constrained to move so that $\dot{x} - t\dot{y} = \alpha(t)$ where the function $\alpha(t)$ is given. The initial conditions are prescribed so that at time $t = t_0$ the particle satisfies this constraint. What are the equations of motion for the system for $t \geq t_0$? What are the constraint forces needed to "guide" the particle so that it satisfies this constraint?

Differentiating the constraint equation once, we get $\ddot{x} - t\ddot{y} = \dot{\alpha}(t) + \dot{y}$. From this we see that $A = [1 \quad -t]$ and $\mathbf{b} = \dot{\alpha}(t) + \dot{y}(t)$, a scalar. The matrix $M = Diag\{m, m\}$is a constant diagonal matrix, the vector $\mathbf{a} = \begin{bmatrix} \frac{F_x}{m} & \frac{F_y}{m} \end{bmatrix}^T$ and $A\mathbf{a} = \begin{bmatrix} \frac{F_x}{m} - t\frac{F_y}{m} \end{bmatrix}$. The equation of motion (3.71) for this constrained system then becomes

$$\begin{bmatrix} \ddot{x} \\ \ddot{y} \end{bmatrix} = \begin{bmatrix} \frac{F_x}{m} \\ \frac{F_y}{m} \end{bmatrix} + A^+\left\{\dot{\alpha} + \dot{y} - \frac{F_x}{m} + t\frac{F_y}{m}\right\}. \tag{3.104}$$

Since

$$A^+ = \frac{1}{(t^2+1)}\begin{bmatrix} 1 \\ -t \end{bmatrix}, \tag{3.105}$$

Example 3.10 (continued)

we get

$$\begin{bmatrix} \ddot{x} \\ \ddot{y} \end{bmatrix} = \begin{bmatrix} \frac{F_x}{m} \\ \frac{F_y}{m} \end{bmatrix} + \frac{1}{(t^2+1)} \begin{bmatrix} 1 \\ -t \end{bmatrix} \left\{ \dot{\alpha} + \dot{y} - \frac{F_x}{m} + t\frac{F_y}{m} \right\}, \tag{3.106}$$

which may be rewritten as

$$\begin{bmatrix} \ddot{x} \\ \ddot{y} \end{bmatrix} = \begin{bmatrix} \frac{t^2}{(1+t^2)} & \frac{t}{(1+t^2)} \\ \frac{t}{(1+t^2)} & \frac{1}{(1+t^2)} \end{bmatrix} \begin{bmatrix} \frac{F_x}{m} \\ \frac{F_y}{m} \end{bmatrix} + \frac{(\dot{\alpha}+\dot{y})}{(t^2+1)} \begin{bmatrix} 1 \\ -t \end{bmatrix}. \tag{3.107}$$

Noting that the Pfaffian form of the constraint equation is $(1)dx + (-t)dy = \alpha(t)$, we can express equation (3.106) as

$$\begin{bmatrix} m\ddot{x} \\ m\ddot{y} \end{bmatrix} = \begin{bmatrix} F_x \\ F_y \end{bmatrix} + \lambda(t) \begin{bmatrix} 1 \\ -t \end{bmatrix} \tag{3.108}$$

where the multiplier $\lambda(t)$ is given by

$$\lambda(t) = \frac{(m\dot{\alpha} + m\dot{y} - F_x + tF_y)}{(1+t^2)}. \tag{3.109}$$

The constraint force can again be written as

$$\begin{bmatrix} F_x^c \\ F_y^c \end{bmatrix} = mA^+(\mathbf{b} - A\mathbf{a}). \tag{3.110}$$

Using equation (3.105) we get

$$\begin{aligned} \begin{bmatrix} F_x^c \\ F_y^c \end{bmatrix} &= \frac{(m\dot{\alpha} + m\dot{y} - F_x + tF_y)}{(1+t^2)} \begin{bmatrix} 1 \\ -t \end{bmatrix} \\ &= \frac{1}{(1+t^2)} \begin{bmatrix} -1 & t \\ t & -t^2 \end{bmatrix} \begin{bmatrix} F_x \\ F_y \end{bmatrix} + \frac{m(\dot{\alpha}+\dot{y})}{(1+t^2)} \begin{bmatrix} 1 \\ -t \end{bmatrix}. \end{aligned} \tag{3.111}$$

Notice that the forces of constraint involve the impressed forces F_x and F_y.

Another Look at the Fundamental Equation

The examples above show the simplicity of application of the fundamental equation. It is now time to go back and have another look at it and gather deeper insights.

We notice that the fundamental equation (3.37) involves the MP-inverse of the constraint matrix $B = AM^{-1/2}$. Since the MP-inverse of any matrix is a unique matrix, for a given set of matrices M and A and a given set of vectors **a** and **b**, the right hand side of equation (3.37) is therefore uniquely obtained. Were this not so, it would have given us some consternation, for one expects the accelerations of the point masses (at some specific time) of a well-modelled constrained system to have unique values.

Furthermore, we know that the MP-inverse of any matrix X can always be written as (see equation (2.88)) $X^T(XX^T)^+$, and hence equation (3.37) can also be expressed as

$$\boxed{\ddot{\mathbf{x}} = \mathbf{a} + M^{-1}A^T(AM^{-1}A^T)^+(\mathbf{b} - A\mathbf{a})}, \tag{3.112}$$

where we have made use of the fact that the matrix $M^{-1/2}$ is symmetric. Premultiplying both sides of this equation by the matrix M yields

$$\boxed{M\ddot{\mathbf{x}} = \mathbf{F} + A^T(AM^{-1}A^T)^+(\mathbf{b} - AM^{-1}\mathbf{F})}. \tag{3.113}$$

To obtain the last equation, we have used equation (3.34) once in each of the two terms on the right hand side of equation (3.112). Notice that we have got rid of $M^{-1/2}$ in this form of the fundamental equation.

Let us now consider the special situation when the m by n matrix A has rank m, i.e., when the constraint equation set $A\ddot{\mathbf{x}} = \mathbf{b}$ constitutes a mutually independent system of equations. Since the matrix M is positive definite, the rank of the m by m matrix $(AM^{-1}A^T)$ in equation (3.113) is also m (see Example 2.18, part (b)). Thus the matrix $(AM^{-1}A^T)$ is nonsingular, and so its MP-inverse is simply its regular inverse! Thus if the rank of A is m, then the fundamental equation (3.112) simplifies to

$$\ddot{\mathbf{x}} = \mathbf{a} + M^{-1}A^T(AM^{-1}A^T)^{-1}(\mathbf{b} - A\mathbf{a}), \tag{3.114}$$

and equation (3.113) simplifies to

$$M\ddot{\mathbf{x}} = \mathbf{F} + A^T(AM^{-1}A^T)^{-1}(\mathbf{b} - AM^{-1}\mathbf{F}). \tag{3.115}$$

Note that equations (3.114) and (3.115) would *only* be valid, in general, when the constraints are independent, or more exactly, when the rank of the matrix A

equals m, the number of its rows; for only then can we replace the MP-inverses in equations (3.112) and (3.113) with their regular inverses.

The reader will realize that most of the examples in this chapter involve one constraint equation, and so the rank of the matrix A equals the number of its rows, i.e., unity. We could then have got away with using equation (3.114) (which uses regular inverses in favor of generalized inverses) instead of using equation (3.37). But conceptually, there are advantages in utilizing equation (3.37) as the fundamental equation not simply because it is more generally applicable, but because it leads us on towards deeper thinking, as we shall presently see.

We return again to the point which we began this discussion with – that the fundamental equation (3.37) utilizes the MP-inverse. We know that the MP-inverse is, in a sense, the most restricted generalized inverse of a matrix, since it has to satisfy *all* four of the MP conditions listed on page 45. It may have perhaps already crossed the reader's mind to inquire if one could relax this situation by requiring the generalized inverse involved in equation (3.37) to satisfy only a subset of the four MP conditions. That is, would some other generalized inverse (which is not as restrictive as the MP-inverse) work in equation (3.37)? For the full answer to this question, the reader will have to wait until Chapter 8, where we take up this issue in some detail. Here we provide some insights, and a flavor of the situation through a partial response.

Going over the validation of the fundamental equation, we notice that the properties of generalized inverses are used at only three places: (1) We needed to use the fact that $(AM^{-1/2})(AM^{-1/2})^+(AM^{-1/2}) = (AM^{-1/2})$ in the first member on the right hand side of the equation (3.40). (2) In the second member on the right hand side of equation (3.40), we needed to use the fact that when the equation $A\ddot{\mathbf{x}} = \mathbf{b}$ is consistent, then $AM^{-1/2}(AM^{-1/2})^+\mathbf{b} = \mathbf{b}$. (3) In the third member on the right hand side of equation (3.43), we needed the fact that for all nonzero vectors $\mathbf{v}$ such that $(AM^{-1/2})(M^{1/2}\mathbf{v}) = \mathbf{0}$, $(M^{1/2}\mathbf{v})^T(AM^{-1/2})^+ = \mathbf{0}$.

We notice that in the first two usages above of the properties of the generalized inverse, one does *not* need the inverse to be an MP-inverse. As pointed out in equations (2.105) and (2.101), these statements would be true even if the MP-inverse were replaced by the {1}-inverse (or G-inverse). But the last usage, as seen from the proofs of equations (2.91) and (2.70), requires (among other things) the last three MP-conditions. At first blush then, it appears that the validation of equation (3.37) requires all four MP conditions and that we may indeed be stuck with requiring that all four be satisfied, and hence stuck with the need for using the MP-inverse in the fundamental equation! This even appears plausible, because the MP-inverse is unique and its use in the fundamental equation will therefore uniquely determine the acceleration $\ddot{\mathbf{x}}$ of the constrained system, as is to be expected.

The reader will be relieved to know that the three conditions (2.141) satisfied by the {1,2,4}-inverse of $(AM^{-1/2})$ are exactly the three conditions which we have enumerated above, and which we used in the verification of the fundamental equation. In view of this, if we were to replace the MP-inverse in equation (3.37) by the {1,2,4}-inverse, our verification of the fundamental equation would go through without a hitch! Indeed then we could replace the MP-inverse in the fundamental equation with the {1,2,4}-inverse to yield

$$\ddot{\mathbf{x}} = \mathbf{a} + M^{-1/2}(AM^{-1/2})^{\{1,2,4\}}(\mathbf{b} - A\mathbf{a}). \tag{3.116}$$

But our job has not yet ended, because, by equation (2.139), the {1,2,4}-inverse of a matrix is not unique! Would the use of *every* specific {1,2,4}-inverse in equation (3.116) leave the right hand side invariant, as it must, if the equation is to have any meaning for a well-modeled mechanical system? Moreover, could we write, by analogy with equation (3.112),

$$M\ddot{\mathbf{x}} = \mathbf{F} + A^T(AM^{-1}A^T)^{\{1,2,4\}}(\mathbf{b} - AM^{-1}\mathbf{F})? \tag{3.117}$$

For the moment, we leave the reader to think about and answer this question. We will revisit this in Chapter 8 where we will show that even more may be done.

PROBLEMS

3.1 Consider a bead on a wire lying in a plane, subjected to the impressed forces $F_x(t) = \sin(\omega t)$ and $F_y(t) = \cos(\omega t)$ (see Figure 1.3). The angle that the wire makes with the X-axis is $\alpha(t) = \sin(\omega t)$, where ω is a fixed frequency. Write the equations of motion for this constrained system.

3.2 In Example 3.6, the length of the pendulum is constrained to be a function of time given by $L(t) = L_0\{1 - \mu\sin(\omega t)\}$, where $|\mu| < 1$. The quantities L_0, μ and ω are constants. Write the equations of motion of the pendulum, and determine the forces of constraint (see Figure 1.5).

3.3 Verify that the result in equation (3.75) obtained in Example 3.7 agrees with that obtained from vector mechanics. What is the force of constraint if impressed forces such as gravity act on the particle? Is the force vector always normal to the circle?

3.4 Write the equations of motion for the bead shown in Problem 1.7 if the only impressed forces on it are those caused by gravity acting downwards (see Figure 1.6). Identify each of the entities in the resulting equation in terms of concepts from elementary dynamics, such as Coriolis accelerations, centripetal accelerations, etc.

3.5 In Example 3.6, instead of confining the motion of the pendulum bob to the XY-plane, suppose that the pendulum bob could move in 3-dimensional space. Write the equations describing the motion of the bob now.

3.6 Write the equations of motion of the pendulum discussed in Example 3.2 (see Figure 3.3) whose supports move in the YZ-plane as shown. The mass of the pendulum bob is m. Find the forces of constraint and show how they depend on the functions $f(t)$ and $g(t)$.

3.7 A particle moves in contact with a smooth surface, given by the equation $z = f(x,y)$, with gravity acting in the negative Z-direction (see Figure 3.6). Find the equation of motion of the particle. What is this equation in the special case when the function $f(x,y)$ is $L_0 \sin \frac{n\pi x}{L_1} \sin \frac{n\pi x}{L_2}$? Provide suitable initial conditions at $t = 0$ so that the problem is well posed. How is your solution dependent on the initial conditions?

3.8 Write the equations of motion for the system discussed in Example 1.3 (see Figure 1.4). Write the forces of constraint on each particle. Show that these forces are equal and opposite. The two-particle system is solely acted on by the force of gravity. What if the two particles are "reeled out" so that the length L is no longer a constant but is given by $L(t) = L_0 - pe^{-\alpha t}$? The parameters L_0, p and α are all positive constants, with $p < L_0$. Write the equations of motion of this constrained system. Provide suitable initial conditions so that the problem is well posed.

3.9 Consider a particle moving in a gravitational field (Z-axis pointing downwards), subjected to the nonholonomic constraint described by equation (3.31). Write the equations of motion of the particle.

3.10 In equation (3.36), if the matrix A is $3n$ by $3n$ and has rank $3n$, then show that equation (3.37) yields the acceleration $\ddot{\mathbf{x}} = A^{-1}\mathbf{b}$. The motion is unaffected by the impressed forces on the system.

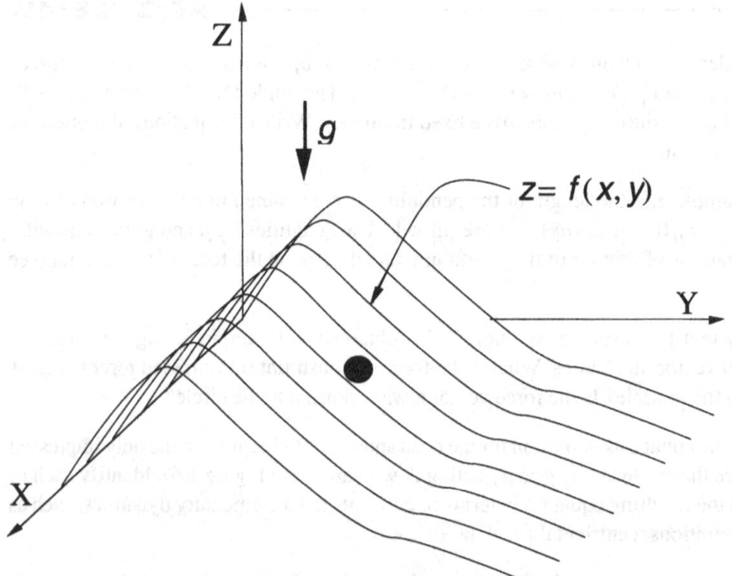

Figure 3.6. Particle moving on the smooth surface $z = f(x,y)$

3.11 Show that the constraint that the time rate of change of the sector area FCD be constant in Figure 3.5 does translate into the constraint equation (3.92).

3.12 Consider a point particle of mass m which moves in 3-dimensional space. Let the Cartesian coordinates which specify the particle's position be x, y and z. Let the components of the given forces which are impressed on the particle in the three coordinate directions be $F_x(x,y,z)$, $F_y(x,y,z)$ and $F_z(x,y,z)$ respectively. The particle is subjected to the constraint $\dot{x}^2(t) + \dot{y}^2(t) - \dot{z}^2(t) = \alpha g(x,y,z,t)$, where α is a given constant, and g is a known function of x, y, z and t. (1) Write the equation of motion for this constrained particle. (2) If α is zero show that the conditions required for the particle to be in equilibrium are the same, irrespective of whether the particle is constrained or not! (3) Think of a mechanism which might (theoretically speaking) provide a nonlinear constraint of the type described.

3.13 Go back to problem 2.5 and solve it, realizing now that the MP-inverse of the matrix $(AM^{-1/2})$ arises in the fundamental equation (3.37).

3.14 Consider the following substitutions. Set $\ddot{\mathbf{x}}_s = M^{1/2}\ddot{\mathbf{x}}$, $\mathbf{a}_s = M^{1/2}\mathbf{a}$, $B = AM^{-1/2}$. Show that the fundamental equation (3.47) then becomes

$$\ddot{\mathbf{x}}_s = \mathbf{a}_s + B^+(\mathbf{b} - B\mathbf{a}_s). \tag{3.118}$$

Notice how this "scaling" of the accelerations $\ddot{\mathbf{x}}$ and $\mathbf{a}$ causes the fundamental equation to take a simple form similar to (3.71). How would the vector $\mathbf{b}$ be altered by the scaling?

3.15 Starting with equation (3.116) verify that this equation minimizes the Gaussian G while satisfying the constraint equation $A\ddot{\mathbf{x}} = \mathbf{b}$.

3.16 What are the objections one might raise regarding the validity of equation (3.117)?

For Further Reading

1.A1 Problem 3.12 is taken from a famous paper by Appell, one of the masters of analytical dynamics. In this paper Appell also describes a mechanism which can theoretically generate a constraint of the type described. The paper is in French.

2.R1 In Chapter 4 of this book the reader will find a detailed description of the various types of constraints on pages 29–53. The author also discusses the necessary and sufficient conditions for the constraints to be integrable. We do not stress this in our book because our method can be uniformly applied to both these types of constraints with equal ease; no special treatment of nonintegrable constraints is required in our approach and hence the need to find out if a constraint is integrable or not is obviated.

3.P1 There is an excellent discussion of Gauss's principle on pages 42–44. The interested reader must read this to obtain a deeper understanding of Gauss's principle. The reader may want to wait until chapter 5 before he reads this though, so that he may have sufficient background to digest this material. Pages 12–19 and 22–27 also provide a short introduction to constrained motion in dynamical systems.

4 Further Applications

In this chapter, we will primarily concentrate on further understanding the fundamental equation and on providing a variety of examples so that the reader can get a deeper feel for the principles and its applications in the context of more complex situations. In the previous chapter we stated the fundamental equation and then we verified it using Gauss's principle. Though such a verification constitutes a complete and rigorous proof of the equation, the perceptive reader must wonder how we magically arrived at this equation in the first place. We show how we did that in this chapter. This chapter begins with a constructive proof of the fundamental equation. We then go on to look at systems with two or more constraints, and then show the ease with which the fundamental equation can be used in computer environments like *Mathematica* and MATLAB.

The Fundamental Equation

As before, we consider a system of n particles whose unconstrained motion is described by the equation

$$M\mathbf{a}(t) = \mathbf{F}(t) \tag{4.1}$$

where the matrix M is as usual the $3n$ by $3n$ diagonal matrix with the masses of the particles down the diagonal in sets of threes, the $3n$ by 1 acceleration vector $\mathbf{a}$ gives the accelerations of the unconstrained system, and the vector $\mathbf{F}$ is the $3n$ by 1 vector of forces which are known, and impressed on the system. These forces may be prescribed functions of the positions of the particles, their velocities and time. In the presence of the constraints, at any time t, the particles will have accelerations denoted by the vector $\ddot{\mathbf{x}}$, which will differ from $\mathbf{a}$ because additional forces will be brought into play by the presence of the constraints. Let the constraints be expressed by the m equations $\varphi_i(\mathbf{x}, \dot{\mathbf{x}}, t) = 0$,

$i = 1, 2, \ldots, m$. Differentiating these m equations with respect to time, as before, yields

$$A(\mathbf{x}, \dot{\mathbf{x}}, t)\ddot{\mathbf{x}} = \mathbf{b}(\mathbf{x}, \dot{\mathbf{x}}, t), \tag{4.2}$$

where the m by n matrix A and the m-vector $\mathbf{b}$ are known functions of $\mathbf{x}$, $\dot{\mathbf{x}}$ and time. In general, the matrix A may have rank $r \le m$, so the m constraint equations need not be independent. The accelerations $\ddot{\mathbf{x}}(t)$ of the constrained system at each instant of time t must satisfy equation (4.2). In what follows, for clarity, we will usually suppress the arguments of the various matrices and vectors.

Starting with equation (4.2), we use the substitution (note that the matrix M is positive definite; see equation (2.32) for the meaning of $M^{1/2}$ and $M^{-1/2}$)

$$\ddot{\mathbf{r}}(t) = M^{1/2}\ddot{\mathbf{x}}(t), \tag{4.3}$$

so that the equation of constraint at time t, equation (4.2), can be expressed as

$$AM^{-1/2}\ddot{\mathbf{r}} = \mathbf{b} \tag{4.4}$$

or equivalently as,

$$B\ddot{\mathbf{r}} = \mathbf{b}, \tag{4.5}$$

where the m by n matrix $B = AM^{-1/2}$. We shall call this matrix B, for reasons which will be clear later, as the Constraint Matrix.

Now we know from equation (2.107) that the general solution for equation (4.5) is simply

$$\ddot{\mathbf{r}} = B^{+}\mathbf{b} + (I - B^{+}B)\mathbf{y} = B^{+}\mathbf{b} + R\mathbf{y}, \tag{4.6}$$

where the n by m matrix B^{+} stands, as usual, for the MP-inverse of the matrix B, and $\mathbf{y}$ is an arbitrary n-vector. Since the vector $\mathbf{y}$ is arbitrary, there are an arbitrarily large number of *possible* acceleration vectors all of which satisfy the constraint (4.2). We will denote the n by n matrix $(I - B^{+}B)$ by R.

We assume that at the time t, we know the positions and the velocities of the particles of the system. Thus the vectors $\mathbf{x}(t)$ and $\dot{\mathbf{x}}(t)$ are known. Since the matrix A and the vector $\mathbf{b}$ are known functions of $\mathbf{x}$, $\dot{\mathbf{x}}$ and t, they too are then known at time t. Hence the first term on the right hand side of equation (4.6) is completely determined. All we need do now is determine the second term $R\mathbf{y}$ on the right hand side to obtain the acceleration of the constrained system. It is the determination of this second term that is dictated by the principles of mechanics, i.e., by Gauss's principle. The principle states that at each instant of time t, the acceleration of the constrained system must be such that while satisfying the constraint equation at that time, it minimizes the Gaussian G given by

$$\begin{aligned} G(\ddot{\mathbf{x}}) &= (\ddot{\mathbf{x}} - \mathbf{a})^T M(\ddot{\mathbf{x}} - \mathbf{a}) = (M^{1/2}\ddot{\mathbf{x}} - M^{1/2}\mathbf{a})^T(M^{1/2}\ddot{\mathbf{x}} - M^{1/2}\mathbf{a}) \\ &= (\ddot{\mathbf{r}} - M^{1/2}\mathbf{a})^T(\ddot{\mathbf{r}} - M^{1/2}\mathbf{a}). \end{aligned} \tag{4.7}$$

Substituting (4.6) in (4.7), we therefore need to find the vector $\mathbf{y}$ such that

$$G(\mathbf{y}) = [R\mathbf{y} - (M^{1/2}\mathbf{a} - B^+\mathbf{b})]^T[R\mathbf{y} - (M^{1/2}\mathbf{a} - B^+\mathbf{b})] \tag{4.8}$$

is minimized.

Now by equation (2.123), the vector $\mathbf{y}$ which minimizes (4.8) is given by

$$\mathbf{y} = R^+(M^{1/2}\mathbf{a} - B^+\mathbf{b}) + (I - R^+R)\mathbf{z} \tag{4.9}$$

where $\mathbf{z}$ is an arbitrary n-vector. But since $R = (I - B^+B)$, by equation (2.75), $R^+ = R$. Moreover, by (2.71) we know that R is idempotent, i.e., $RR = R$. Using these relations in equation (4.9), we get

$$\mathbf{y} = R^+(M^{1/2}\mathbf{a} - B^+\mathbf{b}) + (I - R)\mathbf{z}, \tag{4.10}$$

so that

$$\begin{aligned} R\mathbf{y} &= RR^+(M^{1/2}\mathbf{a} - B^+\mathbf{b}) + R(I - R)\mathbf{z} \\ &= RR(M^{1/2}\mathbf{a} - B^+\mathbf{b}) + (R - RR)\mathbf{z} \\ &= R(M^{1/2}\mathbf{a} - B^+\mathbf{b}). \end{aligned} \tag{4.11}$$

We have thus determined the second term on the right hand side of equation (4.6). Substituting in equation (4.6), we then get

$$\begin{aligned} \ddot{\mathbf{r}} &= B^+\mathbf{b} + R(M^{1/2}\mathbf{a} - B^+\mathbf{b}) = B^+\mathbf{b} + (I - B^+B)(M^{1/2}\mathbf{a} - B^+\mathbf{b}) \\ &= M^{1/2}\mathbf{a} + B^+BB^+\mathbf{b} - B^+BM^{1/2}\mathbf{a} = M^{1/2}\mathbf{a} + B^+\mathbf{b} - B^+BM^{1/2}\mathbf{a} \\ &= M^{1/2}\mathbf{a} + B^+(\mathbf{b} - BM^{1/2}\mathbf{a}) \\ &= M^{1/2}\mathbf{a} + (AM^{-1/2})^+(\mathbf{b} - A\mathbf{a}), \end{aligned} \tag{4.12}$$

where we have made use of the fact that $B^+ = B^+BB^+$, and $B = AM^{-1/2}$. Since by equation (4.3) $\ddot{\mathbf{x}}(t) = M^{-1/2}\ddot{\mathbf{r}}$, we obtain, by premultiplying both sides of equation (4.12) by $M^{-1/2}$,

$$\ddot{\mathbf{x}}(t) = \mathbf{a} + M^{-1/2}(AM^{-1/2})^+(\mathbf{b} - A\mathbf{a}), \tag{4.13}$$

which is the fundamental equation.

Noting the expression for Ry in equation (4.11), we may also write the acceleration $\ddot{\mathbf{x}}$ of the constrained system using equation (4.6) as

$$\begin{aligned}\ddot{\mathbf{x}}(t) &= M^{-1/2}B^{+}\mathbf{b} + M^{-1/2}R\mathbf{y} \\ &= M^{-1/2}B^{+}\mathbf{b} + M^{-1/2}R(M^{1/2}\mathbf{a} - B^{+}\mathbf{b}) \\ &= M^{-1/2}B^{+}\mathbf{b} + (\mathbf{a} - M^{-1/2}B^{+}A\mathbf{a}).\end{aligned} \tag{4.14}$$

From equation (4.14) we observe that the total acceleration $\ddot{\mathbf{x}}(t)$ is made up of the sum of two vectors. The first is the vector $M^{-1/2}B^{+}\mathbf{b}$, which is primarily determined by the constraints imposed on the system. The second is the vector $(\mathbf{a}-M^{-1/2}B^{+}A\mathbf{a})$, which arises because the acceleration must satisfy Gauss's principle; it is therefore determined by the principles of mechanics.

We can premultiply both sides of equation (4.13) by M to yield

$$M\ddot{\mathbf{x}}(t) = M\mathbf{a} + M^{1/2}(AM^{-1/2})^{+}(\mathbf{b} - A\mathbf{a}). \tag{4.15}$$

Using equation (4.1), we can write equation (4.15) in the form

$$M\ddot{\mathbf{x}}(t) = \mathbf{F}(t) + \mathbf{F}^{c}(t), \tag{4.16}$$

where

$$\mathbf{F}^{c}(t) = M^{1/2}(AM^{-1/2})^{+}(\mathbf{b} - A\mathbf{a}). \tag{4.17}$$

The force $\mathbf{F}(t)$ is the given force which is impressed on the system; the vector $\mathbf{F}^{c}(t)$ is the force of constraint brought into play because the system must satisfy the constraint equation (4.2). We have thus *explicitly* obtained the general equation for the force of constraint.

As before, in equations (4.13), (4.15) and (4.17) we may rewrite the term containing the generalized inverse, as $B^{+} = (AM^{-1/2})^{+} = M^{-1/2}A^{T}(AM^{-1}A^{T})^{+}$. In addition, when the rank of A is m, this further simplifies, as before, to $B^{+} = M^{-1/2}A^{T}(AM^{-1}A^{T})^{-1}$.

Also, as pointed out in Chapter 2, the column spaces of the matrices $(AM^{-1/2})^{T}$ and $(AM^{-1/2})^{+}$ are identical. Hence, using the result in Example 2.33, at each instant of time t, the product $(AM^{-1/2})^{+}(\mathbf{b} - A\mathbf{a})$ can always be expressed as $(AM^{-1/2})^{T}\boldsymbol{\lambda}$ where $\boldsymbol{\lambda}$ is a suitable m-vector. The vector $\boldsymbol{\lambda}$ is called the Lagrange multiplier, after J. Lagrange, who used it first in the context of describing the motion of constrained systems. Thus, at each instant of time t, the constraint force $\mathbf{F}^{C}(t)$ can always be expressed through the use of a suitable vector $\boldsymbol{\lambda}$ so that

$$\mathbf{F}^{c}(t) = M^{1/2}(AM^{-1/2})^{+}(\mathbf{b} - A\mathbf{a}) = M^{1/2}(AM^{-1/2})^{T}\boldsymbol{\lambda} = A^{T}\boldsymbol{\lambda}. \tag{4.18}$$

Equation (4.16), which is pertinent to the motion of the constrained system, can therefore always be expressed, using equation (4.18), at each instant of time t, as

$$M\ddot{\mathbf{x}}(t) = \mathbf{F}(t) + A^T\boldsymbol{\lambda}, \tag{4.19}$$

where $\boldsymbol{\lambda}(t)$ is the Lagrange multiplier m-vector. Note that this result is true even when the rank of A is less than m, and the m constraint equations $\varphi_i(\mathbf{x}, \dot{\mathbf{x}}, t) = 0$ are not independent.

Thus we have, through the fundamental equation, proved the existence of a multiplier m-vector $\boldsymbol{\lambda}(t)$, which when multiplied by the matrix A^T, yields the force of constraint $\mathbf{F}^c(t)$ at time t pertinent to the system constrained by the constraint equation (4.2). We shall have more to say later on about the Lagrange multiplier in Chapter 5. However, here we point out to the reader that though equation (4.19) appears as a simple by-product of the fundamental equation, it has occupied a central position in analytical dynamics since its introduction by Lagrange, more than two centuries ago. We thus see that the fundamental equation is in easy conformity with Lagrangian mechanics.

Two General Principles of Analytical Mechanics

We next express equation (4.13) in a manner that lends considerable physical insight into the nature of constrained motion. Let us say that at some time t, we know the position $\mathrm{x}(t)$ and the velocity $\dot{\mathbf{x}}(t)$ of the particles of the system, and that these two vectors are compatible with the m constraint equations $\varphi_i(\mathbf{x}, \dot{\mathbf{x}}, t) = 0,\ i = 1, 2, \ldots, m$. Since $\mathbf{x}(t)$ and $\dot{\mathbf{x}}(t)$ are known, the impressed force at time t, $\mathbf{F}(\mathbf{x}, \dot{\mathbf{x}}, t)$, is therefore known. The acceleration $\mathbf{a}(t)$ of the unconstrained system can then be determined from equation (4.1), as noted before, and equals $M^{-1}\mathbf{F}$. Determining the acceleration of the constrained system at time t then simply amounts to finding out in what way, and by how much, it differs from the known acceleration, $\mathbf{a}(t) = M^{-1}\mathbf{F}$, of the unconstrained system.

To understand this difference, we rewrite equation (4.13) as

$$\ddot{\mathbf{x}}(t) - \mathbf{a}(t) = M^{-1/2}(AM^{-1/2})^+(\mathbf{b} - A\mathbf{a}) = K_1(\mathbf{b} - A\mathbf{a}) \tag{4.20}$$

where the n by m matrix $K_1 = M^{-1/2}(AM^{-1/2})^+$ is just the weighted MP-inverse of the Constraint Matrix, the "weighting" being done by the matrix $M^{-1/2}$, which premultiplies the MP-inverse.

The left hand side of equation (4.20) is now the deviation of the acceleration of the constrained system from that of the unconstrained system at time t; we shall denote this deviation by the vector $\Delta\ddot{\mathbf{x}}(t)$. The quantity $(\mathbf{b} - A\mathbf{a})$ on the right hand side of equation (4.20) is the extent to which the acceleration **a** which corresponds to the acceleration of the unconstrained system does not satisfy the constraint equation (4.2) at time t; we shall denote this by the vector **e**. The fundamental equation (4.13) can therefore be rewritten as

$$\boxed{\Delta\ddot{\mathbf{x}}(t) = K_1\mathbf{e}}\,. \qquad (4.21)$$

Equation (4.21), somewhat surprisingly, exposes the linear relationship between $\Delta\ddot{\mathbf{x}}(t)$ and $\mathbf{e}$. This leads to the following fundamental principle of analytical mechanics:

*The motion of a discrete dynamical system subjected to constraints evolves at each instant of time in such a manner that the deviation of its acceleration from that which it would have had at that instant if there were no constraints on it, is directly proportional to the extent to which the acceleration corresponding to the unconstrained motion, at that instant, does not satisfy the constraints; the matrix of proportionality K_1 is the weighted Moore–Penrose generalized inverse of the Constraint Matrix, and the measure of dissatisfaction of the constraints is the vector **e**.*

In a similar fashion one can interpret equation (4.17), which explicitly gives the constraint force $\mathbf{F}^c$ by noting that this force is also linearly proportional to the vector **e**, which represents the extent to which the acceleration of the unconstrained motion does not satisfy the constraint equation (4.2). This can be written as

$$\boxed{\mathbf{F}^c = K\mathbf{e}} \qquad (4.22)$$

where the n by m matrix $K = M^{1/2}(AM^{-1/2})^+ = A^T(AM^{-1}A^T)^+$. Again the matrix K can be thought of as a weighted MP-inverse of the Constraint Matrix B, this time the "weighting" being done through a premultiplication by the matrix $M^{1/2}$. This leads to the following equivalent principle of analytical mechanics.

*The motion of a discrete dynamical system subjected to constraints evolves, at each instant of time in such a manner that the force of constraint is directly proportional to the extent to which the acceleration corresponding to the unconstrained motion, at that instant, does not satisfy the constraints. The matrix of proportionality K is the weighted Moore–Penrose inverse of the Constraint Matrix, and the measure of dissatisfaction of the constraints is the vector **e**.*

Further Applications

In the last chapter we introduced the reader to the fundamental equation and eased our way into studying the motion of systems that had one kinematical constraint. We found that the equation of constrained motion was obtained in a routine manner by evaluating the matrices M and A, and the vectors **a** and **b**. This routine procedure works for any system with any number of kinematical constraints. However, since we had only one constraint, the matrix A was a 1 by n matrix, and so we used equation (2.82) to obtain in a simple and straightforward manner the MP-inverse of the Constraint Matrix. We now show some examples where we have more than one constraint, and a general approach to handling constrained motion through the use of computational tools like *Mathematica* and MATLAB.

Example 4.1

We consider a particle of mass m moving in 3-dimensional configuration space subjected to no impressed forces, but subjected to the two nonholonomic constraints,

$$\dot{y} = z\dot{x}, \text{ and,} \tag{4.23}$$

$$\dot{y} = z^2\dot{x} + x. \tag{4.24}$$

We intend to find the equations of motion for this particle. Note that this example is an extension of Example 3.9.

Differentiating these equations once with respect to time, we get the constraint on the possible accelerations as

$$\ddot{y} = z\ddot{x} + \dot{z}\dot{x}, \text{ and,} \tag{4.25}$$

$$\ddot{y} = z^2\ddot{x} + 2z\dot{z}\dot{x} + \dot{x}. \tag{4.26}$$

Thus the matrix $A = \begin{bmatrix} -z & 1 & 0 \\ -z^2 & 1 & 0 \end{bmatrix}$, and the vector $\mathbf{b} = \dot{z}\dot{x}\begin{bmatrix} 1 \\ 2z + \frac{1}{\dot{z}} \end{bmatrix}$. To determine the MP-inverse of the matrix $B = AM^{-1/2}$ we use relation (2.88). The matrix $M = Diag\{m, m, m\}$. Hence,

$$B = AM^{-1/2} = \begin{bmatrix} -m^{-1/2}z & m^{-1/2} & 0 \\ -m^{-1/2}z^2 & m^{-1/2} & 0 \end{bmatrix} \text{ and}$$

$$BB^T = \frac{1}{m}\begin{bmatrix} z^2+1 & z^3+1 \\ z^3+1 & z^4+1 \end{bmatrix}, \tag{4.27}$$

so that (note that since BB^T is of rank 2 for z different from zero and unity) $(BB^T)^+ = (BB^T)^{-1}$; thus,

$$(BB^T)^{-1} = \frac{m}{z^2(z-1)^2}\begin{bmatrix} z^4+1 & -(z^3+1) \\ -(z^3+1) & z^4+1 \end{bmatrix} \tag{4.28}$$

and

$$B^+ = B^T(BB^T)^{-1} = \frac{m^{1/2}}{z^2(z-1)}\begin{bmatrix} z & -z \\ z^3 & -z^2 \\ 0 & 0 \end{bmatrix}. \tag{4.29}$$

Since the acceleration vector **a** corresponding to the unconstrained motion of the system is zero, the equations of motion of the constrained system become

$$\begin{bmatrix} \ddot{x} \\ \ddot{y} \\ \ddot{z} \end{bmatrix} = \ddot{\mathbf{x}} = \mathbf{a} + M^{-1/2}B^+\mathbf{b} = \mathbf{0} + M^{-1/2}\frac{m^{1/2}\dot{z}\dot{x}}{z^2(z-1)}\begin{bmatrix} z & -z \\ z^3 & -z^2 \\ 0 & 0 \end{bmatrix}\begin{bmatrix} 1 \\ 2z + \frac{1}{\dot{z}} \end{bmatrix} \tag{4.30}$$

or

$$\begin{bmatrix} \ddot{x} \\ \ddot{y} \\ \ddot{z} \end{bmatrix} = \frac{\dot{x}}{z(z-1)}\begin{bmatrix} \dot{z}(1-2z)-1 \\ -z(1+z\dot{z}) \\ 0 \end{bmatrix}, \tag{4.31}$$

with z not equal to zero or unity. The impressed forces on the system are zero; hence, as promised by equation (4.19), we can express the right hand side of equation (4.31), which is just the constraint force $\mathbf{F}^c(t)$, in the form of $A^T\boldsymbol{\lambda}$. In fact, we can write

Example 4.1 (continued)

$$\begin{bmatrix} m\ddot{x} \\ m\ddot{y} \\ m\ddot{z} \end{bmatrix} = A^T\boldsymbol{\lambda} = \lambda_1(t)\begin{bmatrix} -z \\ 1 \\ 0 \end{bmatrix} + \lambda_2(t)\begin{bmatrix} -z^2 \\ 1 \\ 0 \end{bmatrix} = \mathbf{F}_1^c + \mathbf{F}_2^c, \tag{4.32}$$

where the components $\lambda_1(t)$ and $\lambda_2(t)$ of the Lagrange multiplier 2-vector $\boldsymbol{\lambda}$ are given by

$$\lambda_1(t) = -m\frac{\dot{x}\dot{z}(z^4 + 2z - 1) + \dot{x}(1 + z^3)}{z^2(z-1)^2} \tag{4.33a}$$

and

$$\lambda_2(t) = m\frac{\dot{x}\dot{z}(z^3 + 2z - 1) + \dot{x}(1 + z^2)}{z^2(z-1)^2}. \tag{4.33b}$$

Were we to write the constraint equations (4.23) and (4.24) in Pfaffian form as

$$(-z)dx + (1)dy + (0)dz = 0 \text{ and} \tag{4.34}$$

$$(-z^2)dx + (1)dy + (0)dz = 0, \tag{4.35}$$

we would observe that for the constraints in this problem the matrix A can be inferred from the coefficients of the terms in the above equations.

By equation (4.32), we see that the total force of constraint may then be thought of as being composed of two separate vectors $\mathbf{F}_1^c$ and $\mathbf{F}_2^c$. The first involves the multiplier λ_1(t) and a vector (the first column of the matrix A^T) whose components are the coefficients of dx, dy and dz in the first constraint equation; the second involves the multiplier $\lambda_2(t)$, and a vector (the second column of the matrix A^T) whose components are the coefficients of dx, dy and dz in the second constraint equation.

Equation (4.31) represents the equation of motion in the presence of the two constraints (4.23) and (4.24). The reader should compare this equation with equation (3.100), which was obtained when only the first constraint, i.e., when equation (4.23), is satisfied. Thus the effect that the addition of successive constraints would have on the final equations of motion of the constrained system, or on the forces of constraint which would be required to engender these additional constraints, can be easily seen.

Example 4.2

Consider a pendulum bob of unit mass suspended by a weightless rod of length L_1 which is attached to the origin of the rectangular coordinate system. The bob is also connected to another weightless rod of length L_2 whose other end is pinned to the point $(L_1,0,L_2)$. See Figure 4.1. We want to write the equations of motion describing this system.

The equations describing the constraints can be written as:

$$x^2 + y^2 + z^2 = L_1^2, \text{ and,} \tag{4.36}$$

$$(x - L_1)^2 + y^2 + (z - L_2)^2 = L_2^2. \tag{4.37}$$

Differentiating these equations twice, we can rewrite them as

$$A\ddot{\mathbf{x}} = \mathbf{b}, \tag{4.38}$$

where

$$A = \begin{bmatrix} x & y & z \\ x - L_1 & y & z - L_2 \end{bmatrix}, \text{ and } \mathbf{b} = -\alpha\begin{bmatrix} 1 \\ 1 \end{bmatrix}, \tag{4.39}$$

with $\alpha = (\dot{x}^2 + \dot{y}^2 + \dot{z}^2)$. The matrix $M = I_3$, and the acceleration of the unconstrained system is given by $\mathbf{a} = [g \quad 0 \quad 0]^T$. Using equation (2.88) we obtain

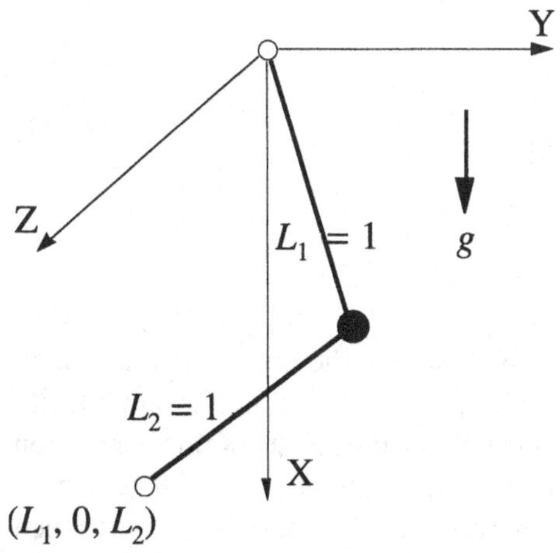

Figure 4.1. Two-link pendulum

Example 4.2 (continued)

$$
\begin{aligned}
B^{+} &= (AM^{-1/2})^{+} = B^{T}(BB^{T})^{-1} \\
&= \frac{1}{\Delta}\begin{bmatrix} L_2^2 x + L_1 y^2 + L_1^2 z^2 - L_1 L_2 z - L_2 xz & L_2 xz - L_1 y^2 - L_1 z^2 \\ y(L_1^2 + L_2^2 - L_1 x - L_2 z) & y(L_1 x + L_2 z) \\ L_2 x^2 + L_2 y^2 - L_1 L_2 x + L_1^2 z - L_1 xz & L_1 xz - L_2 x^2 - L_2 y^2 \end{bmatrix}
\end{aligned}
\tag{4.40}
$$

where $\Delta = [L_2^2 x^2 + (L_1^2 + L_2^2) y^2 - 2L_1 L_2 xz + L_1^2 z^2]$. The vector

$$
\mathbf{b} - A\mathbf{a} = \begin{bmatrix} -\alpha - gx \\ -\alpha - g(x - L_1) \end{bmatrix}, \tag{4.41}
$$

so that

$$
\begin{aligned}
\begin{bmatrix} \ddot{x} \\ \ddot{y} \\ \ddot{z} \end{bmatrix} &= \begin{bmatrix} g \\ 0 \\ 0 \end{bmatrix} + B^{+} \begin{bmatrix} -\alpha - gx \\ -\alpha - g(x - L_1) \end{bmatrix} \\
&= \frac{1}{\Delta}\begin{bmatrix} -\alpha L_2(L_2 x - L_1 z) + g L_2^2 y^2 \\ -y(\alpha L_1^2 + \alpha L_2^2 + g L_2^2 x - g L_1 L_2 z \\ \alpha L_1 (L_2 x - L_1 z) - g L_1 L_2 y^2 \end{bmatrix}.
\end{aligned}
\tag{4.42}
$$

For $L_1 = L_2 = 1$, equation (4.42) reduces to

$$
\begin{bmatrix} \ddot{x} \\ \ddot{y} \\ \ddot{z} \end{bmatrix} = \frac{1}{\Delta_1}\begin{bmatrix} -\alpha(x - z) + gy^2 \\ -y(2\alpha + gx - gz) \\ \alpha(x - z) - gy^2 \end{bmatrix} \tag{4.43}
$$

with $\Delta_1 = [x^2 + 2y^2 - 2xz + z^2]$.

Using the initial conditions $x(0) = L_1$, $y(0) = 0$, $z(0) = 0$, $\dot{x}(0) = 0$, $\dot{y}(0) = 4$, $\dot{z}(0) = 0$, which satisfy the constraints described by equations (4.36) and (4.37) with $L_1 = L_2 = 1$ we can now numerically integrate equation (4.43). We show bel†ow the results of this integration with $g = 10$ units. Notice that were the second constraint (equation (4.36)) absent, the motion of the pendulum would have been constrained to lie, for all time, in the XY-plane. However, constraint (4.36) causes the

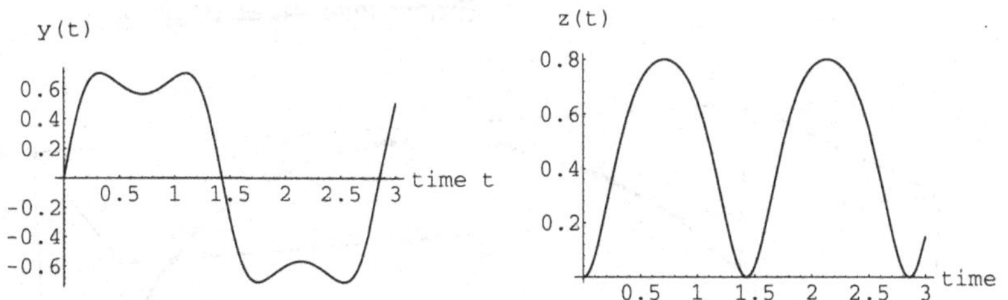

Figure 4.2. Response of doubly constrained pendulum showing nonlinear behavior during large oscillations

bob to move out in the Z-direction causing an out-of-plane motion. Figure 4.2 shows the displacements y and z of the bob as functions of time.

Figure 4.3 shows a parameteric plot of $y(t)$ versus $z(t)$; time is the parameter that is varied. We see here that the motion in the Z-direction is significant.

Figure 4.4 shows a 3-dimensional plot of the trajectory of the bob in configuration space. The bob starts at the top in the center of the box in the Y-direction where $y=0, x=1$ and $z=0$ We see how the two holonomic constraints reduce the accessible configuration space to just a curve.

The equations of motion (4.42) already include in them the constraints (4.36) and (4.37). It is interesting to see the extent of error created through the numerical integration of these differential equations by

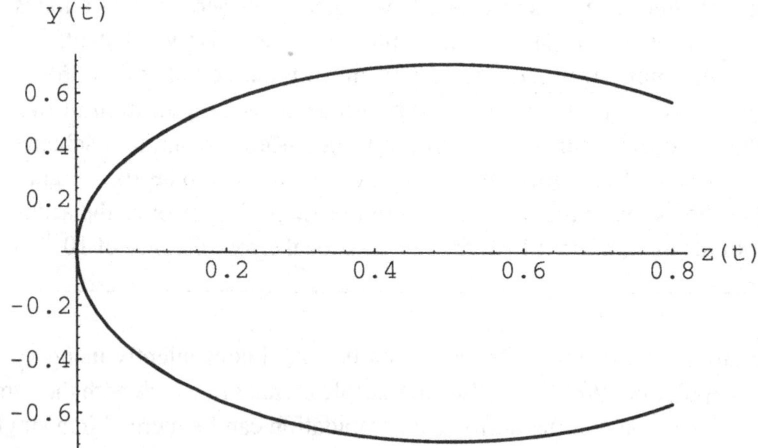

Figure 4.3. Out-of-plane motion of doubly constrained pendulum

Example 4.2 (continued)

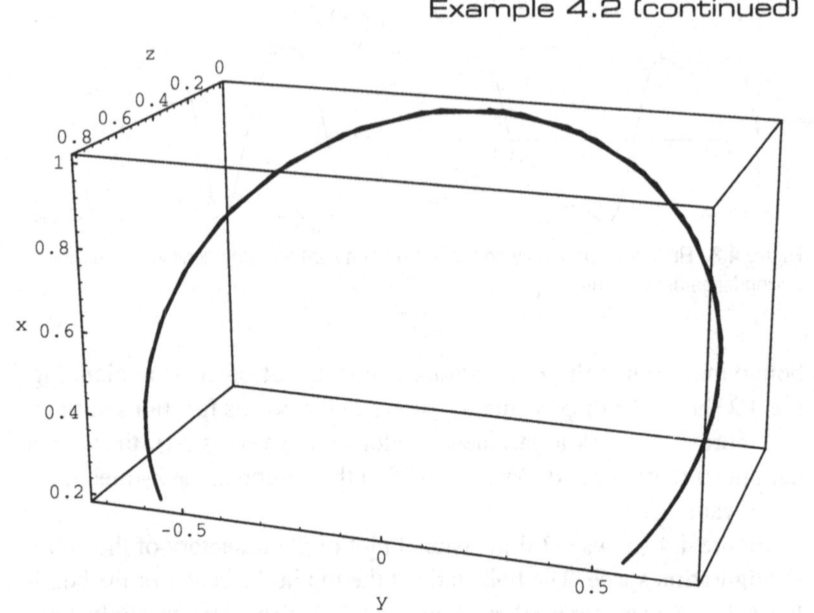

Figure 4.4. Trajectory of the constrained motion of the pendulum bob

checking the extent to which the constraints (4.36) and (4.37) are numerically satisfied. Theoretically speaking, when the numerical integration is to be done with infinite precision, they must be exactly satisfied. We define the numerical error in satisfaction of the constraint (4.36) as $\text{error}(t) = x^2 + y^2 + z^2 - L_1^2$. Figure 4.5(a) gives an indication of this numerical error incurred through the Runge–Kutta integration scheme which was used. The local error tolerance was chosen to be 10^{-6}. Figure 4.5(b) shows the dramatic reduction in this numerical error as the accuracy of the integration is increased using a local error tolerance of 10^{-12}.

In many instances the MP-inverse can be found conveniently using symbolic manipulation. We show below the simple manner in which both the symbolic manipulation and the numerical computation can be merged in a single program using a tool like *Mathematica*. The complete problem is specified through a description of: (1) the matrices A and M, and (2) the vectors **a** and **b**.

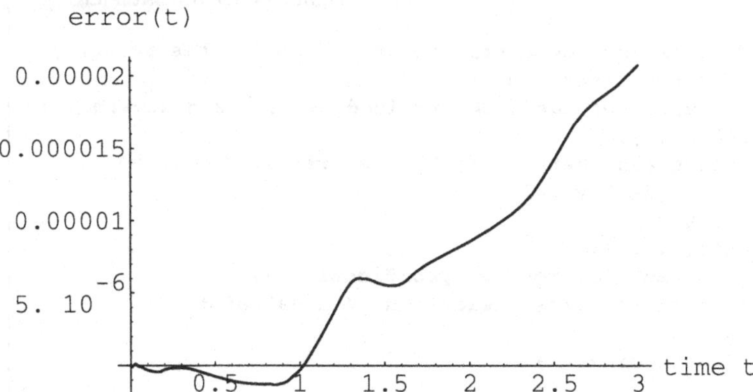

Figure 4.5(a). Numerical integration error in integrating equation (4.43) local error tolerance 10^{-6}

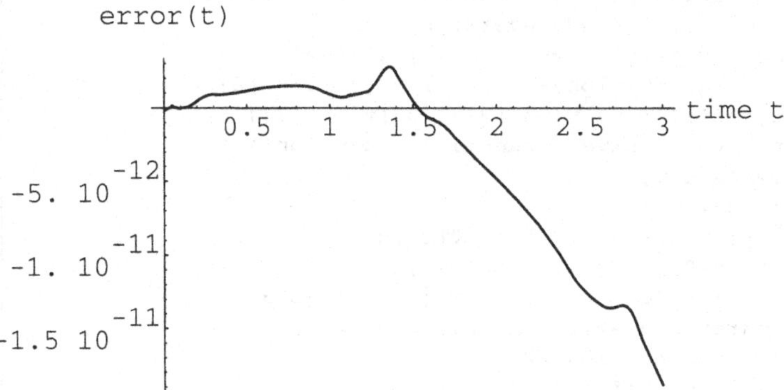

Figure 4.5(b). Numerical integration error in integrating equation (4.43) local error tolerance 10^{-12}

Example 4.3

We provide here a sample *Mathematica* program used to obtain all the results in Example 4.2. The program does the symbolic manipulation to obtain equations (4.39) and (4.41), as well as the numerical computations and the generation of the plots. As seen, the program is rather simple. Even the reader unfamiliar with *Mathematica* will find the program quite understandable.

Example 4.3 (continued)

```
(* In Mathematica, comments are indicated in this manner*)
(* between two asterisks *)
(* acceleration of the unconstrained system, x downwards *)
a={{g},{0},{0}};
(* The Constraint Matrix a1; Mass Matrix is Identity*)
a1={{x,y,z},{x-l1,y,z-l2}};
(* Vector b *)
b=-alpha*{{1},{1}};
(* end of symbolic problem specification *)
(* e is extent of dissatisfaction of constraint *)
e=b-a1.a;
(* MP-inverse of a1 *)
apinverse=
    Simplify[Transpose[a1].Inverse[a1.Transpose[a1]]];
(* Acceleration of constrained system *)
acc=Simplify[a+apinverse.e]
 (* Numerical Calculation of the
    Differential Equation *)
   (* Generation of right hand side of Diff. Eq. *)
    Remove[a,a1,apseudoinverse,b,e];
    u=acc/.{x->x[t],y->y[t],z->z[t],l1->1,l2->1,
            g->10,alpha->x'[t]^2+y'[t]^2+z'[t]^2};
    u1[t]=u[[1]]; u2[t]=u[[2]]; u3[t]=u[[3]];
   (* Call to NDSolve – numerical solution*)
    Remove[acc,u];
    sol=NDSolve[
     {x''[t]==u1[t],y''[t]==u2[t],z''[t]==u3[t],
      x'[0]==0,y'[0]==4,z'[0]==0,
      x[0]==1,y[0]==0,z[0]==0},{x,y,z},{t,3},
      AccuracyGoal->12,PrecisionGoal->12,
      WorkingPrecision->22,
      MaxSteps->Infinity
               ];
(* Plot Results in Various Formats *)
p1=Plot[Evaluate[y[t]/.sol],{t,0,3},
AxesLabel->{"time t", "y(t)"}]
```

Except for the calls to the various plotting routines, this is all that was used to generate the results in the previous example.

Example 4.4

Consider the coupled Duffing's oscillator shown in Figure 4.6 which is subjected to the constraint

$$x_1(t) - x_2(t) = A_0 \exp(-\alpha t)\sin(\omega t). \tag{4.44}$$

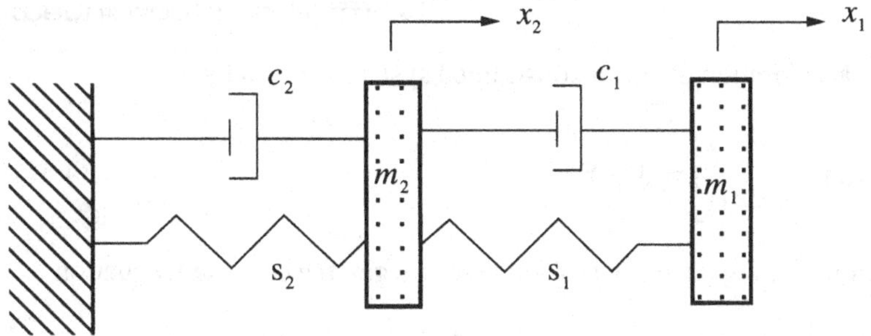

Figure 4.6. Coupled Duffing's oscillator

The two nonlinear springs s_1 and s_2 exert forces denoted by

$$f_i = k_i u_i + k_i^{(nl)} u_i^3, \; i = 1, \; 2, \tag{4.45}$$

where u_i denotes the extension of the ith spring. The second term involving $k_i^{(nl)}$ on the right hand side of equation (4.45) indicates that the spring force has a cubic nonlinearity. The damping elements are linear viscous dampers exerting forces

$$f_i^c = c_i \dot{u}_i, \; i = 1, \; 2. \tag{4.46}$$

The equation of motion of the unconstrained system may be written as

$$M\ddot{\mathbf{x}} = -\left[K\mathbf{x} + C\dot{\mathbf{x}} + \mathbf{g}^{(nl)}\right] = \mathbf{F}, \tag{4.47}$$

where,

$$\mathbf{x} = [x_1 \quad x_2], \text{ and } M = Diag\{m_1, \; m_2\}. \tag{4.48}$$

The matrices

$$K = \begin{bmatrix} k_1 & -k_1 \\ -k_1 & k_1 + k_2 \end{bmatrix} \text{and } C = \begin{bmatrix} c_1 & -c_1 \\ -c_1 & c_1 + c_2 \end{bmatrix}, \tag{4.49}$$

and the vector $\mathbf{g}^{(nl)}$ is given by

$$\mathbf{g}^{(nl)} = \begin{bmatrix} k_1^{(nl)}(x_1 - x_2)^3 \\ k_2^{(nl)} x_2^3 - k_1^{(nl)}(x_1 - x_2)^3 \end{bmatrix}. \tag{4.50}$$

Example 4.4 (continued)

The acceleration of the unconstrained system is given by

$$\mathbf{a}(t) = \begin{bmatrix} a_1(t) \\ a_2(t) \end{bmatrix} = M^{-1}\mathbf{F}. \tag{4.51}$$

Differentiating equation (4.44) twice, we get, the constraint equation

$$\begin{aligned}\ddot{x}_1(t) - \ddot{x}_2(t) = -A_0 \exp(-\alpha t)\{\omega^2 \sin(\omega t) + 2\omega\alpha \cos(\omega t) \\ - \alpha^2 \sin(\omega t)\} = b(t).\end{aligned} \tag{4.52}$$

Hence the matrix $A = [1 \quad -1]$ and $\mathbf{b}$ is a scalar which equals $b(t)$. We then obtain

$$AM^{-1/2} = [m_1^{-1/2} \quad -m_2^{-1/2}], \text{ and } (AM^{-1/2})^+ = \frac{1}{m_1^{-1} + m_2^{-1}} \begin{bmatrix} m_1^{-1/2} \\ -m_2^{-1/2} \end{bmatrix}, \tag{4.53}$$

and the equation of motion of the constrained system becomes

$$\ddot{\mathbf{x}} = \begin{bmatrix} a_1(t) \\ a_2(t) \end{bmatrix} + \frac{m_1 m_2}{m_1 + m_2} \begin{bmatrix} m_1^{-1} \\ -m_2^{-1} \end{bmatrix} \{b(t) - a_1(t) + a_2(t)\}. \tag{4.54}$$

Notice that the effect of the constraint is encapsulated in the second term on the right hand side of equation (4.54).

The force required to be applied to the two masses to "guide" the system so that it satisfies the constraint is explicitly given by

$$\mathbf{F}^c = \begin{bmatrix} F_1^c \\ F_2^c \end{bmatrix} = \frac{m_1 m_2}{m_1 + m_2} \begin{bmatrix} 1 \\ -1 \end{bmatrix} \{b(t) - a_1(t) + a_2(t)\}. \tag{4.55}$$

We next show some numerical results obtained from integrating equation (4.55) using MATLAB. The parameter values describing the system are: $m_1 = 2, m_2 = 1, k_1 = 10, k_2 = 12, k_1^{(nl)} = 1, k_2^{(nl)} = 2, c_1 = 0.1, c_2 = 0.15$. The parameters describing the constraint (4.44) are $A_0 = 1$ and $\omega = 2\pi$. The initial conditions chosen are:

$$x_1(0) = x_2(0) = 1, \text{ and,} \tag{4.56}$$

$$\dot{x}_1(0) = A_0\omega + \dot{x}_2(0),\ \dot{x}_2(0) = 2. \tag{4.57}$$

Regarding the choice of initial conditions which are compatible with the constraint (4.44), we again caution the reader: in the case of holonomic constraints, it might appear at first sight that the satisfaction of the constraint (4.44) requires only that the quantities $x_1(0)$ and $x_2(0)$ be dependent on each other, with no dependence implied on the velocities; yet we know that on differentiating equation (4.44) once, we obtain a constraint that involves the velocities, indicating that these velocities $\dot{x}_1(0)$ and $\dot{x}_2(0)$ cannot be chosen independently. We show this explicitly through equation (4.57). This occurs because the constraint (4.44) must be satisfied at *all* times. Of the four initial conditions, only two can be determined independently.

For different values of the parameter α, the difference $[x_1(t) - x_2(t)]$ can be made to approach zero at different rates. We show the results for $\alpha = 1$. The local error tolerance for the fourth order Runge–Kutta scheme used in MATLAB is set to 10^{-6}.

Figure 4.7(a) shows the response of the system without the constraint (4.44); Figure 4.7(b) shows it with the constraint. The solid line indicates $x_1(t)$; the dashed line represents $x_2(t)$.

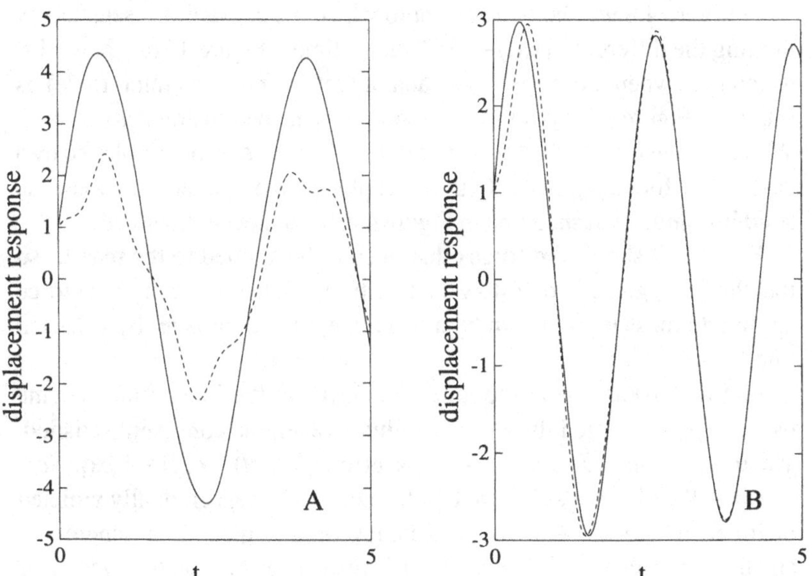

Figure 4.7. Response of the system with and without constraint forces

Example 4.4 (continued)

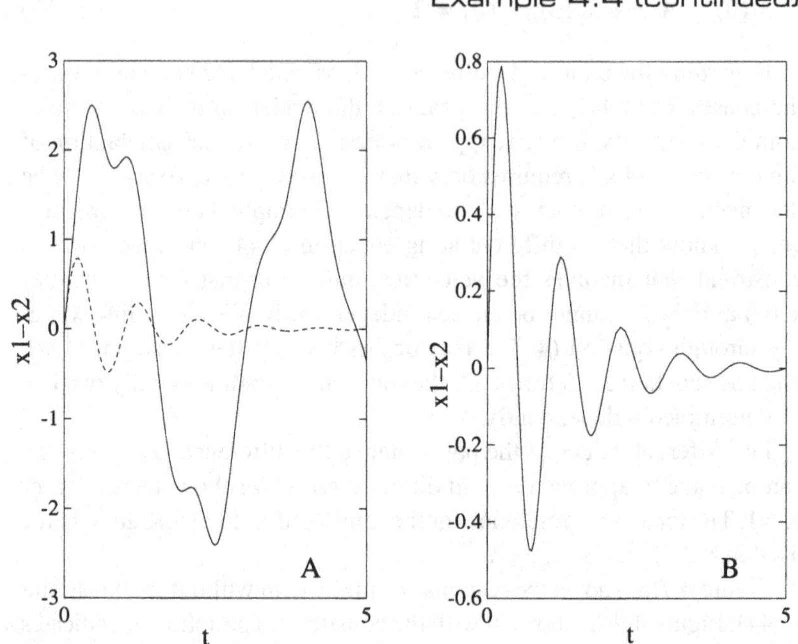

Figure 4.8. Figure illustrating the effect of the constraint force

In Figure 4.8 we show the extent to which the constraint is satisfied by plotting the difference $[x_1(t)-x_2(t)]$ versus time t. Figure 4.8(a) shows this difference when no constraint (and therefore no constraint force) is imposed; Figure 4.8(b) shows the response when the constraint is imposed. We also show in these figures the function $A_0\exp(-\alpha t)\sin(\omega t)$ plotted by a dashed line for comparison. To the scale plotted, the difference between the solid line and the dashed line in Figure 4.8(b) cannot be discerned.

Figure 4.9 shows the forces that need to be applied to the masses so that they are "guided" to follow the constraint (4.44). The constraint force applied to mass m_1 is shown by a solid line, that to mass m_2 by a dashed line.

Figure 4.10(a) shows the numerical error in the integration scheme used, as it is reflected through its ability to keep the constraint satisfied. We denote this numerical error as $\text{error}(t)=x_1(t)-x_2(t)-A_0\exp(-\alpha t)$ $\sin(\omega t)$. We observe as before that the constraint gets gradually violated in time, due to the numerical accuracy of the integration scheme. A smaller local error tolerance of 10^{-12} results in the much lower error shown in Figure 4.10(b).

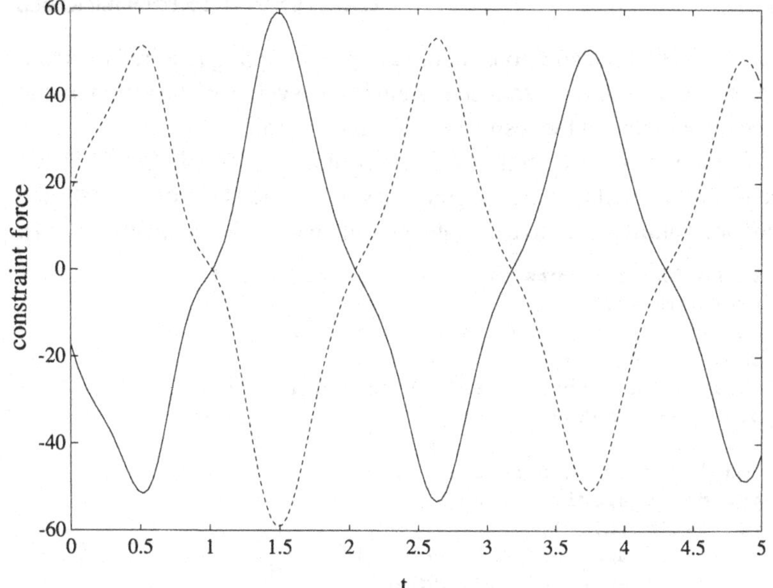

Figure 4.9. The force of constraint acting on the two masses

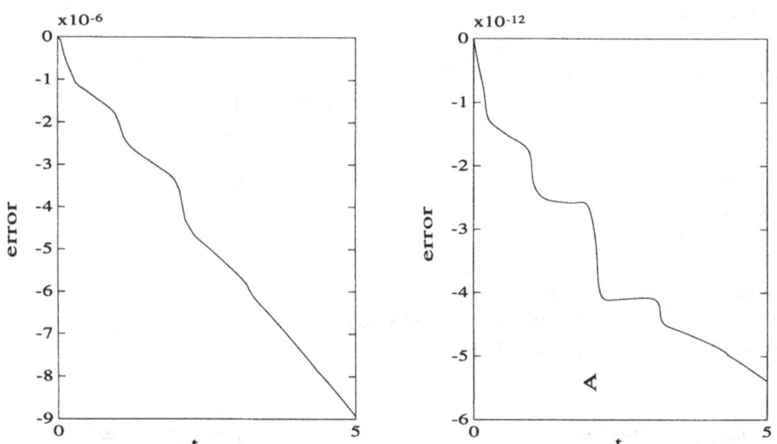

Figure 4.10. Numerical integration error in the satisfaction of the constraints

Example 4.5

We next present here a computer program used to obtain the results in the previous example. This program was used in the MATLAB environment

Example 4.5 (continued)

and shows that the additional computer programming required to obtain the response of the constrained system is but a very small fraction of that necessary to obtain the response of the unconstrained system.

The reader, even though (s)he may be unfamiliar with the MATLAB environment, will find the program easy to follow. The program calls the function "aduffgnew" to calculate the right hand side of equation (4.54).

```
% In MATLAB comments start with the symbol %
format compact;
clear wout error time  time1 l1 f1  f2 con;
global  Am fr alp;
global  alin1 anonlin1 ac1 alin2 anonlin2 ac2;
global MASS  MASSP;

% coupled Duffing's oscillator
% parameter specification
Am=1;fr=2*pi;alp=1;
alin1=10.0; anonlin1=1.0; ac1=0.10;
alin2=12.0; anonlin2=2.0; ac2=0.15;
mass1=2.0; mass2=1.0;
%

MASS=[sqrt(mass1)  0
      0            sqrt(mass2)];
MASSP=[1/(sqrt(mass1))   0
        0                1/(sqrt(mass2)) ];

tol=1.0e-12;
trace=0.;
tstart=0;
tend=5.0

% initial conditions
w01=1.0; w02=1.0; w04=2.0; w03=Am*fr+w04;
w0=[w01  w02 w03  w04]';
%

[time,wout]=ode45('aduffgnew',tstart,tend,w0,tol,trace);
l1=length(time)

% Calculation of error in maintaining constraints
for ka=1:l1,
tte=time(ka);
error(ka)=wout(ka,1)-wout(ka,2)-Am*sin(fr*tte)*exp(-
     alp*tte);
con(ka)=Am*sin(fr*tte)*exp(-alp*tte);
end;
```

```
% Calculation of Constraint Force
a1=[1 -1];
for ka=1:11,
tte=time(ka);
bs=-(fr^2)*sin(fr*tte)-2*fr*alp*cos(fr*tte) +...
    (alp^2)*sin(fr*tte);
b=bs*Am*exp(-alp*tte);
av1=alin1*(wout(ka,1)-wout(ka,2)) +
anonlin1*((wout(ka,1)...
    -wout(ka,2))^3) + ac1*(wout(ka,3)-wout(ka,4));
av2=alin2*(wout(ka,2)) - alin1*(wout(ka,1)-wout(ka,2)) ...
  + anonlin2*(wout(ka,2)^3) -anonlin1*((wout(ka,1)
wout(ka,2))^3)...
- ac1*(wout(ka,3)-wout(ka,4)) + ac2*wout(ka,4);
fgiven=[-av1  -av2]';
acc=(MASSP^2)*fgiven;
force=MASS*pinv(a1*MASSP)*(b-a1*acc);
f1(ka)=force(1);
f2(ka)=force(2);
end;

% plotting results
plot(time,wout(:,1),time,wout(:,2));pause;
plot(time,wout(:,3),time,wout(:,4));pause;
plot(time,(wout(:,1)-wout(:,2)),time,con,'--');pause;
xlabel('Constraint');ylabel('Time');
plot(time,error);
plot(time,f1,time,f2,'--');

% function 'aduffgnew'
function yprime=aduffgnew(tt,w);
yprim(1)=w(3);
yprim(2)=w(4);
% Calculation of the 'given forces', denoted by fgiven
av1=alin1*(w(1)-w(2)) + anonlin1*((w(1)-w(2))^3) +...
    ac1*(w(3)-w(4));
av2=alin2*(w(2)) - alin1*(w(1)-w(2)) +
    anonlin2*(w(2)^3) -...
    anonlin1*((w(1)-w(2))^3) - ac1*(w(3)-w(4)) + ac2*w(4);
fgiven=[-av1  -av2]';
% acceleration of unconstrained system, denoted by acc
acc=(MASSP^2)*fgiven;

% The following lines pertain to constrained motion
 % Matrix A is denoted by a1
 a1=[1  -1];
 % vector b
 bs=-(fr^2)*sin(fr*tt)-2*fr*alp*cos(fr*tt) +...
    (alp^2)*sin(fr*tt);
```

Example 4.5 (continued)

```
b=bs*Am*exp(-alp*tt);
 er=b-a1*acc;
 % Matrix B=A*(M^-.5)
 B=a1*MASSP;
 % Equation (4.13) being used
 beta=1;
 % if beta =0, the motion is unconstrained
% This is the end of the modification needed to take
% account of the constraint
 xxz=acc+beta*MASSP*pinv(B)*er;
yprim(3)=xxz(1);
yprim(4)=xxz(2);
yprime=yprim';
```

PROBLEMS

4.1 Using MATLAB determine the response of the two-bar pendulum described in Example 4.2 using the initial conditions provided. Find the effect of changing the length L_2 by conducting your simulations for $L_2 = 1$, 0.5 and 0.25.

4.2 Consider the system described in Figure 4.6. Use the program provided to determine the control forces required to be applied to the system when the value of α in the constraint equation (4.44) equals 4. Plot the function $[x_1(t) - x_2(t)]$ to verify that the constraint is satisfied when this control force is applied.

4.3 Solve numerically Example 4.4 using *Mathematica* and verify the numerical results obtained in Figures 4.7–4.9.

4.4 After choosing appropriate initial conditions, numerically integrate the equations of motion obtained in Problems 3.1, 3.2, 3.4 and 3.8, using MATLAB and/or *Mathematica*. Do the same for Example 4.1.

4.5 Consider the system shown in Figure 4.11(a), comprising two linear springs k_1 and k_2
this unconstrained system using the coordinates x_1 and x_2, which are shown in the figure. The motion is assumed to occur only in the X-direction. Now, we constrain the system, as shown in Figure 4.11(b), by connecting the two masses by a weightless rigid bar of length L. Write the equation of motion of the constrained system. Find the equilibrium position of the constrained system. Is this equilibrium position unique?

4.6 Consider two particles of masses m_1 and m_2 moving in the XY-plane. The particles are constrained to move in such a way that the line connecting them always passes through a fixed point (say, the origin) in the plane. Assuming that there are no impressed forces on the particles, write the equations of motion for the constrained system.

4.7 A bead of mass m moves on a smooth helical wire defined by the equations $x^2 + y^2 = L^2$; $z = \mu \tan^{-1}\left\{\frac{y}{x}\right\}$ under the action of gravity. The parameters μ and L are

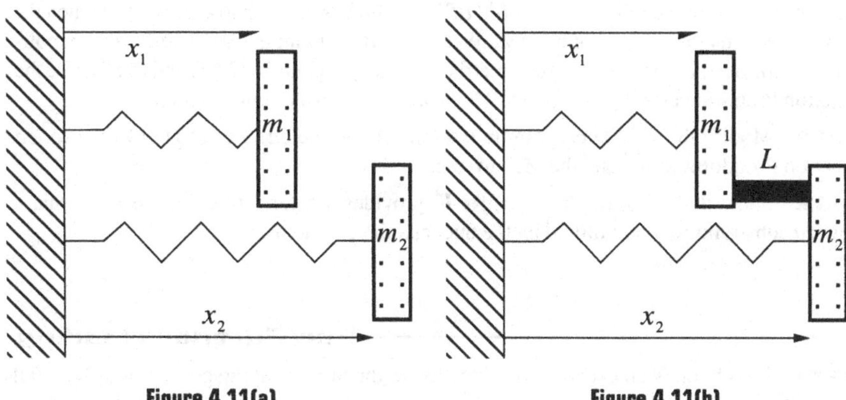

Figure 4.11(a) **Figure 4.11(b)**

positive constants. Write the equations of motion for the particle. Find the forces of constraint. What if the pitch of the helix varies in the vertical direction so that $\mu(t) = \mu_0 + \varepsilon \sin(\omega t)$ where μ_0, ε and ω are constants, with $|\varepsilon| < \mu_0$? See Figure 4.12.

4.8 Consider the second order differential equation governing the motion of an undamped oscillator, $\ddot{x} = -kx$. The energy in the system remains a constant and may be expressed as $\dot{x}^2 + x^2 = 2C$, where C depends on the initial conditions. Look upon this relation as a constraint that must be satisfied in the numerical solution of the second order differential equation. Take $k = 100$, $x(0) = 5$, $\dot{x}(0) = 10$. Use the Runge–Kutta (RK) integration method with different local tolerances and verify how well this constraint is satisfied for long (compared to the period of the system) times. How would you modify the (RK) integration scheme so that this constraint is better satisfied?

4.9 In Figures 4.5 and 4.10 we observe that the Runge–Kutta integration procedure introduces numerical inaccuracies. This causes the error in the satisfaction of the constraints to increase numerically with time. Devise an integration scheme which improves this situation.

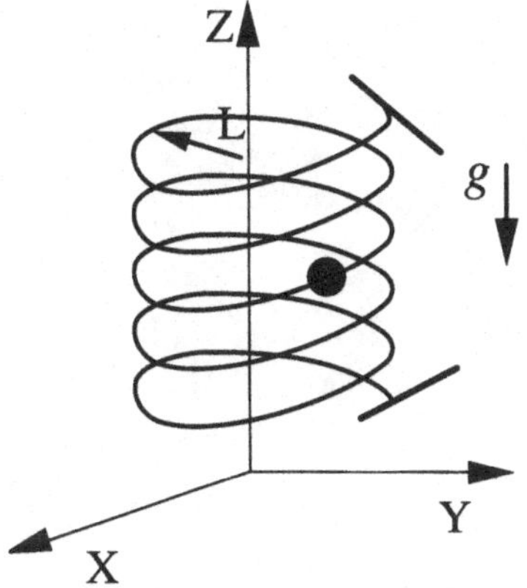

Figure 4.12. Motion of a particle down a helix

4.10 In Example 4.3 we use *Mathematica* to symbolically determine the MP-inverse of the matrix B, using the relation $B^+ = B^T(BB^T)^{-1}$ which is, as we know, only valid when the rank of B equals the number of its rows. Write a short program in *Mathematica* to symbolically determine the MP-inverse of *any* matrix B by using equations (2.86) and (2.87). Use the function PseudoInverse[B] available in *Mathematica* to verify your program.

4.11 Modify the sample program in Example 4.3 to include a call to the *Mathematica* function PseudoInverse to get the MP-inverse.

4.12 Show that the formula (4.17) for $\mathbf{F}^c$ provides a control force for ensuring that a *desired* constraint will be fulfilled by the unconstrained system.

For Further Reading

PFTV This book is an excellent introduction to the numerical integration of differential equations and it explains the Runge–Kutta procedure that is employed in the integration of the examples in this chapter. Chapter 15 is devoted to numerical integration of differential equations. It explains the concepts of local error tolerance as well as other integration schemes which the reader might want to use to integrate the fundamental equation.

W1 This is a book written on the symbolic language *Mathematica* which we have used in Example 4.2. The reader may want to consult the book if some of the statements in Example 4.3 are unclear.

M1 This book, put out by The Math Works, Inc., gives a brief tutorial and a summary of commands which are used in the MATLAB environment. Those not familiar with MATLAB might want to refer to the book to better understand the structure and statements of the program exhibited in Example 4.5.

5 Elements of Lagrangian Mechanics

In this chapter we will introduce the reader to the fundamentals of Lagrangian mechanics. The reader has by now had a fair exposure to the fundamental equation, and has seen how to describe constrained motion in the framework of Cartesian coordinates. Here we will delve deeper into the basics of mechanics and introduce some of the key concepts: the principle of virtual work, and the concept of generalized coordinates. We will be taking on a more rigorous approach and amplifying on several of the issues which we had to gloss over in earlier chapters so that the reader could grasp the big picture better. The first part of this chapter is in a sense going back to better appreciate the issues involved. The second part of the chapter deals with Lagrange's equations.

The chapter introduces, in a rigorous way, what we mean by an "unconstrained" system described in terms of the Lagrangian coordinates chosen to describe the system's configuration. We show that, for a given physical situation, the analyst may exercise considerable choice in what (s)he considers to be an unconstrained system. For any specific choice of an unconstrained system, we then show how to obtain the explicit equations of motion of the constrained mechanical system in terms of the chosen Lagrangian coordinates.

Virtual Displacements

The concept of virtual displacements plays a central role in mechanics. We have already seen that equality constraints of interest to analytical dynamics are of the holonomic type or the nonholonomic type. Thus in a Cartesian coordinate frame of reference consider a system of n particles which are subjected to the h holonomic constraints

$$f_i(\mathbf{x}, t) = 0, \; i = 1, 2, \ldots, h, \tag{5.1}$$

and the r nonholonomic constraints

$$\sum_{j=1}^{j=3n} d_{ij}(\mathbf{x},t)\dot{x}_j + g_i(\mathbf{x},t) = 0, \; i = 1, 2, \ldots, r, \tag{5.2}$$

where we have, as usual, denoted the $3n$-vector describing the positions of each of the particles by $\mathbf{x} = [x_1 \; x_2 \cdots x_{3n-1} \; x_{3n}]^T$. We could replace the finite constraint (5.1) by the differential constraint as we did before in Chapter 3 to give

$$\sum_{j=1}^{j=3n} \frac{\partial f_i(\mathbf{x},t)}{\partial x_j}\dot{x}_j + \frac{\partial f_i(\mathbf{x},t)}{\partial t} = 0, \; i = 1, 2, \ldots, h. \tag{5.3}$$

Equations (5.2) and (5.3) have the same form and can be written concisely as the $m = h + r$ equations

$$\sum_{j=1}^{j=3n} d_{ij}(\mathbf{x},t)\dot{x}_j + g_i(\mathbf{x},t) = 0, \; i = 1, 2, \ldots, m, \tag{5.4}$$

or, as

$$D(\mathbf{x},t)\dot{\mathbf{x}} = \mathbf{g} \tag{5.5}$$

where $d_{ij}(\mathbf{x},t)$ is the i–jth element of the m by $3n$ matrix D, the ith row of the vector $\mathbf{g}$ is $-g_i(\mathbf{x},t)$ and $\dot{\mathbf{x}}(t) = [\dot{x}_1 \; \dot{x}_2 \cdots \dot{x}_{3n-1} \; \dot{x}_{3n}]^T$. In terms of infinitesimal displacements, equation (5.4) can be expressed as

$$\sum_{j=1}^{j=3n} d_{ij}(\mathbf{x},t)dx_j + g_i(\mathbf{x},t)dt = 0, \; i = 1, 2, \ldots, m. \tag{5.6}$$

We will assume that of the set of m equations (5.6), k of them with $k < 3n$, are linearly independent or, more precisely, that the rank of the matrix D is k.

We note that we could have differentiated equation (5.5) once with respect to time to obtain, as before (see equation (3.28)), the constraint equation

$$A(\mathbf{x},t)\ddot{\mathbf{x}} = \mathbf{b}(\mathbf{x},\dot{\mathbf{x}},t), \tag{5.7}$$

where the m by $3n$ matrix A is identical to D, i.e.,

$$A(\mathbf{x},t) = D(\mathbf{x},t). \tag{5.8}$$

For any given position $\mathbf{x}(t)$ of the system at any time t, if the velocity $\dot{\mathbf{x}}(t)$ satisfies the m equations (5.4), then this $3n$-vector of velocity $\dot{\mathbf{x}}(t)$ is called a *possible* velocity. For any given position $\mathbf{x}(t)$ that the system occupies at the instant of time t, there are infinitely many such *possible* velocity vectors. Only one of these is realized in the actual motion of the mechanical system.

Alternatively, we can think of a *possible* infinitesimal displacement $d\mathbf{x}(t) = \dot{\mathbf{x}}dt$ whose $3n$ components satisfy the constraint equations (5.6) which are written in terms of infinitesimal displacements.

Given the position vector $\mathbf{x}(t)$ at a time t, consider now two different *possible* velocity vectors, at that instant, say, $\dot{\mathbf{x}}_1(t)$ and $\dot{\mathbf{x}}_2(t)$. Since the components of both these $3n$-vectors must satisfy the equations (5.4), their difference $\Delta\dot{\mathbf{x}}(t) = \dot{\mathbf{x}}_1(t) - \dot{\mathbf{x}}_2(t)$ must satisfy the homogeneous equations

$$\sum_{j=1}^{j=3n} d_{ij}(\mathbf{x}, t)\Delta\dot{x}_j = 0, \; i = 1, 2, \ldots, m \tag{5.9}$$

where the jth component of the vector $\Delta\dot{\mathbf{x}}(t)$ is

$$\Delta\dot{x}_j = \dot{x}_{1_j} - \dot{x}_{2_j} = \frac{d(x_{1_j} - x_{2_j})}{dt} = \frac{\delta x_j}{dt}. \tag{5.10}$$

Using (5.10) in equations (5.9) and multiplying throughout by dt we get

$$\sum_{j=1}^{j=3n} d_{ij}(\mathbf{x}, t)\delta x_j = 0, \; i = 1, 2, \ldots, m. \tag{5.11}$$

The $3n$-vector $\delta\mathbf{x} = d(\mathbf{x}_1 - \mathbf{x}_2) = \Delta\dot{\mathbf{x}}dt$ is called the vector of virtual displacements. Note that a virtual displacement is just the difference between two *possible* displacements of the system at time t, both *possible* displacements being considered from the same position $\mathbf{x}(t)$ of the system. Since the *actual* displacement of a mechanical system is also a *possible* displacement, a virtual displacement at time t can also be defined as the difference between any *possible* displacement at time t and the actual displacement at time t. We shall use this central idea quite a bit in this book.

More generally, *any* $3n$-vector $\delta\mathbf{x}$ whose components satisfy the homogeneous equations

$$\sum_{j=1}^{j=3n} d_{ij}(\mathbf{x}, t)\,\delta x_j = 0, \; i = 1, 2, \ldots, m \tag{5.12}$$

or, equivalently, *any* $3n$-vector $\delta\mathbf{x}$ which satisfies the matrix equation

$$D\,\delta\mathbf{x} = A\,\delta\mathbf{x} = 0 \tag{5.13}$$

is called a virtual displacement vector. Note that in arriving at equation (5.13) we have used equation (5.8). Thus, by (5.9), $\Delta\dot{\mathbf{x}}(t)$ qualifies for being a virtual displacement vector.

The effect of the m constraints represented by equations (5.4) is thus encapsulated in equations (5.12). These relations say that in the presence of constraints the components of the virtual displacement vector $\delta\mathbf{x}$ *cannot* be chosen arbitrarily; for, these components must satisfy equations (5.12). On the other hand, were there no constraints on the system, then these components could be chosen arbitrarily for they would not be required to satisfy any such relationships.

Notice that equations (5.12) differ from equations (5.6), which define *possible* displacements, by the absence of the term $g_i(\mathbf{x}, t)dt$. For holonomic constraints this term corresponds to $\dfrac{\partial f_i(\mathbf{x}, t)}{\partial t}dt$. It is for this reason that virtual displacements are said to coincide with *possible* displacements *in the case of "frozen" constraints*. Indeed, when we freeze time t, which enters in the equations describing the finite constraints (5.1), these constraints congeal in the configuration that they had at time t. Then the terms $\dfrac{\partial f_i(\mathbf{x}, t)}{\partial t}$ do not appear when differentiating the functions $f_i(\mathbf{x}, t)$, and the h equations (5.3) are of the form (5.12). For the nonintegrable Pfaffian constraints of the form (5.6), "freezing" signifies eliminating the term $g_i(\mathbf{x}, t)dt$ and fixing the t which enters explicitly in the coefficients $d_{ij}(\mathbf{x}, t)$. Then the r equations (5.2) again will coincide in form with equations (5.12). The following example illustrates the concept of virtual displacements.

Example 5.1

a. A wire placed horizontally has a bead moving on it. Describe a virtual displacement of the bead.

Here, let us consider two possible velocities of the bead (see Figure 5.1). The horizontal velocity $\dot{x}_1$ is a possible velocity, and the velocity $\dot{x}_2$ is a possible velocity.

Thus $dx_1 = \dot{x}_1 dt$ and $dx_2 = \dot{x}_2 dt$. Hence $\delta x = dx_1 - dx_2$, which might be considered as a certain vector along the wire denoted by dx.

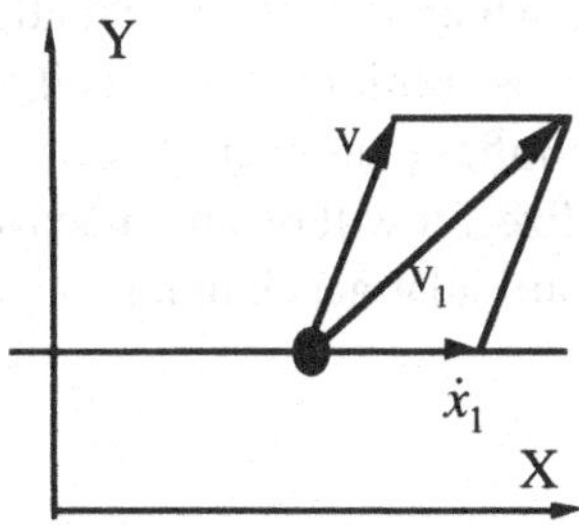

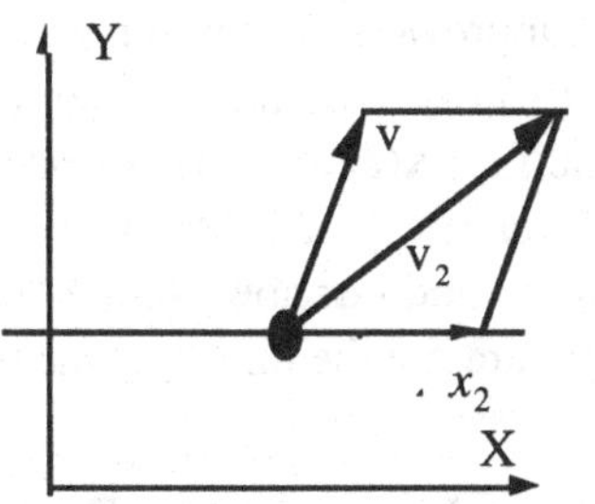

Figure 5.1. A bead moving on a horizontal wire. The wire translates with velocity v. Two *possible* velocities of bead, υ_1 and υ_2, are shown.

b. The wire in part (a) above while remaining horizontal is translating with a velocity $\mathbf{v} = [\dot{x}_w \;\; \dot{y}_w]^T$. In this case, the possible velocity $\mathbf{v}_1$, of the bead is obtained by adding to $\mathbf{v}$ the velocity $\dot{x}_1$ so that $\mathbf{v}_1 = [\dot{x}_w + \dot{x}_1 \;\; \dot{y}_w]^T$. Similarly another possible velocity is $\mathbf{v}_2 = [\dot{x}_w + \dot{x}_2 \;\; \dot{y}_w]^T$. The difference in these two velocities qualifies as a virtual displacement vector with components $[\dot{x}_1 - \dot{x}_2 \;\; 0]^T$.

The virtual displacement $\delta x = (\dot{x}_1 - \dot{x}_2)dt$ constitutes a *possible* displacement for the wire "frozen" in its position at time t.

If k of the equations (5.4) are linearly independent, then of the $3n$-components of the vector $\delta\mathbf{x}$, there will be $d = 3n - k$ independent virtual displacement, where d is called the number of degrees of freedom of the system.

We remind the reader that for a given initial configuration of the system, $\mathbf{x}(t_0)$, at some time t_0, and the m constraint equations of the form (5.4) the generalized virtual displacement vectors at time t_0, are defined as *any* vectors $\mathbf{v}$ which satisfy the relation

$$A(\mathbf{x}, t_0)\mathbf{v} = \mathbf{0} \tag{5.14}$$

where the matrix A is the m by n matrix whose elements are $d_{ij}(\mathbf{x}, t)$.

The concept of the virtual displacement vector at time t_0 can be further extended. We shall show in Chapter 7 that given $\mathbf{x}(t_0)$ and $\dot{\mathbf{x}}(t_0)$, even for systems with more general constraints of the form $A(\mathbf{x}, \dot{\mathbf{x}}, t)\ddot{\mathbf{x}} = \mathbf{b}(\mathbf{x}, \dot{\mathbf{x}}, t)$, a generalized virtual displacement vector is *any* vector which satisfies the relation $A(\mathbf{x}, \dot{\mathbf{x}}, t)\mathbf{v} = \mathbf{0}$. This extension, which accrues under suitable assumptions, will be crucial for our purposes, as we will see later on.

For the system to move in a manner compatible with the constraints, its acceleration $\ddot{\mathbf{x}}$ at every instant of time must satisfy equation (5.7). Analogous to our previous discussion we can call any vector $\ddot{\mathbf{x}}$ which satisfies the equation set (5.7) as a *possible* acceleration. For a system whose position and velocity at time t are $\mathbf{x}(t)$ and $\dot{\mathbf{x}}(t)$ respectively, there are an infinite number of *possible* accelerations, of which only *one* actually materializes. It will be our endeavor now to find out how Nature "chooses" that one particular acceleration vector from among the infinitely many *possible* ones.

The Lagrangian Formulation

Let the $3n$-vector of force $\mathbf{F}$ impressed on the system of particles be known; the vector $\mathbf{F}$ may be prescribed as a known function of $\mathbf{x}$, $\dot{\mathbf{x}}$ and t. If the m constraints (5.4) were absent, then clearly we can obtain the equations of motion of the unconstrained system at each instant of time t as

$$M(t)\mathbf{a} = \mathbf{F}(t), \tag{5.15}$$

where the diagonal $3n$ by $3n$ mass matrix M has the masses of the particles down the diagonal, as usual, in sets of threes. But because of the m constraint equations (5.4), acceleration $\mathbf{a}$ may not be compatible with the constraints. If the acceleration vector $\mathbf{a}$ corresponding to the unconstrained system does not satisfy equation (5.7), then additional forces of constraint $\mathbf{F}^c$ will be exerted on the particles of the system so that

$$M\ddot{\mathbf{x}} = \mathbf{F} + \mathbf{F}^c. \tag{5.16}$$

Notice that unlike the constraint force $\mathbf{F}^c$, the force vector $\mathbf{F}$ is a preassigned vector and is often termed the *impressed* force. Now let us say that we know the position and the velocity of each of the particles of the system at some time t. Furthermore, let them be compatible with the constraints at time t. Our aim is to determine the motion of the system for subsequent times. More specifically, we want to find the acceleration of each of the particles of the system at time t.

If no further information is provided about the nature of the constraint force $\mathbf{F}^c$, the problem becomes indeterminate, because we need to know each of the $3n$-vectors $\ddot{\mathbf{x}}(t)$ and $\mathbf{F}^c$ at each instant of time. This involves $6n$ unknowns. But we have k independent equations (5.4), with $k < 3n$ and an additional $3n$ equations given by (5.16). Thus the number of unknowns exceeds the number of scalar relations ($6n > 3n + k$, since $k < 3n$) and the problem is indeterminate.

To render the basic problem of analytical dynamics determinate, we need an additional $6n - (3n + k) = 3n - k = d$ independent relations. One way of obtaining these relations is to consider the class of *ideal* constraints. This class of constraints, it turns out, is important in many practical situations and has wide applicability.

Constraints are termed ideal if the work done by the force of constraint $\mathbf{F}^c$ in *every* virtual displacement $\delta\mathbf{x}$ is zero, i.e., if

$$\delta\mathbf{x}^T\mathbf{F}^c = (\mathbf{F}^c)^T\delta\mathbf{x} = 0. \tag{5.17}$$

This relation can also be thought of as a relation defining the nature of the forces of constraint that we will consider in this book. It is a generalization both of the principle of Virtual Work in statics first stated by Johann Bernoulli, and of D'Alembert's principle for a single rigid body. More generally, this principle states that $\delta\mathbf{x}^T\mathbf{F}^c \geq 0$, the inequality being considered when we have "one-sided" constraints. The reader may refer to the section "For Further Reading" in this regard. In this book we shall not use one-sided constraints, and equation (5.17) will be utilized.

But how do we manage to obtain d independent equations using (5.17) so as to make the basic problem of mechanics determinate? As we said earlier, the virtual displacements satisfy the equation $A\delta\mathbf{x} = 0$. Since k of these equations are independent, we can express, say, k of the components of $\delta\mathbf{x}$ in terms of the remaining $3n - k = d$ components. Thus k of the components of the vector $\delta\mathbf{x}$ in equation (5.17) can be expressed in terms of the d remaining components, which are then independent. Since equation (5.17) must be satisfied for every generalized virtual vector, we can set to zero the coefficients of each of these d independent components in equation (5.17), thereby obtaining the additional d equations needed to make the problem determinate! We shall revisit this in Chapter 8.

Substituting from equation (5.16) for $\mathbf{F}^c$, equation (5.17) can be rewritten as

$$\delta\mathbf{x}^T(M\ddot{\mathbf{x}} - \mathbf{F}) = 0, \text{ for all } \delta\mathbf{x} \text{ such that } A\delta\mathbf{x} = 0. \tag{5.18}$$

This is the *basic equation of analytical mechanics*. It states that given a system in motion, at each instant of time, the system evolves so that the *scalar* quantity $\delta\mathbf{x}^T(M\ddot{\mathbf{x}} - \mathbf{F})$ equals zero for all generalized virtual displacements $\delta\mathbf{x}$. Its significance was first realized by J. Lagrange, who stated this equation in about 1760.

Note again that the requirement that $A\delta\mathbf{x} = 0$ means that the components of the virtual displacement vector $\delta\mathbf{x}$ *cannot obviously be assigned independently of each other*, for they must satisfy the relation $A\delta\mathbf{x} = 0$.

However, if all the constraints vanish, then the requirement that the virtual displacement vector satisfies the relation $A\delta\mathbf{x} = 0$ disappears, and relation (5.18) would require that

$$\delta\mathbf{x}^T(M\ddot{\mathbf{x}} - \mathbf{F}) = 0 \text{ for all (and any) vectors } \delta\mathbf{x}. \tag{5.19}$$

Now the components of the vector $\delta\mathbf{x}$ could be chosen independently of each other. Hence we could choose all but the jth component of $\delta\mathbf{x}$ to be zero. This would thus lead us to require that $M\ddot{\mathbf{x}} - \mathbf{F} = 0$, or that the acceleration $\ddot{\mathbf{x}} = M^{-1}\mathbf{F}$, which is exactly the acceleration of the unconstrained system as seen from equation (5.15)!

That the components of the virtual displacement vector can be chosen independently of each other in the absence of any constraints is an important insight that we will use later on. When this happens, the components are often said to be independent of each other.

It must have crossed the reader's mind to inquire if the constraints we have considered so far are all the possible types of constraints that arise in mechanics. Indeed they are not. The most conspicuous example is provided by a particle moving on a rough surface so that there is a frictional force acting between the particle and the surface. Such constraints, though important from a practical standpoint, do not lend themselves easily to the methods developed in this book. This is because such constraints are not *ideal* constraints. We have assumed that all the forces acting on a representative particle of a system can be categorized as given forces or constraint forces (see equation (5.16)). The constraint forces satisfy equation (5.17); the given forces are specified a priori. But the reaction of the rough surface on the particle described above is neither a given force nor a constraint force in the sense just mentioned. While such constraints can indeed be handled, Lagrangian formulations for such systems usually turn out to be rather cumbersome. Such constraints will therefore be excluded from consideration in this book. The interested reader may look at the end of this chapter in the "For Further Reading" list.

Example 5.2

Consider a particle of mass m whose position in 3-dimensional space is specified by its Cartesian coordinates (x, y, z). Let the particle be subjected to the effective forces F_x, F_y, F_z in the X-, Y- and Z-directions. Furthermore, let it be subjected to the constraints

$$\dot{y} = z\dot{x} + \dot{z}, \text{ and} \tag{5.20}$$

$$\dot{y} = z^2\dot{x}. \tag{5.21}$$

We want to write the equations of motion of the system in the presence of these constraints.

Equations (5.20) and (5.21) can be expressed in Pfaffian form as

$$dy = zdx + dz \tag{5.22}$$

$$dy = z^2 dx. \tag{5.23}$$

The matrix $A = \begin{bmatrix} z & -1 & 1 \\ z^2 & -1 & 0 \end{bmatrix}$ has rank 2 even at $z = 0$ and $z = 1$, for the determinant of the 2 by 2 matrix $A_1 = \begin{bmatrix} z & -1 \\ z^2 & -1 \end{bmatrix}$ is nonzero.

A virtual displacement is any vector $\mathbf{u} = [u_1\ u_2\ u_3]^T$ such that

$$A\mathbf{u} = \begin{bmatrix} z & -1 & 1 \\ z^2 & -1 & 0 \end{bmatrix}\mathbf{u} = 0 \tag{5.24}$$

or

$$\begin{aligned} zu_1 - u_2 + u_3 &= 0 \text{ and} \\ z^2 u_1 - u_2 + 0u_3 &= 0. \end{aligned} \tag{5.25}$$

These are two equations relating the three components of the virtual displacement vector $\mathbf{u}$. These three components therefore cannot be independent. Alternately put, we can always solve for the dependent quantities u_1 and u_2 given the independent quantity u_3 from the equations

$$A_1\begin{bmatrix} u_1 \\ u_2 \end{bmatrix} = \begin{bmatrix} z & -1 \\ z^2 & -1 \end{bmatrix}\begin{bmatrix} u_1 \\ u_2 \end{bmatrix} = u_3\begin{bmatrix} -1 \\ 0 \end{bmatrix} \tag{5.26}$$

noting that A_1 is nonsingular (except at $z = 0$ and $z = 1$). Thus we can think of the vector $\mathbf{u}$ as composed of the components u_1 and u_2, which are dependent on the quantity u_3; the quantity u_3 may be chosen at will.

In the presence of the constraints, at each instant of time t, the forces of constraints must satisfy relation (5.17) so that

$$F_x^c u_1 + F_y^c u_2 + F_z^c u_3 = 0 \tag{5.27}$$

for all vectors $\mathbf{u}$ which satisfy equation (5.24).

Let us multiply the first of equations (5.25) with a scalar $-\lambda_1$ and the second with the scalar $-\lambda_2$. Adding these equations on to equation (5.27) we get

$$F_x^c u_1 + F_y^c u_2 + F_z^c u_3 - \lambda_1(zu_1 - u_2 + u_3) - \lambda_2(z^2 u_1 - u_2 + 0u_3) = 0, \tag{5.28}$$

Example 5.2 (continued)

or,

$$(F_x^c - \lambda_1 z - \lambda_2 z^2)u_1 + (F_y^c + \lambda_1 + \lambda_2)u_2 + (F_z^c - \lambda_1 - 0\lambda_2)u_3 = 0. \tag{5.29}$$

Now we can choose the multipliers λ_1 and λ_2 to be such that the terms in the first two brackets of (5.29) vanish. This would require that we solve the equations

$$\lambda_1 z + \lambda_2 z^2 = F_x^c, \tag{5.30a}$$

and

$$-\lambda_1 - \lambda_2 = F_y^c, \tag{5.30b}$$

or,

$$A_1^T \begin{bmatrix} \lambda_1 \\ \lambda_2 \end{bmatrix} = \begin{bmatrix} z & z^2 \\ -1 & -1 \end{bmatrix} \begin{bmatrix} \lambda_1 \\ \lambda_2 \end{bmatrix} = \begin{bmatrix} F_x^c \\ F_y^c \end{bmatrix}. \tag{5.31}$$

As already pointed out, $Det(A_1^T) = Det(A_1) \neq 0$ except at $z = 0$ and $z = 1$; hence, we can determine from (5.31), at least in principle, the values of λ_1 and λ_2 that satisfy equations (5.30).

Thus we have arranged for the first two brackets in equation (5.29) to be zero. Hence equation (5.29) reduces to

$$(F_z^c - \lambda_1 - 0\lambda_2)u_3 = 0. \tag{5.32}$$

But equation (5.29) must be valid for *all* virtual displacement vectors **u**, and, as we said before, we may choose u_3 at will. Since (5.32) is true for arbitrary values of u_3, we must have

$$(F_z^c - \lambda_1 - 0\lambda_2) = 0. \tag{5.33}$$

Notice that though the quantities u_1, u_2 and u_3 are *not* independent, through a proper choice of the multipliers λ_1 and λ_2, we have deduced that each of the brackets in equation (5.29) individually equals zero! Furthermore, since equation (5.29) must be true at each instant of time, the values of these multipliers will change with time. They will thus be functions of t.

Equations (5.30) and (5.33) can also be expressed as

$$\begin{bmatrix} F_x^c \\ F_y^c \\ F_z^c \end{bmatrix} = \begin{bmatrix} z & z^2 \\ -1 & -1 \\ 1 & 0 \end{bmatrix} \begin{bmatrix} \lambda_1(t) \\ \lambda_2(t) \end{bmatrix} = A^T \begin{bmatrix} \lambda_1(t) \\ \lambda_2(t) \end{bmatrix}. \tag{5.34}$$

Now the equations of motion for the constrained system, by (5.16), are

$$\begin{bmatrix} m & 0 & 0 \\ 0 & m & 0 \\ 0 & 0 & m \end{bmatrix} \begin{bmatrix} \ddot{x} \\ \ddot{y} \\ \ddot{z} \end{bmatrix} = \begin{bmatrix} F_x \\ F_y \\ F_z \end{bmatrix} + \begin{bmatrix} F_x^c \\ F_y^c \\ F_z^c \end{bmatrix} = \begin{bmatrix} F_x \\ F_y \\ F_z \end{bmatrix} + A^T \begin{bmatrix} \lambda_1 \\ \lambda_2 \end{bmatrix}. \tag{5.35}$$

To these equations we must add the constraint equations

$$\dot{y} = z\dot{x} + \dot{z}, \tag{5.36a}$$

and

$$\dot{y} = z^2 \dot{x}. \tag{5.36b}$$

The equation set (5.35) along with (5.36) constitutes a set of five equations in the five unknowns x, y, z, λ_1 and λ_2. This procedure was first developed by J. Lagrange, and the scalars λ_1 and λ_2 are called Lagrange multipliers. Equation (5.35) is nothing short of remarkable, for we have been able to express the force of constraint in terms of an auxiliary 2-vector $\boldsymbol{\lambda}$. Solution of this system of equations may still be a nontrivial task, for the constraint equations (5.36) are nonlinear in the variables.

We can generalize (for a proof of this the reader will have to wait till Chapter 8) the above example and write the equations of motion of a system subjected to the m linearly independent constraint equations (5.5) (so that the m by n matrix $A = D$ has rank m) in terms of Lagrange multipliers, $\boldsymbol{\lambda}(t) = [\lambda_1(t)\ \lambda_2(t) \cdots \lambda_m(t)]^T$, as

$$M\ddot{\mathbf{x}} = \mathbf{F} + A^T\boldsymbol{\lambda} \tag{5.37}$$

along with the constraint equation (note that $A = D$)

$$A\dot{\mathbf{x}} = \mathbf{g}. \tag{5.38}$$

Here the m-vector $\boldsymbol{\lambda}(t)$ of Lagrange multipliers and the $3n$-vector $\mathbf{x}$ of position are to be determined using the $(3n + m)$ equations (5.37) and (5.38). Along with these two sets of equations we need a set of initial conditions $\mathbf{x}(t_0)$ and $\dot{\mathbf{x}}(t_0)$ which must satisfy the constraint equation (5.38) so that the problem is well-posed and fully specified.

This general formulation of the equations of motion for constrained systems was given by J. Lagrange in his famous treatise called *Mechanique Analytique* in 1788.

Example 5.3

Consider a particle which is subjected to the externally impressed forces F_x, F_y, F_z and which is constrained to remain on the surface

$$f(x, y, z) = 0. \tag{5.39}$$

We can differentiate this equation to yield

$$\frac{\partial f}{\partial x}dx + \frac{\partial f}{\partial y}dy + \frac{\partial f}{\partial z}dz = 0. \tag{5.40}$$

The mechanism by which the actual motion is made to conform to equation (5.39) is as follows. An additional force of reaction created by the surface is brought into play. This force is normal to the surface. This is the only limitation on this force of reaction *a priori*; its magnitude and sign adjust themselves so that the motion of the particle, under the influence of both the applied impressed forces and this reaction force, is confined to the surface (5.39). This force of reaction is called the constraint force.

A generalization of the character of this force of constraint is provided by D'Alembert's principle, which states that the total work done by this force under any and all virtual displacements must sum to zero. From our definition of a virtual displacement, by equation (5.40) the vector $\mathbf{u} = [dx\ dy\ dz]^T$ qualifies as a virtual displacement vector. Thus we require that the components F_x^c, F_y^c, F_z^c, of the force of constraint satisfy the equation

$$F_x^c dx + F_y^c dy + F_z^c dz = 0 \tag{5.41}$$

for all vectors $\mathbf{u}$ satisfying equation (5.40). From equations (5.40) and (5.41) we get

$$\frac{F_x^c}{\partial f/\partial x} = \frac{F_y^c}{\partial f/\partial y} = \frac{F_z^c}{\partial f/\partial z} = \lambda, \text{ say,} \tag{5.42}$$

so that the equations of motion of the constrained system are

$$\begin{aligned} m\ddot{x} &= F_x + F_x^c = F_x + \lambda\frac{\partial f}{\partial x}, \\ m\ddot{y} &= F_y + F_y^c = F_y + \lambda\frac{\partial f}{\partial y}, \text{ and} \\ m\ddot{z} &= F_z + F_z^c = F_z + \lambda\frac{\partial f}{\partial z}. \end{aligned} \tag{5.43}$$

Along with this equation we need to adjoin equation (5.39) so that we can determine the four quantities x, y, z and λ.

Notice that equation (5.42) states that the force of constraint, at each instant of time, is along the direction of the normal to the surface since the vector $\begin{bmatrix} \frac{\partial f}{\partial x} & \frac{\partial f}{\partial y} & \frac{\partial f}{\partial z} \end{bmatrix}^T$ is normal to the surface (5.39). Its magnitude and sign are determined by the magnitude and sign of λ.

Explicit Determination of the Lagrange Multipliers

As stated before, the determination of the m-vector of Lagrange multipliers is usually a difficult job and depends on the specific problem at hand. When the constraints are nonintegrable their determination is by no means a simple or routine matter. However, comparing the fundamental equation of motion, namely,

$$M\ddot{\mathbf{x}} = \mathbf{F} + M^{1/2}(AM^{-1/2})^+(\mathbf{b} - A\mathbf{a}) \tag{5.44}$$

with equations (5.16) and (5.37), we get the force of constraint explicitly as

$$\mathbf{F}^c = A^T\boldsymbol{\lambda} = M^{1/2}(AM^{-1/2})^+(\mathbf{b} - A\mathbf{a}) = A^T(AM^{-1}A^T)^+(\mathbf{b} - A\mathbf{a}). \tag{5.45}$$

If, further, as is commonly assumed in the Lagrangian formulation, the rank of the matrix A is m, i.e., the m constraint equations are linearly independent, then the m by m matrix AA^T has rank m, and so is nonsingular. Premultiplying equation (5.45) first by A, and then by $(AA^T)^{-1}$, we get

$$\boldsymbol{\lambda} = (AM^{-1}A^T)^{-1}(\mathbf{b} - A\mathbf{a}). \tag{5.46}$$

Through the use of the fundamental equation we have thus obtained the explicit equation for the Lagrange multiplier vector $\boldsymbol{\lambda}$! Notice that since the rank of the matrix A is m, the rank of $(AM^{-1}A^T)$ is also m, and hence $(AM^{-1}A^T)^+ = (AM^{-1}A^T)^{-1}$. Also, in this case we find that the vector is unique. In general, of course, when the rank of A is less than m, this is not true; nor is equation (5.46) then valid. See Problem 5.17.

Example 5.4

Consider a bead moving on a wire, as described in Example 3.4. Let us state the problem in the Lagrangian framework and directly determine the Lagrange multiplier $\lambda(t)$.
As before, the equation for the unconstrained motion in the Cartesian coordinate frame is given by (see equation (3.45))

$$\begin{bmatrix} m & 0 \\ 0 & m \end{bmatrix} \mathbf{a} = \begin{bmatrix} F_x \\ F_y \end{bmatrix}, \tag{5.47}$$

the matrix $A = [-\tan\alpha \quad 1]$, and $\mathbf{b} = 0$, a scalar. Hence using equations (5.37) and (5.38) we have

$$\begin{bmatrix} m & 0 \\ 0 & m \end{bmatrix} \begin{bmatrix} \ddot{x} \\ \ddot{y} \end{bmatrix} = \begin{bmatrix} F_x \\ F_y \end{bmatrix} + \lambda(t) \begin{bmatrix} -\tan\alpha \\ 1 \end{bmatrix} \tag{5.48}$$

and

$$y = x\tan\alpha. \tag{5.49}$$

Equations (5.48) and (5.49) provide three equations for the three unknowns $x(t)$, $y(t)$ and $\lambda(t)$. Rather than solve this set of differentio-algebraic equations, we can use equation (5.46) directly to obtain $\lambda(t)$.

Since $(AM^{-1}A^T) = \left[\dfrac{(\tan^2\alpha + 1)}{m}\right]$, we have,

$$\begin{aligned} \boldsymbol{\lambda} &= (AM^{-1}A^T)^{-1}(\mathbf{b} - A\mathbf{a}) \\ &= \left[\frac{(\tan^2\alpha + 1)}{m}\right]^{-1} \cdot \left\{\frac{F_x}{m}\tan\alpha - \frac{F_y}{m}\right\} = \frac{(F_x\tan\alpha - F_y)}{(\tan^2\alpha + 1)}, \end{aligned} \tag{5.50}$$

which is exactly what we had in equation (3.50)! We have explicitly determined the Lagrange multiplier without having to solve the system of equations (5.48) and (5.49).

The reader may now go back to several of the examples in Chapter 3 and see the way the Lagrange multiplier vector $\boldsymbol{\lambda}$ enters the equations of motion.

It is also instructive to see what would happen if two of the constraints were not linearly independent. This is important because in real-life systems with many constraints, it is often difficult to ascertain whether the constraints are independent or not. We would expect that the addition of linearly dependent constraints to the set of constraint equations adds nothing new to the description of the system and therefore the equations of motion would remain unchanged. We show in the following simple example that this is indeed true; what happens to the Lagrange multipliers, though, as we shall see, is a slightly different story.

Example 5.5

Consider the pendulum of Example 3.6 in Chapter 3. We use the (x,y) coordinates of the pendulum bob to describe its position, with the Y-axis pointing downwards. Along with the constraint equation

$$x^2 + y^2 = L^2 \tag{5.51}$$

we consider another constraint equation

$$2x^2 + 2y^2 = 2L^2. \tag{5.52}$$

Clearly the two equations (5.51) and (5.52) are linearly dependent, the second having been obtained by multiplying the first by 2. Differentiating these constraint equations twice we express them in the form $A\ddot{\mathbf{x}} = \mathbf{b}$, as

$$\begin{bmatrix} x & y \\ 2x & 2y \end{bmatrix} \begin{bmatrix} \ddot{x} \\ \ddot{y} \end{bmatrix} = \begin{bmatrix} -\dot{x}^2 - \dot{y}^2 \\ -2\dot{x}^2 - 2\dot{y}^2 \end{bmatrix}. \tag{5.53}$$

The unconstrained motion of the system is described, as before, by the equations

$$\begin{bmatrix} m & 0 \\ 0 & m \end{bmatrix} \begin{bmatrix} \ddot{x} \\ \ddot{y} \end{bmatrix} = \begin{bmatrix} 0 \\ mg \end{bmatrix}, \tag{5.54}$$

and the constrained equations of motion are, as usual, given by

$$\begin{bmatrix} \ddot{x} \\ \ddot{y} \end{bmatrix} = \begin{bmatrix} 0 \\ g \end{bmatrix} + M^{-1/2}(AM^{-1/2})^{+}(\mathbf{b} - A\mathbf{a}). \tag{5.55}$$

Example 5.5 (continued)

Since $(AM^{-1/2})^+ = (Am^{-1/2}I)^+ = m^{1/2}(AI)^+ = m^{1/2}(A)^+$, this reduces to the equation

$$\begin{bmatrix} \ddot{x} \\ \ddot{y} \end{bmatrix} = \begin{bmatrix} 0 \\ g \end{bmatrix} + (A)^+(\mathbf{b} - A\mathbf{a}). \tag{5.56}$$

Using relation (2.100) we obtain the MP-inverse of A to be

$$A^+ = \begin{bmatrix} x & y \\ 2x & 2y \end{bmatrix}^+ = \frac{1}{5(x^2+y^2)}\begin{bmatrix} x & 2x \\ y & 2y \end{bmatrix}, \tag{5.57}$$

so that

$$\begin{aligned} \begin{bmatrix} \ddot{x} \\ \ddot{y} \end{bmatrix} &= \begin{bmatrix} 0 \\ g \end{bmatrix} + \frac{1}{5(x^2+y^2)}\begin{bmatrix} x & 2x \\ y & 2y \end{bmatrix}\begin{bmatrix} (-\dot{x}^2 - \dot{y}^2 - gy) \\ (-2\dot{x}^2 - 2\dot{y}^2 - 2gy) \end{bmatrix} \\ &= \begin{bmatrix} 0 \\ g \end{bmatrix} - \frac{(\dot{x}^2 + \dot{y}^2 + gy)}{(x^2+y^2)}\begin{bmatrix} x \\ y \end{bmatrix}, \end{aligned} \tag{5.58}$$

which is the same as equation (3.66), which was obtained using just the constraint (5.51). Clearly, the addition of the second constraint added no new information to the description of the constrained system and hence resulted in no changes in the equations of motion.

However, if we were to express the equations of motion through the use of Lagrange multipliers, then we would write the constrained equation of motion as

$$M\ddot{\mathbf{x}} = \mathbf{F} + M^{1/2}(AM^{-1/2})^+(\mathbf{b} - A\mathbf{a}) = \mathbf{F} + A^T\boldsymbol{\lambda} \tag{5.59}$$

so that, as in equation (5.45), we get

$$A^T\boldsymbol{\lambda} = M^{1/2}(AM^{-1/2})^+(\mathbf{b} - A\mathbf{a}) = -\frac{m(\dot{x}^2+\dot{y}^2+gy)}{(x^2+y^2)}\begin{bmatrix} x \\ y \end{bmatrix}. \tag{5.60}$$

Notice that $\boldsymbol{\lambda}$ is a 2-vector. But now the matrix A^T has rank unity. The solution $\boldsymbol{\lambda}$ of equation (5.60) can be obtained, using (2.107), as

$$\boldsymbol{\lambda} = -\frac{m(\dot{x}^2+\dot{y}^2+gy)}{(x^2+y^2)}(A^T)^+\begin{bmatrix} x \\ y \end{bmatrix} + [I - (A^T)^+A^T]\mathbf{h} \tag{5.61}$$

where $\mathbf{h}$ is an arbitrary 2-vector. Noting relations (2.65) and (5.57) we get

$$\begin{bmatrix} \lambda_1 \\ \lambda_2 \end{bmatrix} = -\frac{m(\dot{x}^2 + \dot{y}^2 + gy)}{(x^2 + y^2)} \frac{1}{5(x^2 + y^2)} \begin{bmatrix} x & y \\ 2x & 2y \end{bmatrix} \begin{bmatrix} x \\ y \end{bmatrix}$$
$$+ \left\{ I - \begin{bmatrix} 1/5 & 2/5 \\ 2/5 & 4/5 \end{bmatrix} \right\} \begin{bmatrix} h_1 \\ h_2 \end{bmatrix} \quad (5.62)$$
$$= -\frac{m(\dot{x}^2 + \dot{y}^2 + gy)}{5(x^2 + y^2)} \begin{bmatrix} 1 \\ 2 \end{bmatrix} + \begin{bmatrix} 4/5 & -2/5 \\ -2/5 & 1/5 \end{bmatrix} \begin{bmatrix} h_1 \\ h_2 \end{bmatrix}.$$

Thus we see that though the equations of motion are unique as is the force of constraint, the Lagrange multipliers at each instant of time are non-unique, the quantities h_1 and h_2 being arbitrary! Viewed as functions of time, these multipliers are nonunique functions!

Generalized Coordinates and Holonomic Constraints

To introduce the idea of generalized coordinates, let us consider a system of n particles subjected to the h holonomic constraints

$$df_i(\mathbf{x}, t) = 0, \; i = 1, 2, \ldots, h, \quad (5.63)$$

where the $3n$-vector $\mathbf{x}(t)$ describes the position of the system of particles, in an inertial rectangular Cartesian coordinate frame of reference, at the instant of time t. We note that equations (5.63) can be integrated to yield

$$f_i(\mathbf{x}, t) = \alpha_i, \; i = 1, 2, \ldots, h \quad (5.64)$$

where the α_i's are constants. Let us assume, for simplicity, that the h functions f_i are independent.

We can then theoretically express h components of the vector $\mathbf{x}$, in terms of the remaining $(3n - h)$ components (locally), and regard these $(3n - h)$ components as independent quantities that define the position of the system at time t.

However, we need not take Cartesian coordinates for these $(3n - h)$ independent components. All the $3n$ Cartesian coordinate components of the vector $\mathbf{x}$ may be expressed as some functions of the $d = (3n - h)$ independent parameters $q_1, q_2, q_3, \ldots, q_{(3n-h)}$ and of t, so that

$$x_i = x_i(q_1, q_2, q_3, \ldots, q_{(3n-h)}, t), \; i = 1, 2, \ldots, 3n. \quad (5.65)$$

More precisely, we can choose $3n$ parameters q_i, $i = 1, 2, \ldots, 3n$, such that

$$q_r = g_r(\mathbf{x}, t), \; r = 1, 2, \ldots, 3n - h, \quad (5.66)$$

and,

$$q_r = f_{r-(3n-h)}(\mathbf{x}, t), \ r = 3n - h + 1, \ldots, 3n. \tag{5.67}$$

Thus the first $(3n-h)$ q's are suitably chosen functions of $\mathbf{x}$ and t, while the latter h of them are functions that arise from the equations of constraint. If we consider any small neighborhood of a point $\mathbf{x}$ at time t in configuration space, we want the transformation from the x's to the q's to be one-to-one. That is, we want the Jacobian

$$J = \frac{\partial(g_1, g_2, \ldots, g_{3n-h}, f_1, \ldots, f_h)}{\partial(x_1, x_2, \ldots\ldots\ldots\ldots, x_{3n})} \tag{5.68}$$

to be nonvanishing. Thus we can solve equations (5.66) and (5.67) (locally) for the x's as functions of the q's and t. But now the equations of constraint in terms of the new variables take the simple form

$$q_r = \alpha_{r-(3n-h)}, \ r = 3n - h + 1, \ldots, 3n \tag{5.69}$$

where the α_i's are constants. The values of these last h q's being fixed, it is therefore the values of the first $(3n-h)$ q's that determine the configuration of the system. The x_i's are thus expressed as explicit functions of the $(3n-h)$ coordinates $q_1, q_2, \ldots, q_{(3n-h)}$ and we are led to equation (5.65).

Thus any position of the system that is compatible with the constraints at a given instant of time will be obtained from equations (5.65) for certain values of the parameters $q_1, q_2, q_3, \ldots, q_{(3n-h)}$. When these values of x_i are put in the constraint equations (5.64), the latter become identities; the equations of constraint are therefore automatically satisfied.

The minimum number of quantities q_i with which equations (5.65) can encompass all possible positions of the holonomic system coincides with the number of degrees of freedom d of the system and equals $(3n-h)$. The parameters $q_1, q_2, q_3, \ldots, q_{(3n-h)}$ are called the independent generalized coordinates of the system. We now illustrate these concepts by the following examples.

Example 5.6

Consider a pendulum bob of mass m suspended from an inextensible string of length L in the XY-plane. The $2n$-vector of Cartesian coordinates that describes the position of the bob is $[x \quad y]^T$. The system is subjected to the holonomic constraint $x^2 + y^2 = L^2$. Since the motion occurs in the XY-plane, the number of independent coordinates is $2n - h = 2(1) - 1 = 1$.

We can express both the Cartesian coordinates in terms of just one generalized coordinate θ, so that we have

$$x = L\sin\theta\,,\ \text{and}\ \ y = L\cos\theta. \tag{5.70}$$

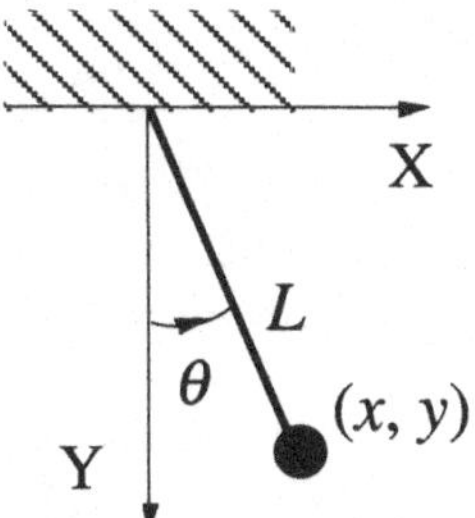

Figure 5.2

Note that these relations when substituted in the constraint equation $x^2 + y^2 = L^2$ reduce it to an identity, and the constraint equation is therefore no longer required since it is automatically satisfied. Thus to specify the position of the bob under the given constraint, we need only one generalized coordinate, namely θ. More precisely, we use the two generalized coordinates $q_1 = \theta$ and $q_2 = L$. The reader can verify that the Jacobian J of the transformation given by equations (5.70) is

$$Det\begin{bmatrix} L\cos\theta & \sin\theta \\ -L\sin\theta & \cos\theta \end{bmatrix} = L \neq 0 \tag{5.71}$$

signifying a one-to-one mapping from the coordinate system (x, y) to the coordinate system (q_1, q_2). Notice that since the coordinate q_2 is a constant, the configuration of the system is entirely determined by the coordinate q_1.

Of course, we could have used two coordinates (x, y) to specify the position of the bob as we did in Chapter 3; however, one of them would be redundant. We would then need to carry the constraint equation $x^2 + y^2 = L^2$ along in our description of the problem.

Example 5.7

Consider now a double pendulum as shown below moving in the XY-plane. Rigid massless bars of lengths L_2 and L_1 respectively connect the two masses, and the mass m_1 to the point of suspension. The position of

Example 5.7 (continued)

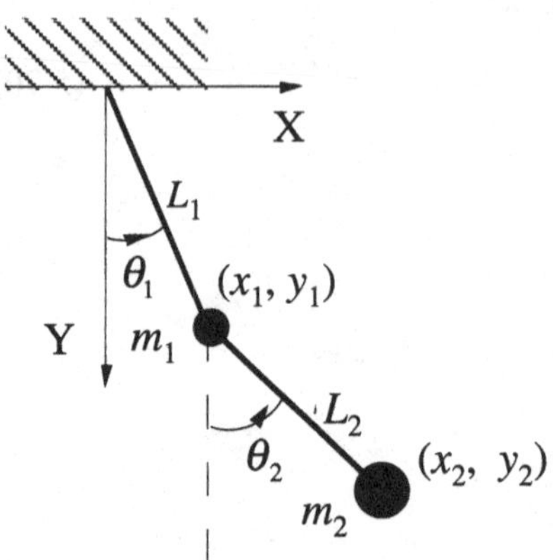

Figure 5.3. The double pendulum

the system is described by the 4-vector of Cartesian components $[x_1 \quad y_1 \quad x_2 \quad y_2]^T$. These four components satisfy the two constraints $x_1^2 + y_1^2 = L_1^2$ and $(x_1 - x_2)^2 + (y_1 - y_2)^2 = L_2^2$.

There are two independent coordinates because among the four Cartesian coordinates x_1, y_1, x_2, y_2 there are two linearly independent constraint relations. We can use the two generalized coordinates θ_1 and θ_2 to specify the configuration of the system. Note that equations (5.65) now take the form

$$x_1 = L_1 \sin\theta_1, y_1 = L_1 \cos\theta_1, \text{ and,} \tag{5.72}$$

$$x_2 - x_1 = L_2 \sin\theta_2, y_2 - y_1 = L_2 \cos\theta_2. \tag{5.73}$$

With these two independent generalized coordinates, relations (5.72) and (5.73) when substituted into the two equations of constraint make them into identities. Thus were we to frame the problem in terms of these two generalized coordinates, we would automatically satisfy these two constraint relations; that is, we do away with the need for specifying these constraints since they are automatically always satisfied!

If we were to denote the vector of these generalized coordinates by $\boldsymbol{\theta} = [\theta_1 \quad \theta_2]$, then the fact that no additional constraints need to be specified implies that the components of the vector $\delta\boldsymbol{\theta} = [\delta\theta_1 \quad \delta\theta_2]$ can be independently varied. Thus *every* vector $\delta\boldsymbol{\theta}$ satisfies the qualification for being a virtual displacement vector.

Again as before, we can use redundant coordinates provided we augment the problem description with the required number of relations between these redundant coordinates. For example we could have used the three coordinates θ_1, x_2, y_2, instead of just the minimum number of two. Now we need to include the additional constraint $(L_1 \sin\theta_1 - x_2)^2 + (L_1 \cos\theta_1 - y_2)^2 = L_2^2$ to complete the description of the system. Notice that this equation involves all three coordinates, not just x_2 and y_2.

We could have also used x_1, y_1, x_2, y_2 as a set of coordinates, with two coordinates in excess of the minimum required; this would then require us, in addition, to specify the two constraint equations $x_1^2 + y_1^2 = L_1^2$ and $(x_1 - x_2)^2 + (y_1 - y_2)^2 = L_2^2$ explicitly, to complete the description of the system.

Example 5.8

Consider a particle constrained to lie on the surface of a moving sphere of constant radius r so that at each instant of time it is constrained by the equation

$$(x - u(t))^2 + (y - v(t))^2 + (z - w(t))^2 = r. \tag{5.74}$$

Since the particle can be described by three Cartesian coordinates and these coordinates must satisfy equation (5.74), we have two independent

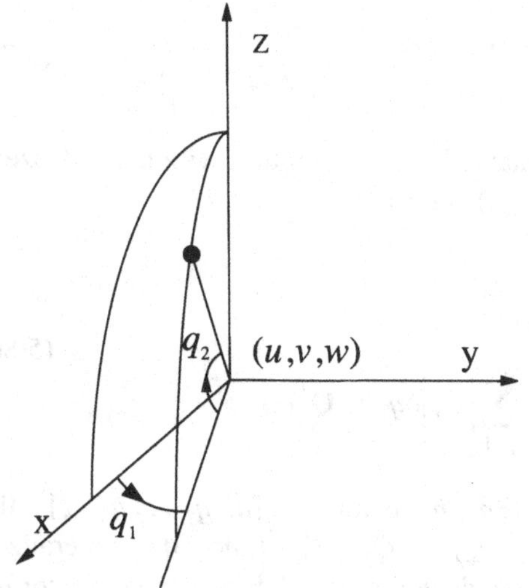

Figure 5.4. Generalized coordinates for a particle on a moving sphere.

Example 5.8 (continued)

coordinates. If we were to choose q_1 and q_2 to be the longitude and the latitude of the particle respectively, we would then be able to state the position at, say, time t using these two independent generalized coordinates. Equations (5.65) that express the Cartesian coordinates in terms of the generalized coordinates as

$$x(t) = u(t) + r\cos q_1 \cos q_2, \tag{5.75}$$

$$y(t) = v(t) + r\sin q_1 \cos q_2, \text{ and} \tag{5.76}$$

$$z(t) = w(t) + r\sin q_2. \tag{5.77}$$

Let the parameters $q_1, q_2, q_3, \ldots, q_{(3n-h)}$ be the independent generalized coordinates describing the holonomic system. Associated with each generalized coordinate q_i is a generalized force Q_i. These generalized forces are determined as follows. Consider the work done by the impressed forces under the virtual displacement vector $\delta\mathbf{x}$,

$$\delta W = \mathbf{F}^T\delta\mathbf{x} = \sum_{i=1}^{3n} F_i\delta x_i. \tag{5.78}$$

But

$$\delta x_i = \sum_{j=1}^{3n-h}\frac{\partial x_i}{\partial q_j}\delta q_j, \quad i = 1, 2, \ldots, 3n. \tag{5.79}$$

Notice that the virtual displacements refer to differentials with fixed "frozen" time t. Using equation (5.79) in (5.78) we get

$$\begin{aligned}\mathbf{F}^T\delta\mathbf{x} &= \sum_{i=1}^{3n} F_i\delta x_i = \sum_{i=1}^{3n} F_i \sum_{j=1}^{3n-h}\frac{\partial x_i}{\partial q_j}\delta q_j \\ &= \sum_{j=1}^{3n-h}\left[\sum_{i=1}^{3n} F_i\frac{\partial x_i}{\partial q_j}\right]\delta q_j = \sum_{j=1}^{3n-h}\left[Q_j\right]\delta q_j = \mathbf{Q}^T\delta\mathbf{q}\end{aligned} \tag{5.80}$$

where we have associated with the $(3n - h)$ vector $\mathbf{q} = [q_1 \; q_2 \cdots q_{3n-h}]^T$, the generalized force vector $\mathbf{Q} = [Q_1 \; Q_2 \ldots Q_{3n-h}]^T$. Since the generalized coordinates are assumed to be independent, the virtual displacement vector $\delta\mathbf{q}$ is such that all its components can be assigned independently of every other.

We could then choose all the components of $\delta\mathbf{q}$ to be zero except for the jth component. In that situation by comparing the first two expressions in the second line of equation (5.80) we would get

$$Q_j = \sum_{i=1}^{3n} F_i \frac{\partial x_i}{\partial q_j}, \; j = 1, 2, \ldots, 3n - h. \tag{5.81}$$

In summary, when we have h independent holonomic constraints we can express h of the Cartesian coordinates in terms of the remaining $(3n - h)$ coordinates. In fact we can use $(3n - h)$ generalized coordinates, and in terms of these generalized coordinates there are no longer any equations of constraint needed to be specified, for they are all identically satisfied; thus these coordinates are independent. Also, $(3n - h)$ is the minimum number of coordinates that we can use to specify the configuration of the system. Were we to decide to use u more coordinates than this minimum number, we would need to add u relevant additional relations (holonomic constraints) to complete the specification of the system.

Hence given a holonomically constrained dynamic system we have alternative ways of describing it. These alternative ways are related to our choice regarding the number of generalized coordinates we may want to use. If we use the minimum number, i.e., $(3n - h)$, and make a proper choice of generalized coordinates, we would do away with the need to specify any of the h constraints for they would automatically be satisfied because of our choice of the generalized coordinates; thus, these coordinates would be independent, and the problem in these $(3n - h)$ coordinates would essentially become an "unconstrained" problem. If we choose to describe the system using u redundant coordinates (i.e., u more coordinates than the minimum number, a total of $3n - h + u$ coordinates), we would need to specify the u relations between these redundant coordinates. We would then have a holonomically constrained system with u holonomic constraints as we illustrated in Examples 5.6 and 5.7.

Generalized Coordinates and Nonholonomic Constraints

Let us now assume that in addition to the h independent holonomic constraints there are also r nonholonomic constraints that a system of n particles satisfies. Because there are h holonomic constraints we can find $(3n - h)$ generalized coordinates so that the h holonomic constraint equations, as before, become identities. We thus have the generalized coordinates given again by equations of the form

$$x_i = x_i(q_1, q_2, q_3, \ldots, q_{(3n-h)}, t), \; i = 1, 2, \ldots, 3n. \tag{5.82}$$

We assume, as before, that these equations that transform the $(3n-h)$ q_i's to the $3n$ x_i's are reversible, i.e., the mapping in a small neighborhood of any point $\mathbf{x}$ is one-to-one. From equation (5.82) it follows that

$$\dot{x}_i = \sum_{j=1}^{(3n-h)} \frac{\partial x_i}{\partial q_j}\dot{q}_j + \frac{\partial x_i}{\partial t},\ i = 1, 2, \ldots, 3n. \tag{5.83}$$

The r nonholonomic equations are nonintegrable and hence we cannot get rid of any more coordinates. But in these r equations,

$$\sum_{j=1}^{j=3n} d_{ij}(\mathbf{x}, t)\dot{x}_j + g_i(\mathbf{x}, t) = 0,\ i = 1, 2, \ldots, r, \tag{5.84}$$

we can substitute for x_i and $\dot{x}_i$ from equations (5.82) and (5.83) so that we obtain

$$\sum_{j=1}^{j=3n-h} \hat{d}_{ij}(\mathbf{q}, t)\dot{q}_j + \hat{g}_i(\mathbf{q}, t) = 0,\ i = 1, 2, \ldots, r. \tag{5.85}$$

Thus in the presence of nonholonomic constraints the $\dot{q}_i$'s are related and must satisfy equations (5.85). The $(3n-h)$-vector of virtual displacement is the vector $\delta\mathbf{q}$ whose components are δq_j. These components, as before, satisfy the equations

$$\sum_{j=1}^{j=3n-h} \hat{d}_{ij}(\mathbf{q}, t)\delta q_j = 0,\ i = 1, 2, \ldots, r, \tag{5.86}$$

or

$$\hat{\mathrm{D}}(\mathbf{q}, t)\,\delta\mathbf{q} = 0, \tag{5.87}$$

where the i–jth element of the r by $(3n-h)$ matrix $\hat{\mathrm{D}}$ is $\hat{d}_{ij}(\mathbf{q}, t)$.

Thus for the system which has h independent holonomic and r nonholonomic constraints, the minimum number of coordinates that need to be used to describe the system is $(3n-h)$. The generalized coordinates are never independent. They are linked by the r nonholonomic constraints described by equations (5.85).

We note in passing here that equations (5.85) can also be expressed as

$$\hat{\mathrm{D}}(\mathbf{q}, t)\dot{\mathbf{q}} = \hat{\mathbf{g}}(\mathbf{q}, t) \tag{5.88}$$

where the ith component of the $(3n-h)$-vector $\hat{\mathbf{g}}(\mathbf{q},t)$ is $-\hat{g}_i(\mathbf{q},t)$. This equation, on being differentiated once with respect to time, yields

$$\hat{\mathrm{A}}(\mathbf{q},t)\,\ddot{\mathbf{q}} = \hat{\mathbf{b}}(\mathbf{q},\dot{\mathbf{q}},t) \tag{5.89}$$

where, as usual, we have denoted the matrix $\hat{\mathrm{A}}(\mathbf{q},t) = \hat{\mathrm{D}}(\mathbf{q},t)$ so as to express the constraint in our standard form. The relation defining a virtual displacement $\delta\mathbf{q}$ can then be expressed as

$$\hat{\mathrm{A}}(\mathbf{q},t)\delta\mathbf{q} = \mathbf{0}. \tag{5.90}$$

We shall see in Chapter 7 that even when the constraints are of a more general form, namely,

$$\hat{\mathrm{A}}(\mathbf{q},\dot{\mathbf{q}},t) = \hat{\mathbf{b}}(\mathbf{q},\dot{\mathbf{q}},t), \tag{5.91}$$

the virtual displacement vector is still defined in a form similar to (5.90) as

$$\hat{\mathrm{A}}(\mathbf{q},\dot{\mathbf{q}},t)\delta\mathbf{q} = \mathbf{0}. \tag{5.92}$$

Having understood how equations (5.85)–(5.92) come about, in what follows we will, for convenience, drop the hats over the quantities d_{ij}, D, A, $\mathbf{g}$ and $\mathbf{b}$ in these equations.

In summary then with nonholonomic constraints we cannot, by means of a proper choice of generalized coordinates, have the constraints automatically satisfied, as we did in the case of holonomic constraints. The nonholonomic constraints are there to stay. We of course could do away with all the h independent holonomic constraints by a proper choice of generalized coordinates as we did earlier; or else, we could choose u redundant coordinates (u coordinates more than $(3n-h)$) as before, and include u appropriate relations each of the holonomic form

$$f_i(q_1, q_2, \ldots, q_{3n-h+u}, t) = 0,\ i = 1, 2, \ldots, u \tag{5.93}$$

to complete the description of the system.

Example 5.9

In this example we present some basic results related to transformation of coordinates which we will use later on.

Consider the transformation equations

Example 5.9 (continued)

$$x_i = x_i(q_1, q_2, q_3, \ldots, q_s, t),\ i = 1, 2, \ldots, 3n, \tag{5.94}$$

so that

$$\dot{x}_i = \sum_{j=1}^{s} \frac{\partial x_i}{\partial q_j}\dot{q}_j + \frac{\partial x_i}{\partial t},\ i = 1, 2, \ldots, 3n. \tag{5.95}$$

Then we have the following relations:

$$\text{a. } \frac{\partial \dot{x}_i}{\partial \dot{q}_j} = \frac{\partial x_i}{\partial q_j}, \tag{5.96}$$

$$\text{b. } \frac{\partial \dot{x}_i}{\partial q_j} = \frac{d}{dt}\left(\frac{\partial x_i}{\partial q_j}\right), \text{ and,} \tag{5.97}$$

$$\text{c. } \ddot{x}_i \frac{\partial x_i}{\partial q_j} = \frac{d}{dt}\left(\dot{x}_i \frac{\partial \dot{x}_i}{\partial \dot{q}_j}\right) - \dot{x}_i \frac{\partial \dot{x}_i}{\partial q_j}. \tag{5.98}$$

The first relation follows directly from equations (5.95) by taking partial derivatives on both sides with respect to $\dot{q}_j$. The second relation follows by partially differentiating (5.95) with respect to q_j because

$$\begin{aligned}\frac{\partial \dot{x}_i}{\partial q_j} &= \sum_{m=1}^{s} \frac{\partial^2 x_i}{\partial q_j \partial q_m}\dot{q}_m + \frac{\partial^2 x_i}{\partial q_j \partial t} \\ &= \sum_{m=1}^{s} \frac{\partial^2 x_i}{\partial q_m \partial q_j}\dot{q}_m + \frac{\partial^2 x_i}{\partial t \partial q_j} = \frac{d}{dt}\left(\frac{\partial x_i}{\partial q_j}\right).\end{aligned} \tag{5.99}$$

The third relation can be proved as follows. We start with the identity

$$\ddot{x}_i \frac{\partial x_i}{\partial q_j} = \frac{d}{dt}\left(\dot{x}_i \frac{\partial x_i}{\partial q_j}\right) - \dot{x}_i \frac{d}{dt}\left(\frac{\partial x_i}{\partial q_j}\right). \tag{5.100}$$

Using equation (5.97), in the right hand side of equation (5.100), we obtain equation (5.98).

We define the kinetic energy of a system of n particles to be

$$T = \frac{1}{2}\sum_{i=1}^{3n} m_i \dot{x}_i^2 \tag{5.101}$$

where for convenience of notation we have described the masses by $m_1, m_2, \ldots, m_{3n}$, though we know that each set of three consecutive masses, starting from m_1, has the same value. In this section and the next we will use this simpler notation for the masses corresponding to the $3n$ coordinates. The $\dot{x}_i$'s are, as usual, the components of the $3n$-vector of velocity of the particles in an inertial Cartesian coordinate frame of reference, i.e., $\dot{\mathbf{x}} = [\dot{x}_1\ \dot{x}_2 \ldots \dot{x}_{3n}]^T$. Were we to substitute for the $\dot{x}_i$'s from (5.95) then the kinetic energy would be a polynomial of second degree in the $\dot{q}_i$'s.

Example 5.10

Find the kinetic energy of the double pendulum system shown in Figure 5.3 in terms of the independent coordinates θ_1 and θ_2 and their derivatives.

The kinetic energy of mass m_1 is obtained by realizing that $x_1 = L_1 \sin\theta_1$ and $y_1 = L_1 \cos\theta_1$. Thus $\dot{x}_1 = L_1\dot{\theta}_1 \cos\theta_1$, and $\dot{y}_1 = -L_1\dot{\theta}_1 \sin\theta_1$, so that $\frac{1}{2}m_1(\dot{x}_1^2 + \dot{y}_1^2) = \frac{1}{2}m_1 L_1^2\dot{\theta}_1^2$.

Similarly, the kinetic energy of mass m_2 is obtained by noting that $x_2 = x_1 + L_2 \sin\theta_2$, and $y_2 = y_1 + L_2\cos\theta_2$, so that $\dot{x}_2 = \dot{x}_1 + L_2\dot{\theta}_2\cos\theta_2$ and $\dot{y}_2 = \dot{y}_1 - L_2\dot{\theta}_2\sin\theta_2$. Hence we get $\frac{1}{2}m_2(\dot{x}_2^2 + \dot{y}_2^2) = \frac{1}{2}m_2(L_1\dot{\theta}_1\cos\theta_1 + L_2\dot{\theta}_2\cos\theta_2)^2 + \frac{1}{2}m_2(L_1\dot{\theta}_1\sin\theta_1 + L_2\dot{\theta}_2\sin\theta_2)^2$. This simplifies to $\frac{1}{2}m_2(L_1^2\dot{\theta}_1^2 + L_2^2\dot{\theta}_2^2) + m_2L_1L_2\dot{\theta}_1\dot{\theta}_2\cos(\theta_1 - \theta_2)$. Thus the total kinetic energy of the two masses is given by

$$T = \frac{1}{2}(m_1 + m_2)L_1^2\dot{\theta}_1^2 + m_2L_1L_2\dot{\theta}_1\dot{\theta}_2\cos(\theta_1 - \theta_2) + \frac{1}{2}m_2L_2^2\dot{\theta}_2^2. \tag{5.102}$$

Example 5.11

In this example we present some basic results related to the derivatives of the kinetic energy with respect to the generalized coordinates. In the next section, these relations will be used in deriving Lagrange's equations.

We have, using the notation of equation (5.101),

a. $$\frac{\partial T}{\partial \dot{q}_j} = \sum_{i=1}^{3n} m_i\left(\dot{x}_i\frac{\partial \dot{x}_i}{\partial \dot{q}_j}\right) \tag{5.103}$$

Example 5.11 (continued)

b. $$\frac{d}{dt}\left(\frac{\partial T}{\partial \dot{q}_j}\right) = \sum_{i=1}^{3n} m_i \frac{d}{dt}\left(\dot{x}_i \frac{\partial \dot{x}_i}{\partial \dot{q}_j}\right) \tag{5.104}$$

c. $$\frac{\partial T}{\partial q_j} = \sum_{i=1}^{3n} m_i \left(\dot{x}_i \frac{\partial \dot{x}_i}{\partial q_j}\right) \tag{5.105}$$

d. $$\sum_{i=1}^{3n} m_i \ddot{x}_i \frac{\partial x_i}{\partial q_j} = \frac{d}{dt}\left(\frac{\partial T}{\partial \dot{q}_j}\right) - \frac{\partial T}{\partial q_j} \tag{5.106}$$

where equations (5.94) transform the q_j's to the x_i's.

These results for parts (a)–(c) follow directly by differentiating equation (5.101). We show part (d) as follows. First using equation (5.98) we have

$$\sum_{i=1}^{3n} m_i \ddot{x}_i \frac{\partial x_i}{\partial q_j} = \sum_{i=1}^{3n} m_i \frac{d}{dt}\left(\dot{x}_i \frac{\partial \dot{x}_i}{\partial \dot{q}_j}\right) - \sum_{i=1}^{3n} m_i \dot{x}_i \frac{\partial \dot{x}_i}{\partial q_j}. \tag{5.107}$$

Substituting from (5.104) and (5.105), equation (5.106) follows.

Lagrange's Equations for Unconstrained Systems

How do we write the equations of motion corresponding to the unconstrained system when we describe the position of the n particles of a system using generalized coordinates? This is the question which we will answer in this section.

First we must be clear what we mean by an unconstrained system described in a given set of generalized coordinates. By this we mean that the virtual displacements corresponding to those coordinates are independent. Alternately stated, the coordinates are treated as being independent of each other.

Let $\mathbf{x}(t)$ be the $3n$-vector describing the position of the n particles in an inertial Cartesian frame of reference. Let us say that the system is described by the s generalized coordinates $q_1, q_2, q_3, \ldots, q_s$ where the $3n$ equations (5.94) transform the Lagrangian components q_i of the s-vector $\mathbf{q}$ of generalized coordinates to the Cartesian coordinate components x_i.

By the *basic equation of analytical mechanics* we know that

$$\delta\mathbf{x}^T(M\ddot{\mathbf{x}} - \mathbf{F}) = \sum_{i=1}^{3n} (m_i\ddot{x}_i - F_i)\delta x_i = 0, \tag{5.108}$$

for all virtual displacement vectors $\delta\mathbf{x}$. But as in equation (5.79),

$$\delta x_i = \sum_{j=1}^{j=s} \frac{\partial x_i}{\partial q_j}\delta q_j, \tag{5.109}$$

so that equation (5.108) now becomes

$$\sum_{j=1}^{j=s} \frac{\partial x_i}{\partial q_j}\delta q_j \sum_{i=l}^{3n}(m_i\ddot{x}_i - F_i) = \sum_{j=1}^{s}\left\{\sum_{i=l}^{3n}(m_i\ddot{x}_i - F_i)\frac{\partial x_i}{\partial q_j}\right\}\delta q_j = 0. \tag{5.110}$$

Now by equation (5.81), we know that

$$\sum_{i=1}^{3n} F_i \frac{\partial x_i}{\partial q_j} = Q_j, \tag{5.111}$$

the generalized force corresponding to the jth generalized coordinate. Also, by equation (5.106), we have

$$\sum_{i=1}^{3n} m_i\ddot{x}_i \frac{\partial x_i}{\partial q_j} = \frac{d}{dt}\left(\frac{\partial T}{\partial \dot{q}_j}\right) - \frac{\partial T}{\partial q_j}. \tag{5.112}$$

Using equations (5.111) and (5.112) in (5.110) we thus get

$$\sum_{j=l}^{s}\left\{\frac{d}{dt}\left(\frac{\partial T}{\partial \dot{q}_j}\right) - \frac{\partial T}{\partial q_j} - Q_j\right\}\delta q_j = 0. \tag{5.113}$$

Now if s equals the minimum number of generalized coordinates for the problem (i.e., $(3n-h)$, where h is the number of independent holonomic constraints) and if there are no nonholonomic constraints, we know that the constraint equations are automatically satisfied by a proper choice of the $(3n-h)$ generalized coordinates. The components of the virtual displacement vector $\delta\mathbf{q}$ are therefore then independent and so, from equation (5.113), we get the equations

$$\frac{d}{dt}\left(\frac{\partial T}{\partial \dot{q}_j}\right) - \frac{\partial T}{\partial q_j} - Q_j = 0, \quad j = 1, 2, \ldots, s. \tag{5.114}$$

There are two groups of impressed forces which are of special significance – those that arise from a "potential function" and those that arise from a "dissipation function." We consider these two groups next.

Were the generalized forces Q_j to be expressible as

$$Q_j = -\frac{\partial V}{\partial q_j} + \frac{d}{dt}\left(\frac{\partial V}{\partial \dot{q}_j}\right) + Q_j^{nc} = Q_j^c + Q_j^{nc},\ j = 1, 2, \ldots, s \tag{5.115}$$

where the quantity V is a given function of $q_1, q_2, \ldots, q_s, \dot{q}_1, \dot{q}_2, \ldots, \dot{q}_s$, then V is called the V generalized potential energy function of the system. When V is only a function of the generalized coordinates $q_1, q_2, \ldots, q_s$ in equation (5.115), then it is usually called the potential energy of the system in the configuration $q_1, q_2, \ldots, q_s$. The force Q_j^c is sometimes referred to as the conservative part of the total generalized force Q_j, and Q_j^{nc} is referred to as the non-conservative part. Using this expression for Q_j in the equation set (5.114) we can recast equation set (5.114) as

$$\frac{d}{dt}\left(\frac{\partial L}{\partial \dot{q}_j}\right) - \frac{\partial L}{\partial q_j} - Q_j^{nc} = 0,\ \ j = 1,\ 2, \ldots, s \tag{5.116}$$

where we have denoted the quantity $(T - V)$ as L, the "Lagrangian" or "kinetic potential" of the system. The forces Q_j^{nc} are usually a consequence of externally impressed forces and/or forces which cannot be engendered from a generalized potential energy.

The second group of forces whose existence it is advantageous to recognize is those arising from friction or viscosity. If we suppose that each particle of the system is retarded by forces proportional to its component velocities, then

$$F_i = -c_i\dot{x}_i,\ i = 1,\ 2, \ldots, 3n, \tag{5.117}$$

and the quantity $\sum_{i=1}^{3n} F_i \delta x_i$, which is the second term on the left in equation (5.108), becomes

$$\begin{aligned}\sum_{i=1}^{3n} F_i\delta x_i &= -\sum_{i=1}^{3n} c_i\dot{x}_i\delta x_i = -\sum_{i=1}^{3n} c_i\dot{x}_i\left[\sum_{j=1}^{j=s}\frac{\partial x_i}{\partial q_j}\delta q_j\right] \\ &= -\sum_{j=1}^{j=s}\left[\sum_{i=1}^{3n} c_i\dot{x}_i\frac{\partial x_i}{\partial q_j}\right]\delta q_j,\end{aligned} \tag{5.118}$$

where we have used equation (5.109) to express δx_i in generalized coordinates. Noting equation (5.96), the last expression on the right hand side then becomes

$$\sum_{i=1}^{3n} F_i \delta x_i = -\sum_{j=1}^{j=s}\left[\sum_{i=1}^{3n} c_i \dot{x}_i \frac{\partial \dot{x}_i}{\partial \dot{q}_j}\right]\delta q_j = \sum_{j=1}^{j=s} Q_j \delta q_j. \tag{5.119}$$

Thus the generalized force corresponding to such a group of forces is simply

$$Q_j = -\sum_{i=1}^{3n} c_i \dot{x}_i \frac{\partial \dot{x}_i}{\partial \dot{q}_j},\ j = 1, 2, \ldots, s. \tag{5.120}$$

Now consider the function

$$R_D = \frac{1}{2}\sum_{i=1}^{3n} c_i \dot{x}_i^2, \tag{5.121}$$

where, the c_i's are constants. We shall refer to R_D as the dissipation function. Using equations (5.95) in the right hand side of equation (5.121), we can think of R_D as a function of the q_i's and the $\dot{q}_i$'s. Then differentiating R_D partially with respect to $\dot{q}_j$'s yields,

$$\frac{\partial R_D}{\partial \dot{q}_j} = \sum_{i=1}^{3n} c_i \dot{x}_i \frac{\partial \dot{x}_i}{\partial \dot{q}_j} = -Q_j, \tag{5.122}$$

where the last equality follows from equations (5.120). Hence in the presence of dissipative forces of the type described by equations (5.117), Lagrange's equations become

$$\frac{d}{dt}\left(\frac{\partial L}{\partial \dot{q}_j}\right) - \frac{\partial L}{\partial q_j} + \frac{\partial R_D}{\partial \dot{q}_j} - Q_j^{nc} = 0,\ j = 1, 2, \ldots, s. \tag{5.123}$$

The above discussion refers to retarding forces proportional to the absolute velocities; but it is equally important to consider such retarding forces which depend on the components of the *relative* velocities between the particles. It is a simple matter to show that in such a situation one can again define an appropriate dissipation function. We leave this as an exercise for the reader (see Problem 5.9).

Recalling now the general situation, let us say that we choose s generalized coordinates which may *not* necessarily be independent. Determining the equations of motion of the "unconstrained" system corresponding to this choice of our s coordinates is a simple matter. We simply always assume that the components of the virtual displacement s-vector $\boldsymbol{\delta q}$ are independent and hence that equations (5.114) are valid even when s > $(3n - h)$. Say $u = s - (3n - h)$. All we need do in that situation is to append the appropriate u holonomic constraints (in terms of the q's) as we did earlier (see equations (5.93)) and, as always, the nonholonomic constraints, to complete the description of the system. The following example illustrates the meaning of this paragraph.

Example 5.12

Consider the double pendulum of Example 5.7 shown in Figure 5.3. We will indicate how we can generate the equations of motion corresponding to the "unconstrained" system when using different sets of coordinates.

a. Let us use the two generalized coordinates θ_1 and θ_2 as shown in Figure 5.3. The virtual work done by the force of gravity is $m_1 g\delta y_1 + m_2 g\delta y_2$. But $y_1 = L_1 \cos\theta_1$ so that $\delta y_1 = -L_1 \sin\theta_1 \delta\theta_1$, and $y_2 = L_1 \cos\theta_1 + L_2 \cos\theta_2$ so that $\delta y_2 = -(L_1 \sin\theta_1 \delta\theta_1 + L_2 \sin\theta_2 \delta\theta_2)$. Using these expressions in the expression for virtual work, we get

$$\begin{aligned} Q_1\delta\theta_1 + Q_2\delta\theta_2 &= -\{m_1 g L_1 \sin\theta_1\delta\theta_1 + m_2 g(L_1 \sin\theta_1\delta\theta_1 \\ &\quad + L_2 \sin\theta_2\delta\theta_2)\} \\ &= -(m_1 + m_2)gL_1 \sin\theta_1\delta\theta_1 - m_2 g L_2 \sin\theta_2\delta\theta_2 \end{aligned} \tag{5.124}$$

from which it follows that $Q_1 = -(m_1 + m_2)gL_1 \sin\theta_1$ and $Q_2 = -m_2 g L_2 \sin\theta_2\delta\theta_2$. The kinetic energy can be written as (see equation (5.102))

$$T = \frac{1}{2}(m_1 + m_2)L_1^2\dot{\theta}_1^2 + m_2 L_1 L_2 \dot{\theta}_1\dot{\theta}_2 \cos(\theta_1 - \theta_2) + \frac{1}{2}m_2 L_2^2\dot{\theta}_2^2. \tag{5.125}$$

The first Lagrange equation now becomes

$$\frac{d}{dt}\left(\frac{\partial T}{\partial \dot{\theta}_1}\right) - \frac{\partial T}{\partial \theta_1} = Q_1, \tag{5.126}$$

which yields

$$\begin{aligned} &\frac{d}{dt}[(m_1 + m_2)L_1^2\dot{\theta}_1 + m_2 L_1 L_2 \dot{\theta}_2 \cos(\theta_1 - \theta_2)] \\ &\quad + m_2 L_1 L_2 \dot{\theta}_1\dot{\theta}_2 \sin(\theta_1 - \theta_2) = -(m_1 + m_2)gL_1 \sin\theta_1. \end{aligned} \tag{5.127}$$

The second equation

$$\frac{d}{dt}\left(\frac{\partial T}{\partial \dot{\theta}_2}\right) - \frac{\partial T}{\partial \theta_2} = Q_2 \tag{5.128}$$

becomes

$$\frac{d}{dt}[m_2 L_2^2 \dot{\theta}_2 + m_2 L_1 L_2 \dot{\theta}_1 \cos(\theta_1 - \theta_2)] - m_2 L_1 L_2 \dot{\theta}_1 \dot{\theta}_2 \sin(\theta_1 - \theta_2) = -m_2 g L_2 \sin\theta_2 . \tag{5.129}$$

Differentiating with respect to time the left hand members of equations (5.127) and (5.129), we obtain the equations of motion of the system – two equations each involving $\ddot{\theta}_1$ and $\ddot{\theta}_2$. The reader may want to determine these equations explicitly, and place them in vector-matrix form to verify that the matrix multiplying the vector $[\ddot{\theta}_1 \;\; \ddot{\theta}_2]^T$ is symmetric, positive definite, and a function of θ_1 and θ_2.

With this choice of generalized coordinates, the two distance-constraints have already been automatically satisfied, and equations (5.127) and (5.129) are the equations of motion corresponding to the unconstrained system as these coordinates are independent.

The reader will observe that upon differentiation with respect to time, a system of two equations linear in the quantities $\ddot{\theta}_1$ and $\ddot{\theta}_2$ is obtained. These equations can be solved algebraically for $\ddot{\theta}_1$ and $\ddot{\theta}_2$.

b. Let us next consider using the three coordinates θ_1, x_2 and y_2 as the generalized coordinates. We now have the virtual work expressed as

$$Q_1 \delta\theta_1 + Q_2 \delta x_2 + Q_3 \delta y_2 = -m_1 g L_1 \sin\theta_1 \delta\theta_1 + m_2 g \delta y_2 \tag{5.130}$$

so that $Q_1 = -m_1 g L_1 \sin\theta_1$, $Q_2 = 0$ and $Q_3 = m_2 g$. Notice that to arrive at the last three equations from the single equation (5.130) *we treat the three virtual displacements as though they were independent*, i.e., as though the three generalized coordinates θ_1, x_2 and y_2 are independent. Similarly, the kinetic energy is now given by

$$T = \frac{1}{2}(m_1 L_1^2 \dot{\theta}_1^2 + m_2 \dot{x}_2^2 + m_2 \dot{y}_2^2), \tag{5.131}$$

as opposed to equation (5.125).

Treating the three coordinates as though they are independent is, physically speaking, tantamount to assuming that there is no constraint

Example 5.12 (continued)

between the mass m_1 and the mass m_2, i.e., the bar L_2 does not exist! The three Lagrange equations corresponding to the "unconstrained" system in terms of these three generalized coordinates now become, through the use of equations (5.114),

$$m_1 L_1^2 \ddot{\theta}_1 = -m_1 g L_1 \sin\theta_1, \tag{5.132}$$

$$m_2 \ddot{x}_2 = 0, \text{ and} \tag{5.133}$$

$$m_2 \ddot{y}_2 = m_2 g. \tag{5.134}$$

These are thus the equations of the "unconstrained" system.

However, while the three coordinates θ_1, x_2 and y_2 are treated as though they are independent, we know that they really are in fact not independent; one of them is redundant. Thus to describe the actual constrained motion, we need to append the constraint equation

$$(x_2 - L_1 \sin\theta_1)^2 + (y_2 - L_1 \cos\theta_1)^2 = L_2^2. \tag{5.135}$$

Thus equations (5.132)–(5.134), which correspond to the equations of the unconstrained system, along with the corresponding constraint equation (5.135), complete the description of the dynamical system in terms of the three coordinates θ_1, x_2 and y_2.

Notice that the expression for the kinetic energy now is substantially simpler than equation (5.125). In fact, the determination of the kinetic energy of the "unconstrained" system becomes easier and easier, in general, as we assume that more and more of the coordinates are independent.

c. Let us now consider the coordinates x_1, y_1, x_2, y_2 as the generalized coordinates and treat them as though they are independent coordinates to obtain the unconstrained motion of the system. Now,

$$Q_1 \delta x_1 + Q_2 \delta y_1 + Q_3 \delta x_2 + Q_4 \delta y_2 = m_1 g \delta y_1 + m_2 g \delta y_2 \tag{5.136}$$

so that $Q_1 = Q_3 = 0$, $Q_2 = m_1 g$ and $Q_4 = m_2 g$. The kinetic energy is simply given by $T = \frac{1}{2} m_1 (\dot{x}_1^2 + \dot{y}_1^2) + \frac{1}{2} m_2 (\dot{x}_2^2 + \dot{y}_2^2)$. Notice that since we are treating all four coordinates as though they are independent, as promised, the expression for the kinetic energy can be effortlessly written down, and is very simple in form compared to equation (5.125).

For this "unconstrained" system where all four coordinates are considered independent, Lagrange's equations now become

$$m_1\ddot{x}_1 = 0, \tag{5.137}$$

$$m_1\ddot{y}_1 = m_1 g, \tag{5.138}$$

$$m_2\ddot{x}_2 = 0, \tag{5.139}$$

$$m_2\ddot{y}_2 = m_2 g. \tag{5.140}$$

Along with these equations describing the "unconstrained" system, we must append now the two constraint equations

$$x_1^2 + y_1^2 = L_1^2, \text{ and,} \tag{5.141}$$

$$(x_2 - x_1)^2 + (y_2 - y_1)^2 = L_2^2. \tag{5.142}$$

Notice that we have imagined here that the given system is obtained from: (1) an unconstrained system described by equations (5.137)–(5.140) composed of the two point masses moving freely in space subjected to the given impressed forces (gravity in this case), and (2) the two additional constraints described by equations (5.141) and (5.142).

There is an important feature that emerges from the example presented above. *Depending on the way we choose the generalized coordinates we can describe the same system with different sets of equations corresponding to the "unconstrained" system, and different sets (and numbers) of equations corresponding to the constraints imposed on this "unconstrained" system.* In general, as the number of coordinates used in describing the "unconstrained" system increases, i.e., as we treat more and more of the coordinates as though they were independent, the "unconstrained" equations of motion become simpler, and easier and easier to write; compare, for example, the equations (5.127) and (5.129) on the one hand, with equations (5.137)–(5.140) on the other. Yet, these simpler equations have to be augmented by a larger and larger number of constraint equations to complete the description of the system. This could make the analytical determination of the generalized inverse which is needed to obtain the final equation of motion of the constrained system more cumbersome. However, if one were to use symbolic manipulations (see Problem 4.10), this difficulty is considerably alleviated.

What does the general form of the equations (5.114), in terms of the s generalized coordinates, look like ? Since we know that

$$T = \frac{1}{2}\sum_{i=1}^{3n} m_i\dot{x}_i^2, \text{ and } x_i = x_i(q_1, q_2, \ldots, q_s, t),\ i = 1, 2, \ldots, n, \tag{5.143}$$

substituting for the $\dot{x}$'s in terms of the q's and the $\dot{q}$'s we get

$$T = T_2 + T_1 + T_0, \tag{5.144}$$

where

$$T_2 = \frac{1}{2}\sum_{i=1}^{s}\sum_{j=1}^{s} \rho_{ij}\dot{q}_i\dot{q}_j,\ T_1 = \sum_{i=1}^{s} \mu_i\dot{q}_i, \text{ and } T_0 = \nu_0. \tag{5.145}$$

Here

$$\rho_{ij} = \rho_{ji} = \sum_{r=1}^{3n} m_r \frac{\partial x_r}{\partial q_i}\frac{\partial x_r}{\partial q_j} = \rho_{ij}(\mathbf{q}, t). \tag{5.146}$$

$$\mu_i = \sum_{r=1}^{3n} m_r \frac{\partial x_r}{\partial q_i}\frac{\partial x_r}{\partial t} = \mu_i(\mathbf{q}, t) \text{ and} \tag{5.147}$$

$$\nu_0 = \frac{1}{2}\sum_{r=1}^{3n} m_r\left(\frac{\partial x_r}{\partial t}\right)^2 = \nu_0(\mathbf{q}, t). \tag{5.148}$$

If we use equations (5.114) and if we treat the s generalized coordinates $q_1, q_2, \ldots, q_s$ as though they are independent, the "unconstrained" equation of motion now results in an equation of the form

$$\mathrm{M}(\mathbf{q}, t)\ddot{\mathbf{q}} = \mathrm{F}(\mathbf{q}, \dot{\mathbf{q}}, t) \tag{5.149}$$

where M is the s by s symmetric matrix ($s \le 3n$) whose i–jth element is $\rho_{ij}(\mathbf{q}, t)$ and F is an s-vector whose components are functions of the q's, the $\dot{q}$'s and time. It can be shown that the matrix M in equation (5.149) corresponding to the unconstrained system is always positive definite. We note from equations (5.143) that T is always positive definite; also, by proper scaling, the quadratic term in equation (5.144) can always be made to dominate, thus leading to the positive definiteness of T_2.

The problem formulation, depending on our choice of generalized coordinates, in general, would also include, say, u holonomic constraints of the form

$$f_i(\mathbf{q}, t) = 0,\ i = 1, 2, \ldots, u \tag{5.150}$$

and r nonholonomic constraints of the form (we drop the hats)

$$\sum_{j=1}^{s} d_{ij}(\mathbf{q},t)\dot{q}_j + g_i(\mathbf{q},t) = 0, \; i = 1, 2, \ldots, r. \tag{5.151}$$

Thus equations (5.149)–(5.151) describe the constrained system in the Lagrangian coordinates q_i, $i = 1, 2, \ldots, s$; $s = (3n - h + u)$. Equation (5.149) is obtained from Lagrange's equations; these equations are the "unconstrained" equations of motion. They are determined *as though the s coordinates were independent of one another.* Equations (5.150) and (5.151) provide the additional holonomic constraints and the nonholonomic constraints to characterize the system, a total of $w = u + r$ equations.

Equations (5.150) can be differentiated twice, and equations (5.151) can be differentiated once to obtain the constraint equations in our standard form

$$\mathrm{A}(\mathbf{q},t)\ddot{\mathbf{q}} = \mathbf{b}(\mathbf{q},\dot{\mathbf{q}},t) \tag{5.152}$$

where the matrix A is a w by s matrix whose i–jth element is $d_{ij}(\mathbf{q},t)$. We will assume the rank of A to be less than $3n$.

Thus in the s generalized coordinates q_i, $i = 1, 2, \ldots, s$, we have once again the generic formulation wherein the unconstrained equations of motion are provided by equation (5.149) and the constraints are provided by equation (5.152). One may wish to further reduce the number of generalized coordinates used to describe the configuration of the system. Depending on the generalized coordinates q_i chosen, and their specific number used to describe the "unconstrained" system, some or all of the u holonomic constraints (among the total of w constraints) may then be redundant.

The Fundamental Equation in Lagrangian Coordinates

The motion of the constrained system described by equations (5.149) and (5.152) now needs to be determined. Using these equations the reader can guess what this equation might be in analogy with what we had when we looked at systems described by their Cartesian coordinates. After all, the Cartesian coordinates can be considered as a special set of generalized coordinates. The result is immediate. The generalized acceleration of the constrained system is given by

$$\ddot{\mathbf{q}} = \mathbf{a} + \mathrm{M}^{-1}\mathrm{A}^T(\mathrm{A}\mathrm{M}^{-1}\mathrm{A}^T)^+(\mathbf{b} - \mathrm{A}\mathbf{a}), \tag{5.153}$$

where $\mathbf{a}$ is the acceleration of the unconstrained system, namely $\mathrm{M}^{-1}\mathrm{F}$. We shall prove this equation later in the next chapter. Further on in Chapter 7 we will show that this equation of motion is also valid for our more general constraint equation in standard form, namely $\mathrm{A}(\mathbf{q},\dot{\mathbf{q}},t)\ddot{\mathbf{q}} = \mathbf{b}(\mathbf{q},\dot{\mathbf{q}},t)$.

Example 5.13

Consider the double pendulum shown in Figure 5.3. Let us constrain the mass m_2 to always remain on the horizontal plane, which is at a constant distance y_0, $y_0 < (L_1 + L_2)$, below the point of suspension O.

Observe that with this additional constraint, the system has just one degree of freedom. Its configuration can be completely specified, for example, by either the coordinate θ_1 or the coordinate θ_2. However, the reader can see that were (s)he to try to determine the equation of motion of the system in terms of this single coordinate (either θ_1 or θ_2) directly by determining the kinetic energy and thence the Lagrange equation, it might take a good portion of an evening to do. Instead, the fundamental equation approach is more direct and swifter.

a. The unconstrained equations of motion of this system using θ_1 and θ_2 as the generalized coordinates are given by equations (5.127) and (5.129). The constraint equation is given in terms of these coordinates as $y = L_1 \cos\theta_1 + L_2 \cos\theta_2 = y_0$. On twice differentiation, this equation yields

$$[-L_1 \sin\theta_1 \quad -L_2 \sin\theta_2]\ddot{\mathbf{q}} = L_1 \cos\theta_1 \dot{\theta}_1^2 + L_2 \cos\theta_2 \dot{\theta}_2^2 \tag{5.154}$$

so that $A = [-L_1 \sin\theta_1 \quad -L_2 \sin\theta_2]$ and $\mathbf{b}$ is the scalar $L_1 \cos\theta_1 \dot{\theta}_1^2 + L_2 \cos\theta_2 \dot{\theta}_2^2$. Equations (5.127), (5.129) and (5.154) along with the initial conditions now provide a complete specification of the problem.

b. Alternately, we could have used the vector $\mathbf{q} = [\theta_1\ x_2\ y_2]^T$ and utilized the three equations (5.132)–(5.134) to describe the unconstrained motion of the system. Specification of the constrained motion would now require us to append to these three equations the two equations of constraint

$$(x_2 - L_1 \sin\theta_1)^2 + (y_2 - L_1 \cos\theta_1)^2 = L_2^2, \text{ and,} \tag{5.155}$$

$$y_2 = y_0. \tag{5.156}$$

Differentiating these equations twice we get

$$\begin{aligned}(x_2 - L_1 \sin\theta_1)(\ddot{x}_2 - L_1\ddot{\theta}_1 \cos\theta_1) + (y_2 - L_1 \cos\theta_1)(\ddot{y}_2 + L_1\ddot{\theta}_1 \sin\theta_1) \\ = -(\dot{x}_2 - L_1\dot{\theta}_1 \cos\theta_1)^2 - (\dot{y}_2 + L_1\dot{\theta}_1 \sin\theta_1)^2 \\ -(x_2 - L_1 \sin\theta_1)L_1\dot{\theta}_1^2 \sin\theta_1 - (y_2 - L_1 \cos\theta_1)L_1\dot{\theta}_1^2 \cos\theta_1, \text{ and,}\end{aligned} \tag{5.157}$$

$$\ddot{y}_2 = 0. \tag{5.158}$$

A complete description of the system would now include equations (5.132), (5.133), (5.134), (5.157) and (5.158). The first three are the equations of the "unconstrained" system in these three coordinates, and the last two are the equations of constraint relevant to this unconstrained system. To these equations we need to append a consistent set of initial conditions to specify the problem fully.

c. Were we to use the 4-vector $\mathbf{q} = [x_1 \quad y_1 \quad x_2 \quad y_2]^T$ to represent the positions of the masses, then the "unconstrained equations of motion" would be given by equations (5.137)–(5.140). As before, we would also have the twice differentiated constraints (5.141) and (5.142), yielding

$$x_1\ddot{x}_1 + y_1\ddot{y}_1 = \dot{x}_1^2 - \dot{y}_1^2, \text{ and} \tag{5.159}$$

$$(x_2 - x_1)(\ddot{x}_2 - \ddot{x}_1) + (y_2 - y_1)(\ddot{y}_2 - \ddot{y}_1) = -(\dot{x}_2 - \dot{x}_1)^2 - (\dot{y}_2 - \dot{y}_1)^2. \tag{5.160}$$

In addition, we have the new constraint

$$\ddot{y}_2 = 0. \tag{5.161}$$

The matrix A now becomes

$$A = \begin{bmatrix} x_1 & y_1 & 0 & 0 \\ (x_1 - x_2) & (y_1 - y_2) & (x_2 - x_1) & (y_2 - y_1) \\ 0 & 0 & 0 & 1 \end{bmatrix} \tag{5.162}$$

and the vector $\mathbf{b}$ is given by

$$\mathbf{b} = -\begin{bmatrix} \dot{x}_1^2 + \dot{y}_1^2 \\ (\dot{x}_2 - \dot{x}_1)^2 + (\dot{y}_2 - \dot{y}_1)^2 \\ 0 \end{bmatrix}. \tag{5.163}$$

The matrix $M = Diag\{m_1, m_1, m_2, m_2\}$, and the vector $\mathbf{a} = [0 \ g \ 0 \ g]^T$. Problem 5.14 requires the student to complete this example.

Summary

This long chapter deserves a summary. Having gone into the details of Lagrangian mechanics, for our purposes, we must carry forward the following basic concepts:

1. We can use generalized coordinates to describe the configuration of a system of n particles.

2. The minimum number of such coordinates needed to describe the configuration of the system equals $(3n - h)$, where h is the number of independent holonomic constraints. Were we to use this minimum number of coordinates, a proper choice of coordinates would ensure that all the holonomic constraints would be identically satisfied. We don't need to specify any holonomic constraints then, because the choice of coordinates will cause them to be automatically satisfied.

3. We could use more than this minimum number of coordinates, but then we would need to add additional relations between the constraints – relations which will be holonomic in nature.

4. The nonintegrable constraints are there to stay; no transformation of coordinates will eliminate them. They have to be additionally imposed to complete the description of the system.

5. Having decided on the generalized coordinates we can obtain the "unconstrained" equations of motion in terms of the generalized coordinates by using Lagrange's equations (5.114). We treat each coordinate as though it were independent of the others. We write the kinetic energy in terms of these coordinates, and find the generalized force corresponding to each coordinate. Expressing the Lagrange equations so found, in the form of equations (5.149), we next determine the matrix M and vector F, and from these determine the unconstrained acceleration $\mathbf{a} = \mathrm{M}^{-1}\mathrm{F}$. Note that both M and F are, in general, functions of time.

6. Depending on the choice of coordinates and their number, we will need to append an appropriate number of holonomic constraints. In addition, we impose *all* the nonholonomic constraints. This will now provide a complete description of the system. All these constraints will naturally be expressed in terms of the generalized coordinates. We differentiate these constraint equations and put them in the standard form of equation (5.152); we thus determine the matrix A and the vector **b**.

7. The equation of motion of the constrained system in generalized coordinates can now be found, knowing the matrices M and A, and the vectors **a** and **b**, by simply using the fundamental equation (5.153).

8. A new conceptualization of the way a given constrained system is "put together" emerges. The constrained system can be thought of as an "unconstrained system" to which "appropriate constraints" are then added. *The analyst has a choice. (S)He can visualize the unconstrained system in one of many possible ways; to the particular unconstrained system (s)he selects, the appropriate constraints need to be added so that the given constrained physical system is correctly described.* The more the number of generalized coordinates (s)he chooses as independent in the description of the unconstrained system, the easier will be the task, in general, of writing the unconstrained equations of motion. The more the number of generalized coordinates (s)he chooses as independent for the description of the unconstrained system, the larger the number of additional constraint equations (s)he will require to describe the given constrained system. The larger the number of additional constraint equations, the harder will be the task of analytically obtaining the generalized inverse involved in the fundamental equation of motion, unless of course one is doing this through the use of symbolic manipulation. Computationally, however, there may be little additional burden posed since the generalized inverse can be easily found, assuming that proper attention is paid to the numerical methods used.

PROBLEMS

5.1 Consider a system whose position $\mathbf{x}(t)$ and whose velocity $\dot{\mathbf{x}}(t)$ are known. Show that the difference between any two *possible* acceleration vectors qualifies as a virtual displacement vector.

5.2 In Example 5.2, we have said that the matrix A_1 is nonsingular except when $z = 0$ and $z = 1$. What would happen at these two points? Is relation (5.34) still valid? Hint: Note that the rank of matrix A is 2 for *all* z.

5.3 Use relation (5.46) to obtain directly the Lagrange multipliers for Examples 3.8 and 3.9. Compare your results with those obtained in these examples.

5.4 Use polar coordinates to describe the configuration of the two particles described in Problem 4.6. Determine the equations of motion using Lagrange's equations. Note that by choosing the coordinates that describe the configuration of the system as r_1, r_2 and θ, where r_1 and r_2 are the polar distances from the origin, and θ is the inclination of the line connecting the particles with the X-axis, the constraint is automatically satisfied.

5.5 In Example 5.7, consider the transformation of the coordinates from the Cartesian coordinates (x_1, y_1, x_2, y_2) to the generalized coordinates $(\theta_1, L_1, \theta_2, L_2,)$ as described by equations (5.72) and (5.73). Is this transformation one-to-one, as is required?

5.6 Show that if time t in equation (5.64) does not appear explicitly, then it is always possible to choose the coordinates $q_1, q_2, q_3, \ldots, q_{(3n-h)}$ so that time t does not enter in equations (5.65) either.

5.7 Show that the kinetic energy can be expressed in terms of the generalized coordinates as given by equations (5.144)—(5.148).

5.8 Use Lagrange's equations (5.114) along with the expressions for the kinetic energy in terms of the generalized coordinates given in equations (5.144)–(5.148) to explicitly write the equation (5.149). Determine explicitly the components of the vector $F(\mathbf{q}, \dot{\mathbf{q}}, t)$ as well as the elements of the matrix $M(\mathbf{q}, t)$.

5.9 Consider the group of forces which retard particles and which are proportional to the components of their *relative* velocities. Hence F_{ij}, the force exerted on particle i by particle j, is $-c_{ij}(\dot{x}_i - \dot{x}_j)$. Find the dissipation function for such a retarding set of forces. Write Lagrange's equations in terms of this dissipation function.

5.10 Express the dissipation function given by equation (5.121) in terms of the generalized coordinates.

5.11 A particle whose position is denoted by x, moves along the X-direction in a Cartesian coordinate frame. Its potential energy is given by $V = \frac{1}{2}kx^2$, where k is a constant. If c is a constant, and the dissipation function is given by $R_D = \frac{1}{2}c|\dot{x}|^{\alpha+1}$, with $\alpha = 1$, write the equation of motion for the unconstrained system. Discuss the behavior of such a system for a given set of initial conditions.

5.12 In the above problem, discuss the particle's behavior if the parameter α in the dissipation function, instead of being unity, is given by $\alpha = \frac{\beta}{\beta + 1}$ where β is some odd positive integer. How does the behavior of such a system differ from that found in the previous example? What happens as $\beta \to \infty$?

5.13 Find the vector $\mathbf{b}$ and the matrix A corresponding to equation (5.152) for Example 5.13, part (b).

5.14 Complete Example 5.13 and write the explicit equations of motion for the coordinates chosen in part (a) and then again in part (b). Write a program in MATLAB to integrate these equations. Note that you don't have to find the MP-inverse analytically; it can be done numerically. Compare your numerical results using the two different formulations of the problem. Take $y_0 = (L_1 + L_2)/2$.

5.15 Formulate the problem where the lower mass m_2 in Example 5.13 is constrained to move on a horizontal sinusoidal surface (instead of a flat horizontal surface) which is defined by $y = y_0 - a \sin \omega x$, where $y_0 < L_1 + L_2 - a$.

5.16 In Figure 5.5 mass M is constrained to move horizontally on frictionless rollers. The spring is a linear spring with a spring constant k. A pendulum consisting of a weightless rod of length L with an end mass m is attached to the mass M as shown. Assume that the given system is "arrived at" from two different visualizations of the unconstrained system: (1) use the X-coordinate of mass M and the angle θ as the two independent coordinates describing the unconstrained system; (2) use the X- coordinate of mass M, the X-coordinate of mass m and the Y-coordinate of the mass m as the three independent coordinates to describe the unconstrained system. What is the equation of constraint that the three coordinates in case (2) must satisfy? What does the unconstrained system physically look like? Write the equations of motion of the given system using these two different visualizations.

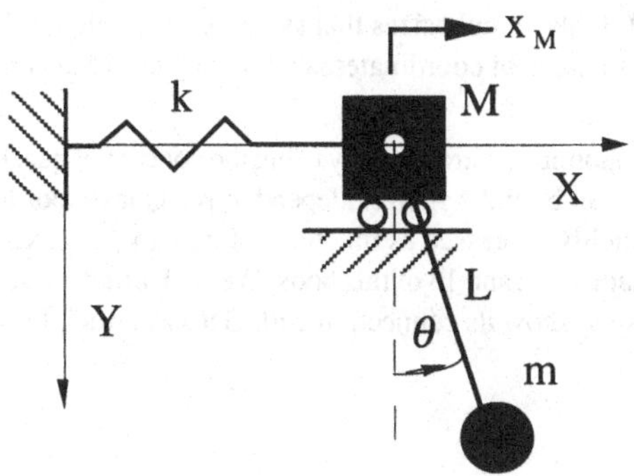

Figure 5.5.

5.17 From relation (5.45) we have determined explicitly the vector $\boldsymbol{\lambda}$ when the matrix A has rank m. What if the matrix A has rank r where $r < m$? Can the vector $\boldsymbol{\lambda}$ now be uniquely found? Find the general solution for $\boldsymbol{\lambda}$ which solves equation (5.45) by using equation (2.107) of Chapter 2. Does this make the force of constraint $\mathbf{F}^c$ nonunique?

5.18 Consider the constrained motion of a system of particles, described in generalized coordinates, by: (a) the equations (5.149) corresponding to the unconstrained system, and, (b) the relations (5.152) which prescribe the constraints. Provide a formulation of this problem in terms of the Lagrange multiplier vector $\boldsymbol{\lambda}$ similar to equations (5.37) and (5.38). What would the dimension of the vector $\boldsymbol{\lambda}$ be? Noting equation (5.153), find the general solution for the vector $\boldsymbol{\lambda}$, when the rank r of the matrix A is less than or equal to w. When would the vector $\boldsymbol{\lambda}$ be unique?

For Further Reading

1. P1 This book has an excellent discussion, though detailed, regarding the basic equation of mechanics, virtual work, Lagrangian coordinates and Lagrange's equations. The material between pages 23 and 89 would be of value to the interested reader.

2. R1 The discussion of virtual work in Chapter 9 is thorough and very well done. Chapters 12–15 of the book deal with the basic equation of analytical mechanics in generalized coordinates, Lagrange's equations and the handling of constraints via Lagrange multipliers. The discussion is lucid and very readable.

3. R1 The constraint imposed by a particle moving on a rough surface is the type of constraint we said is not easily amenable to the methods developed in this book. Chapter 9, Section 9.10 of this book gives an indication of how to handle such problems in the Lagrangian framework.

4. R3 The dissipation function is discussed very well in this book in Article 81. The book is an excellent reference on vibration theory and contains valuable material, especially in Chapter 4, which may not be found even in most modern books on the subject.

5. W2 This book, which is a classic in the field of mechanics, has a nice development of Lagrange's equations in sections 20 to 29. It also emphasizes that systems with holonomic constraints can be treated, through a proper choice of coordinates, as though they are unconstrained.

6. P1 Another way of handling nonholonomic constraints is by using the concept of quasi-coordinates. This method was developed by Gibbs and Appell (independently). Quasi-coordinates are coordinates which cannot be explicitly expressed as functions of $x_1, x_2, x_3, \ldots, x_{3n}$. An excellent treatment may be found in Chapters 12 and 13 of this book. We shall briefly touch upon this method in Chapter 8, where we shall show its connection with Gauss's principle.

6 The Fundamental Equation in Generalized Coordinates

In the last chapter we dealt with the use of generalized coordinates and the fundamental equation in generalized coordinates. While we did not present a proof of this equation (see equation (5.153)) we did provide some plausibility arguments in its favor. In this chapter we will prove this equation starting from the basic equation of analytical mechanics. We will amplify some of our observations related to the use of generalized coordinates, and provide more examples so that the reader acquires a greater familiarity with the concepts developed in the previous chapter through increased exposure.

We will utilize the fundamental equation, now stated in terms of generalized coordinates, to show how simple and elegant the determination of the equations of motion related to complex, constrained mechanical systems turns out to be. For this, we expand our compass of application to include rigid bodies, the description of their configurations and the determination of their kinetic energy through the use of Euler angles. We include in some detail the theory of rotations, and the concepts of infinitesimal rotations and angular velocity. Our treatment of rigid bodies may be somewhat swift for the reader who is completely unfamiliar with this topic; however, our main aim here is not to provide a comprehensive account of the kinematics and dynamics of rigid bodies; for this the reader may look at our "For Further Reading" list at the end of this chapter. It is simply to provide the reader with a flavor for what can be done with the fundamental equation in problems which have so far been difficult to handle, such as those dealing with the constrained motion of rigid bodies.

Before we begin deriving the fundamental equation in generalized coordinates, we will remind ourselves of some of the concepts which we have dealt with in the previous chapter.

To formulate the equation of motion of a constrained system, we can always imagine the system to be initially unconstrained, and then impose the necessary constraints. We saw in the last chapter that what we specifically mean by an "unconstrained" system is simply that the coordinates used to describe the system are treated as though they are independent. Thus for any given physical situation wherein we have a constrained system, we can have different visualizations of the "unconstrained" system corresponding to it and of course the necessary constraints that need to be imposed. The reader should once again review Examples 5.12 and 5.13 to get a more palpable idea of the last three sentences. It should be noted that we do not need the constraints to be independent; indeed independence of the constraints may be a difficult task to check, especially if there are many particles constituting the system and many constraints. All we require is that the constraints be *consistent*: that is, the satisfaction of any one constraint does not preclude the satisfaction of any other constraint. Furthermore, we assume that the constraint equations can be differentiated so that we obtain a set of equations which are linear in the possible accelerations that the particles of the system may have (see equation (5.152)). In addition to these equations that describe the constraints, we have the "unconstrained" equations of motion. These equations are also linear in the accelerations of the particles and describe the motion of the system were the constraint equations nonexistent. Thus, armed with these two pieces of information, the equations corresponding to the "unconstrained" system, and the constraint equations, let us set off in search of the explicit equations of motion of the constrained system.

Gauss's principle can be derived from the principle of virtual work and D'Alembert's principle. In fact that is the way Gauss proved his principle back in 1829. The reader will recall that in Chapter 3, we had used Gauss's principle to prove the fundamental equation in terms of Cartesian coordinates. Rather than appeal to Gauss's principle, as we did earlier, in this chapter we shall use the fundamentals that we have developed in Chapter 5 regarding the *basic principle of analytical mechanics* to directly arrive at the fundamental equation. Later on we will show that Gauss's principle can also be cast in generalized coordinates (something perhaps not well known), and a direct application of the methods of Chapter 3 then will yield the fundamental equation in a trivially obvious fashion.

We now want to collect some of the mathematical machinery which we have developed in the previous chapters to arm ourselves before we embark on the first main objective in this chapter – to prove the fundamental equation using generalized coordinates.

Useful Mathematical Machinery

In this section we will not resort to proving any results; they have already been proved by us in detail in Chapter 2. We will merely collect four important results related to the Moore–Penrose inverse. The reader may want to refresh his/her memory by skimming over Chapter 2, especially the section entitled "Generalized Inverse of a Matrix."

1. The general solution to the consistent equation $\mathbf{Ax} = \mathbf{b}$, where the matrix A is m by n is

$$\mathbf{x} = A^{+}\mathbf{b} + (I - A^{+}A)\mathbf{h}, \tag{6.1}$$

where $\mathbf{h}$ is *any* n by 1 vector. See equation (2.107) for details.

2. The Moore–Penrose inverse of the n-vector $\mathbf{u}$ is the 1 by n vector

$$\mathbf{u}^{+} = \frac{1}{(\mathbf{u}^{T}\mathbf{u})}\mathbf{u}^{T}. \tag{6.2}$$

See equation (2.83) for details.

3. If

$$\mathbf{u} \neq \mathbf{0} \text{ and } A\mathbf{u} = \mathbf{0}, \text{ then } \mathbf{u}^{T}A^{+} = \mathbf{0}. \tag{6.3}$$

See relation (2.91) for details.

4. For any matrix A,

$$(I - A^{+}A) = (I - A^{+}A)^{T}. \tag{6.4}$$

We have $(I-A^{+}A)^{T}=I^{T}-(A^{+}A)^{T}=I-A^{+}A$, where the last equality follows from the MP-condition (2.55).

The Fundamental Equation in Generalized Coordinates

We are now ready to prove the fundamental equation in generalized coordinates. We shall do this using D'Alembert's principle and the basic equation of analytical dynamics. Recall from Chapter 5 that the equations for the "unconstrained" system can always be cast in the form of equation (5.149) as

$$\mathrm{M}(\mathbf{q}, t)\ddot{\mathbf{q}} = \mathrm{F}(\mathbf{q}, \dot{\mathbf{q}}, t) \tag{6.5}$$

and that the equations of constraint (differentiated a suitable number of times) can be cast as

$$\mathrm{A}(\mathbf{q}, t)\ddot{\mathbf{q}} = \mathbf{b}(\mathbf{q}, \dot{\mathbf{q}}, t). \tag{6.6}$$

Here the vector $\mathbf{q}$ is an s-vector and the matrix A is w by s. Of the w equations represented by equation (6.6), h equations come from the holonomic constraints and r from the nonholonomic constraints.

The arguments and results that follow in the rest of this section are actually valid for a more general class of constraints than indicated in equation (6.6); we could expand the set of constraints to include those that can be expressed in our standard form as $A(\mathbf{q}, \dot{\mathbf{q}}, t) = \mathbf{b}(\mathbf{q}, \dot{\mathbf{q}}, t)$, where the matrix A may now be not just a function of $\mathbf{q}$ and t, as in equation (6.6), but also a function of $\dot{\mathbf{q}}$. But in order to do this we need a slightly deeper understanding of what constitutes a virtual displacement; we shall leave this discussion for Chapter 7. The reader may simply, at this stage, make a mental note that the line of thinking in the rest of this section and the form of the final results are exactly applicable to the more general above mentioned form of the constraint equation as well.

In the absence of the constraint equation (6.6), the acceleration of the "unconstrained" system of course is simply given, using equation (6.5), by

$$\mathbf{a}(t) = M^{-1}(\mathbf{q}, t)F(\mathbf{q}, \dot{\mathbf{q}}, t). \tag{6.7}$$

Now, as noted in equation (5.14), a virtual displacement vector $\mathbf{v}$ at time t is any (nonzero) vector which is such that

$$A(\mathbf{q}, t)\mathbf{v} = \mathbf{0}. \tag{6.8}$$

We mention in passing that equation (6.8) remains the defining equation for a virtual displacement vector, $\mathbf{v}$, even when the matrix A is a more general matrix which is a function of $\mathbf{q}, \dot{\mathbf{q}}$ and t. However, this statement requires a deeper appreciation of what constitutes a virtual displacement; we shall defer this discussion to Chapter 7. It is this fact about the nature of virtual displacements that allows us to expand the applicability of all the results obtained in this section to the more general form of the constraint equation referred to earlier.

Let us define the vector $\mathbf{u}$ as

$$\mathbf{u}(t) = M^{1/2}(\mathbf{q}, t)\, \mathbf{v}(t). \tag{6.9}$$

The matrix M is positive definite and hence the matrices $M^{1/2}$ and $M^{-1/2}$ are well defined. Then equation (6.8) is equivalent to the equation

$$\mathbf{Bu} = \mathbf{0}, \tag{6.10}$$

where the w by s matrix $\mathbf{B} = AM^{-1/2}$ is the Constraint Matrix. By equation (6.6), the accelerations of the constrained system satisfy the equation

$$B(\mathbf{q}, t)\ddot{\mathbf{r}} = \mathbf{b}(\mathbf{q}, \dot{\mathbf{q}}, t) \tag{6.11}$$

where

$$\ddot{\mathbf{r}}(t) = \mathbf{M}^{1/2}(\mathbf{q}, t)\ddot{\mathbf{q}}(t). \tag{6.12}$$

Now we know from equation (6.1) that the explicit solution of equation (6.11) is simply

$$\ddot{\mathbf{r}}(t) = \mathbf{B}^{+}\mathbf{b} + (I - \mathbf{B}^{+}\mathbf{B})\mathbf{h} \tag{6.13}$$

where the s by w matrix $\mathbf{B}^{+}$ is, as usual, the MP-inverse of the matrix B. In arriving at equation (6.13) we have not used any mechanics; all we did was use the constraint equation (6.11). Linear algebra provides us with equation (6.13).

We notice that if the vector $\ddot{\mathbf{r}}$ is known, then the acceleration of the system $\ddot{\mathbf{q}}$ can be determined by using equation (6.12), since $\ddot{\mathbf{q}} = \mathbf{M}^{-1/2}\ddot{\mathbf{r}}$. Furthermore, since the matrix B is known, we know $\mathbf{B}^{+}$, and hence $\mathbf{B}^{+}\mathbf{b}$. The first term on the right hand side of equation (6.13) is therefore known. All we need to do now is to determine the vector $(I - \mathbf{B}^{+}\mathbf{B})\mathbf{h}$ in accordance with the principles of mechanics.

In the presence of the constraints (6.11) we know that the equation of motion of the "constrained" system must differ from that of the "unconstrained" system, namely from equation (6.5), by the presence of the constraint forces which are engendered by the presence of the constraints. Hence the equation of motion of the constrained system can be expressed as

$$\mathbf{M}(\mathbf{q}, t)\ddot{\mathbf{q}} = \mathbf{F}(\mathbf{q}, \dot{\mathbf{q}}, t) + \mathbf{F}^{c}(\mathbf{q}, \dot{\mathbf{q}}, t) \tag{6.14}$$

where the vector $\mathbf{F}^{c}(\mathbf{q}, \dot{\mathbf{q}}, t)$ is the additional constraint force vector generated by virtue of the imposition of the constraints (6.11).

Now by the basic equation of analytical mechanics, we know that these constraint forces are such that, at each instant of time, the sum total of the work done by them under any virtual displacement must equal zero.

Hence for any nonzero vector $\mathbf{v}$, such that $\mathbf{A}\mathbf{v} = \mathbf{0}$, we must have

$$\mathbf{v}^{T}\mathbf{F}^{c}(\mathbf{q}, \dot{\mathbf{q}}, t) = \mathbf{v}^{T}\{M(\mathbf{q}, t)\ddot{\mathbf{q}} - \mathbf{F}(\mathbf{q}, \dot{\mathbf{q}}, t)\} = 0. \tag{6.15}$$

We now express this condition in terms of the vectors $\mathbf{u}$ and $\ddot{\mathbf{r}}$. Thus the acceleration vector $\ddot{\mathbf{r}}$ corresponding to the constrained motion must be such that for all nonzero vectors $\mathbf{u}$ for which $\mathbf{B}\mathbf{u} = \mathbf{0}$,

$$\begin{aligned} \mathbf{v}^{T}\mathbf{F}^{c}(\mathbf{q}, \dot{\mathbf{q}}, t) &= \mathbf{u}^{T}\mathbf{M}^{-1/2}\{\mathbf{M}(\mathbf{q}, t)\ddot{\mathbf{q}} - \mathbf{F}(\mathbf{q}, \dot{\mathbf{q}}, t)\} \\ &= \mathbf{u}^{T}\{\ddot{\mathbf{r}} - \mathbf{M}^{-1/2}\mathbf{F}(\mathbf{q}, \dot{\mathbf{q}}, t)\} = 0. \end{aligned} \tag{6.16}$$

But the acceleration vector $\ddot{\mathbf{r}}$ must satisfy the constraints and must therefore have the form given in equation (6.13). We can substitute for $\ddot{\mathbf{r}}$ from (6.13) in the last equality in (6.16). Hence for all nonzero vectors $\mathbf{u}$ such that $\mathrm{B}\mathbf{u} = \mathbf{0}$, we require that

$$\mathbf{u}^T\{\mathrm{B}^+\mathbf{b} + (I - \mathrm{B}^+\mathrm{B})\mathbf{h} - \mathrm{M}^{-1/2}\mathrm{F}(\mathbf{q}, \dot{\mathbf{q}}, t)\} = 0. \tag{6.17}$$

But by relation (6.3), $\mathrm{B}\mathbf{u} = \mathbf{0}$ implies that $\mathbf{u}^T\mathrm{B}^+ = \mathbf{0}$. Condition (6.17) then reduces to requiring that

$$\mathbf{u}^T\{\mathbf{h} - \mathrm{M}^{-1/2}\mathrm{F}(\mathbf{q}, \dot{\mathbf{q}}, t)\} = 0, \text{ for all nonzero vectors } \mathbf{u} \text{ such that } \mathbf{u}^T\mathrm{B}^T = \mathbf{0}. \tag{6.18}$$

But this can only happen if the vector $\{\mathbf{h} - \mathrm{M}^{-1/2}\mathrm{F}(\mathbf{q}, \dot{\mathbf{q}}, t)\}$ belongs to the column space of the matrix B^T. But if a vector belongs to the column space of B^T it must be expressible as $\mathrm{B}^T\mathbf{w}$ for some w-vector $\mathbf{w}$. Hence

$$\mathbf{h} - \mathrm{M}^{-1/2}\mathrm{F}(\mathbf{q}, \dot{\mathbf{q}}, t) = \mathrm{B}^T\mathbf{w}, \tag{6.19}$$

so that

$$\mathbf{h} = \mathrm{M}^{-1/2}\mathrm{F} + \mathrm{B}^T\mathbf{w}. \tag{6.20}$$

Using this expression for $\mathbf{h}$ in equation (6.13) we get the equation of motion for the constrained system to be

$$\ddot{\mathbf{r}}(t) = \mathrm{B}^+\mathbf{b} + (I - \mathrm{B}^+\mathrm{B})(\mathrm{M}^{-1/2}\mathrm{F} + \mathrm{B}^T\mathbf{w}). \tag{6.21}$$

But, $(I - B^+B)$ is symmetric so that we get

$$[(I - \mathrm{B}^+\mathrm{B})\mathrm{B}^T\mathbf{w}]^T = \mathbf{w}^T\mathrm{B}(I - \mathrm{B}^+\mathrm{B}) = \mathbf{w}^T(\mathrm{B} - \mathrm{B}) = \mathbf{0}, \tag{6.22}$$

where we have used the MP-condition $\mathrm{B} = \mathrm{B}\mathrm{B}^+\mathrm{B}$. Using relation (6.22), equation (6.21) simplifies to

$$\ddot{\mathbf{r}}(t) = \mathrm{B}^+\mathbf{b} + (I - \mathrm{B}^+\mathrm{B})\mathrm{M}^{-1/2}\mathrm{F}. \tag{6.23}$$

Noting that $\ddot{\mathbf{q}} = \mathrm{M}^{-1/2}\ddot{\mathbf{r}}$, we get

$$\boxed{\ddot{\mathbf{q}}(t) = \mathbf{a}(t) + \mathrm{M}^{-1/2}(\mathrm{A}\mathrm{M}^{-1/2})^+(\mathbf{b} - \mathrm{A}\mathbf{a})} \tag{6.24}$$

where we have made use of equation (6.7) and the fact that $\mathrm{B} = \mathrm{A}\mathrm{M}^{-1/2}$. Equation (6.24) expresses the acceleration of the constrained system at time t in terms of: (1) the matrix $M(\mathbf{q}, t)$, (2) the acceleration of the unconstrained system $\mathbf{a}(t)$, (3) the matrix $A(\mathbf{q}, t)$ and (4) the vector $\mathbf{b}$. The first two of these quantities are related to the description of the unconstrained motion of the system; the last two are related to the constraints imposed.

Premultiplying both sides of equation (6.24) by the matrix M we get

$$\boxed{M(\mathbf{q},t)\ddot{\mathbf{q}}(t) = F(\mathbf{q},\dot{\mathbf{q}},t) + M^{1/2}(AM^{-1/2})^{+}(\mathbf{b} - A\mathbf{a})} \tag{6.25}$$

where $F(\mathbf{q}, \dot{\mathbf{q}}, t)$ is the *impressed* force on the system which enters into the description of the "unconstrained" motion. Comparing equation (6.25), which describes the motion of the constrained system with equation (6.5), which describes the unconstrained system, we infer that the second term on the right hand side of equation (6.25) is a consequence of the presence of the constraint equation (6.6). Furthermore, comparing equation (6.14) with equation (6.25) we obtain explicitly the constraint force vector F^c as

$$\boxed{F^c(\mathbf{q},\dot{\mathbf{q}},t) = M^{1/2}(AM^{-1/2})^{+}(\mathbf{b} - A\mathbf{a}).} \tag{6.26}$$

As mentioned earlier, equations (6.24)–(6.26) and the method we used to obtain them in this section remain valid when the constraint equations are of the more general form $A(\mathbf{q},\dot{\mathbf{q}},t)\ddot{\mathbf{q}}(t) = \mathbf{b}(\mathbf{q},\dot{\mathbf{q}},t)$ instead of equation (6.6). This result follows as soon as we can prove that equation (6.8) remains valid for defining a virtual displacement even when the matrix A is a function of $\dot{\mathbf{q}}$. This will be done in the following chapter.

The reader may want to obtain alternate forms of equations (6.24)–(6.26), like we did in Chapter 3, by using (see equation (2.88)) the fact that $(AM^{-1/2})^{+} = M^{-1/2}A^T(AM^{-1}A^T)^{+}$, thereby removing the quantity $M^{-1/2}$ from these equations. We now go on to develop two important principles of analytical dynamics.

Two Principles of Analytical Mechanics

The second term on the right hand side of equation (6.24) is a consequence of the presence of the constraints (6.6) and indicates that the acceleration of the constrained system $\ddot{\mathbf{q}}(t)$ differs from the acceleration of the unconstrained system $\mathbf{a}(t)$ by the quantity

$$\ddot{\mathbf{q}}(t) - \mathbf{a}(t) = M^{-1/2}(AM^{-1/2})^{+}(\mathbf{b} - A\mathbf{a}). \tag{6.27}$$

In the right hand side of the equation above, we note that the quantity $\mathbf{e}(t) = \mathbf{b}(t) - A(\mathbf{q}, t)\mathbf{a}(t)$ is simply the extent to which the acceleration $\mathbf{a}(t)$ corresponding to the unconstrained system does not satisfy the constraint equation (6.6) at time t. Furthermore, using equation (6.27), we can now write the difference (or deviation) at time t, $\Delta\ddot{\mathbf{q}}(t) = \ddot{\mathbf{q}}(t) - \mathbf{a}(t)$, between the acceleration corresponding to the constrained system and that corresponding to the unconstrained system as

$$\Delta\ddot{\mathbf{q}}(t) = K_1\mathbf{e}\,, \tag{6.28}$$

where the matrix K_1 is defined to be $M^{-1/2}(AM^{-1/2})^+$. The matrix K_1 is nothing but the weighted Moore–Penrose inverse of the Constraint Matrix, the weighting being done through premultiplication by the matrix $M^{-1/2}$. Thus at each instant in time, the deviation in acceleration from that corresponding to the unconstrained system is directly proportional to $\mathbf{e}$, the matrix of proportionality being the matrix K_1.

This leads to the following principle of analytical mechanics:

The motion of a discrete dynamical system subjected to constraints evolves, at each instant in time, in such a manner that the deviation of its acceleration from that which it would have had, at that instant, if there were no constraints on it, is directly proportional to the extent to which the acceleration corresponding to the unconstrained motion, at that instant, does not satisfy the constraints; the matrix of proportionality K_1 *is the weighted Moore–Penrose generalized inverse of the Constraint Matrix, and the measure of the dissatisfaction of the constraints is the vector* $\mathbf{e}$.

Alternately, we observe from equation (6.26) that the force of constraint can be explicitly expressed as

$$F^c(\mathbf{q}, \dot{\mathbf{q}}, t) = K\mathbf{e}, \tag{6.29}$$

where the matrix $K = M^{1/2}(AM^{-1/2})^+$. The matrix K is again the weighted Moore–Penrose inverse of the Constraint Matrix, this time the weighting is done through premultiplication by the matrix $M^{1/2}$.

This result can be expressed as a principle of analytical mechanics as follows:

The motion of a discrete dynamical system subjected to constraints evolves, at each instant in time, in such a manner that the force of constraint acting on it, at that instant, is directly proportional to the extent to which the acceleration corresponding to the unconstrained motion, at that instant, does not satisfy the constraints; the matrix of proportionality K *is the weighted Moore-Penrose generalized inverse of the Constraint Matrix, and the measure of the dissatisfaction of the constraints is the vector* $\mathbf{e}$.

The reader will note that these two principles are identical to those developed in Chapter 4 except that we have now been able to obtain them while using generalized coordinates here instead of the Cartesian coordinates which we did in Chapter 4. As stated before, they are also valid for the more general standard form of the constraint equation $A(\mathbf{q}, \dot{\mathbf{q}}, t)\ddot{\mathbf{q}}(t) = \mathbf{b}(\mathbf{q}, \dot{\mathbf{q}}, t)$.

Example 6.1

Consider two masses m_1 and m_2 connected to each other by a rigid weightless rod AB of length 4 units (see Figure 6.1). Let the masses m_1 and m_2 be connected to the pivot O by two weightless rigid rods OA and OB of lengths 3 and 5 units respectively. We want to determine the equations of motion of this system of particles as it moves under the influence of gravity which acts in the positive Y-direction. Notice that this system of particles, as constrained above, could be thought of as a rigid body.

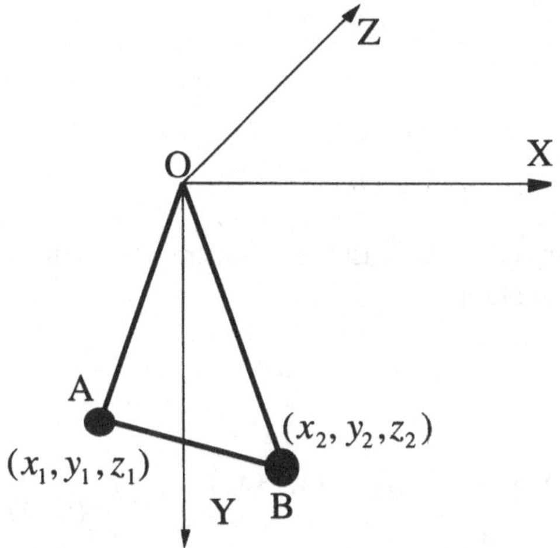

Figure 6.1. A two particle system connected by three weightless rods

Let us describe the configuration of the unconstrained system of particles by the coordinates of each mass. Let the coordinates of the mass m_1 located at A be denoted by (x_1, y_1, z_1) and those of mass m_2 located at B by (x_2, y_2, z_2). We thus use six coordinates to describe the unconstrained configuration of the system. The unconstrained equations of motion of this system are given by

$$m_1\ddot{x}_1 = 0,\ m_1\ddot{y}_1 = m_1 g,\ m_1\ddot{z}_1 = 0, \tag{6.30}$$

Example 6.1 (continued)

and

$$m_2\ddot{x}_2 = 0,\ m_2\ddot{y}_2 = m_2 g,\ m_2\ddot{z}_2 = 0 \tag{6.31}$$

where g denotes the acceleration due to gravity which acts in the positive Y-direction. Let the vector which describes the configuration of the unconstrained system be denoted by $\mathbf{q}(t) = [x_1\ y_1\ z_1\ x_2\ y_2\ z_2]^T$. Hence the vector $\mathbf{a}$ of the acceleration of the system corresponding to the unconstrained system becomes $\mathbf{a} = [0\ g\ 0\ 0\ g\ 0]^T$.

There are three constraints on this system of particles. We require that the lengths OA=3 units, OB=5 units and AB=4 units. These three constraint equations can be written as

$$x_1^2 + y_1^2 + z_1^2 = 9, \tag{6.32}$$

$$x_2^2 + y_2^2 + z_2^2 = 25, \tag{6.33}$$

and,

$$(x_2 - x_1)^2 + (y_2 - y_1)^2 + (z_2 - z_1)^2 = 16. \tag{6.34}$$

Differentiating these equations twice, we cast these constraint equations in the form $A\ddot{\mathbf{q}} = \mathbf{b}$. We then obtain

$$A(t) = \begin{bmatrix} x_1 & y_1 & z_1 & 0 & 0 & 0 \\ 0 & 0 & 0 & x_2 & y_2 & z_2 \\ (x_1 - x_2) & (y_1 - y_2) & (z_1 - z_2) & -(x_1 - x_2) & -(y_1 - y_2) & -(z_1 - z_2) \end{bmatrix} \tag{6.35}$$

and the vector

$$\mathbf{b}(t) = -\begin{bmatrix} \dot{x}_1^2 + \dot{y}_1^2 + \dot{z}_1^2 \\ \dot{x}_2^2 + \dot{y}_2^2 + \dot{z}_2^2 \\ (\dot{x}_2 - \dot{x}_1)^2 + (\dot{y}_2 - \dot{y}_1)^2 + (\dot{z}_2 - \dot{z}_1)^2 \end{bmatrix}. \tag{6.36}$$

Now using the fundamental equation we get the motion of the constrained system directly to be

$$\ddot{\mathbf{q}} = \mathbf{a} + M^{-1/2}(AM^{-1/2})^{+}(\mathbf{b} - A\mathbf{a}), \tag{6.37}$$

where $M^{-1/2} = Diag\{m_1^{-1/2}, m_1^{-1/2}, m_1^{-1/2}, m_2^{-1/2}, m_2^{-1/2}, m_2^{-1/2}\}$. The second term on the right hand side of equation (6.37) constitutes the contribution of the forces of constraint in altering the acceleration of the two masses so that in the presence of gravity their motion satisfies the constraint equations (6.32)–(6.34).

Let us assume that the initial conditions at $t = 0$ are given by

$$\mathbf{q}(0) = [x_1(0) \quad y_1(0) \quad 0 \quad x_2(0) \quad y_2(0) \quad 0]^T, \tag{6.38}$$

and

$$\dot{\mathbf{q}}(0) = [0 \quad 0 \quad \dot{z}_1(0) \quad 0 \quad 0 \quad \dot{z}_2(0)]^T. \tag{6.39}$$

Thus we start the system off when it lies in the XY-plane, and provide a velocity in the Z-direction to each of the two masses. Were we to start the system when the vertical line through O passes through the center of the weightless bar AB, then $x_1(0) = -6/\sqrt{13}$, $x_2(0) = 6/\sqrt{13}$, $y_1(0) = \sqrt{13} - (4/\sqrt{13})$ and $y_2(0) = \sqrt{13} + (4/\sqrt{13})$. Since the initial position of the system lies in the XY-plane we are free to choose the quantities $\dot{z}_1(0)$ and $\dot{z}_2(0)$ arbitrarily.

We remind the reader that the initial conditions for holonomic constraints must be so chosen that these constraints *as well as their once differentiated forms* are both identically satisfied. The reader should verify that this is so for the initial conditions we have chosen while still leaving $\dot{z}_1(0)$ and $\dot{z}_2(0)$ arbitrary.

We are now ready to numerically solve for the trajectories describing the motion of the system as it evolves in time. As in Chapter 4, we will use the fourth order Runge–Kutta scheme to numerically integrate the fundamental equation, numerically determining the MP-inverse as we go along.

Figure 6.2 shows the motion of the system when $\dot{z}_1(0)$ and $\dot{z}_2(0)$ are both zero. The masses m_1 and m_2 are taken to be 1 and 4 units respectively, and the value of g is taken to be 9.8 units. Note that the initial position of the system is not an equilibrium position. The system pendulates in the XY-plane, the Z-component of motion being zero. We show plots

Example 6.1 (continued)

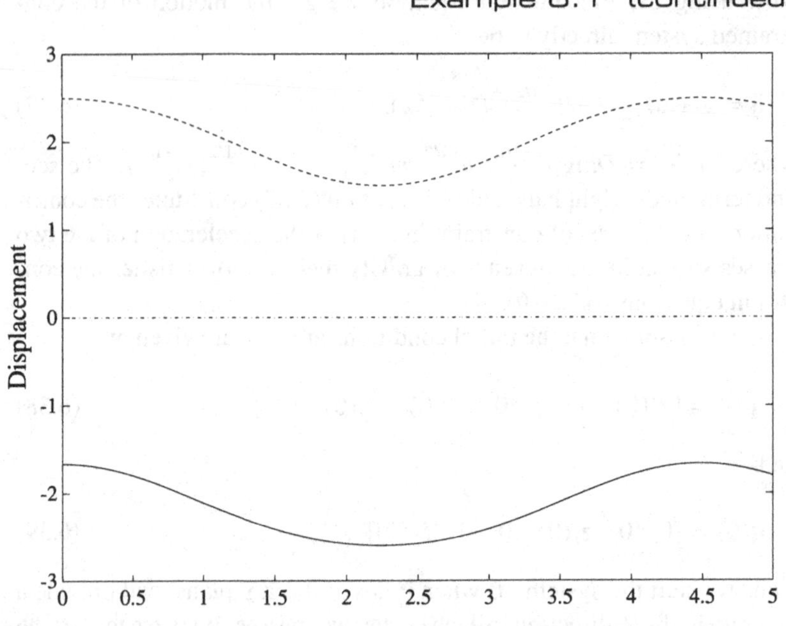

Figure 6.2(a). Motion of the mass m_1. The solid line shows the X-component, the dashed line the Y-component and the dash–dot line the Z-component of motion

of each component as a function of time. The local error tolerance for the Runge–Kutta scheme was set to 10^{-10}.

The errors in the satisfaction of the constraints are defined as

$$e_1(t) = x_1^2 + y_1^2 + z_1^2 - 9, \tag{6.40}$$

$$e_2(t) = x_2^2 + y_2^2 + z_2^2 - 25, \tag{6.41}$$

$$e_3(t) = (x_2 - x_1)^2 + (y_2 - y_1)^2 + (z_2 - z_1)^2 - 16. \tag{6.42}$$

We observe that these errors appear to increase with time and are, over the time interval considered, of the same order of magnitude as the numerical integration error tolerance.

Figure 6.3 shows the motion of the system with masses m_1 and m_2 being 1 and 2 units respectively. Using the same initial position for the two masses as before, we now take the initial velocities to be $\dot{z}_1(0) = 10$ and $\dot{z}_2(0) = 0$.

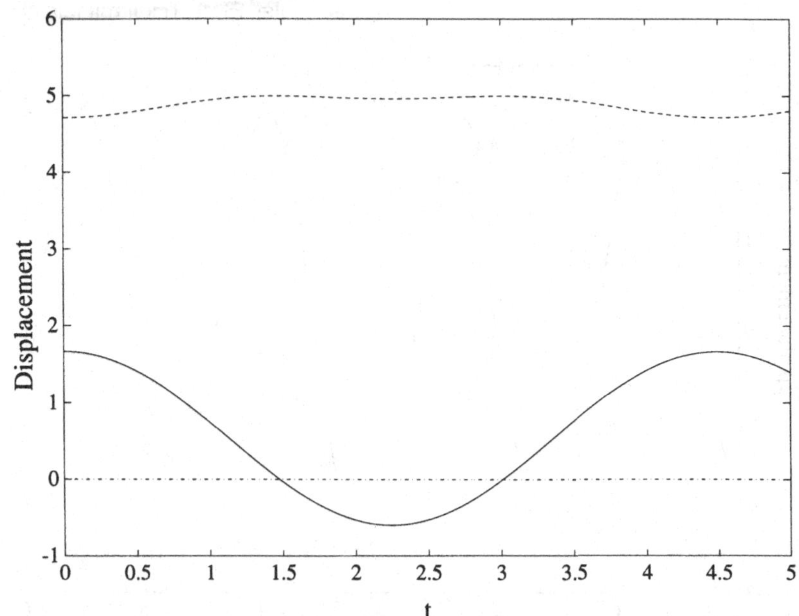

Figure 6.2(b). Motion of mass m_2. The solid line shows the X-component, the dashed line the Y-component and the dash–dot line the Z-component of motion

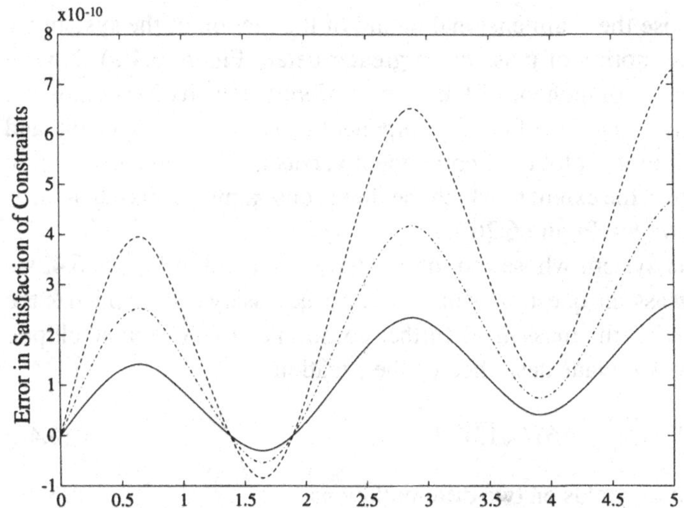

Figure 6.2(c). The error in the dissatisfaction of the constraint equations. The error $e_1(t)$ is shown by a solid line, $e_2(t)$ by a dashed line and $e_3(t)$ by the dash–dot line

Example 6.1 (continued)

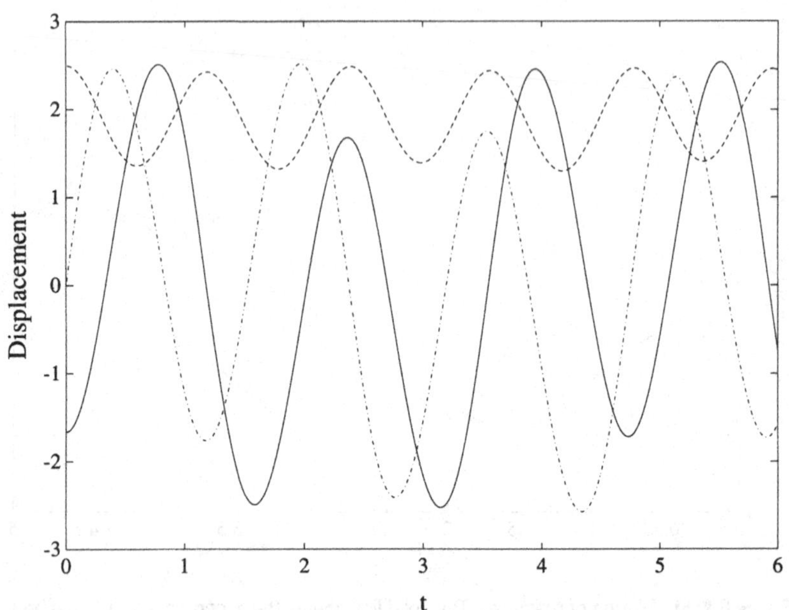

Figure 6.3(a). Motion of the mass m_1. The solid line shows the X-component, the dashed line the Y-component and the dash–dot line the Z-component of motion

To expose the 3-dimensional nature of the motion of the system we exhibit the motion of mass m_1 in greater detail. Figure 6.4(a) shows a plot of the Y-component of the motion of m_1 versus its X-component. Figure 6.4(b) is a plot of the Z-component versus its X-component, and Figure 6.4(c) is a plot of Z-component versus its Y-component. Figure 6.4(d) shows the extent to which the three constraints are dissatisfied, as we did before in Figure 6.2(c).

For this system whose motion we have exhibited in Figure 6.4, we could now ask the question: what forces are necessary to be applied to the system so that the mass m_1 is further constrained to move in an elliptic path in the XZ-plane described by the equation

$$x_1^2 + (1.3)^2 z_1^2 = (6)/\sqrt{13})^2 \text{ ?} \tag{6.43}$$

This equation yields on two differentiations

$$x_1\ddot{x}_1 + (1.3)^2 z_1\ddot{z}_1 = -\dot{x}_1^2 - (1.3)^2 \dot{z}_1^2 . \tag{6.44}$$

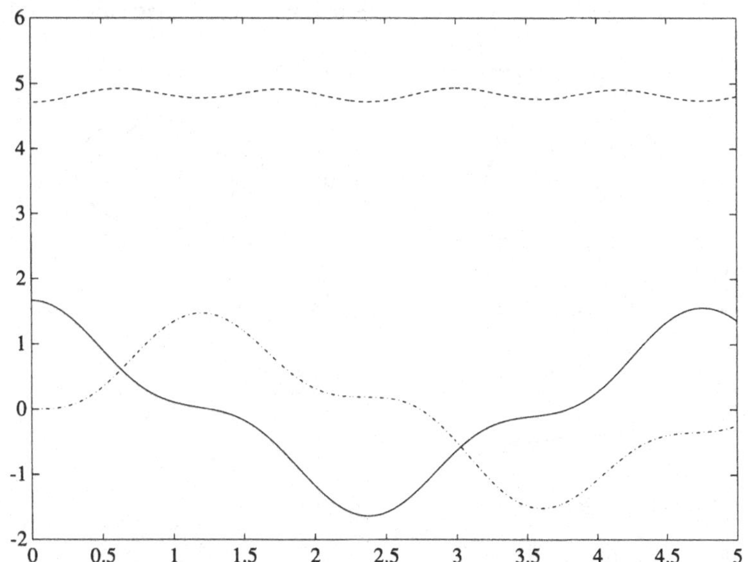

Figure 6.3(b). Motion of mass m_2. The solid line shows the X-component, the dashed line the Y-component and the dash–dot line the Z-component of motion

The matrix A now becomes the 4 by 6 matrix given by

$$A(t) = \begin{bmatrix} x_1 & y_1 & z_1 & 0 & 0 & 0 \\ 0 & 0 & 0 & x_2 & y_2 & z_2 \\ (x_1 - x_2) & (y_1 - y_2) & (z_1 - z_2) & -(x_1 - x_2) & -(y_1 - y_2) & -(z_1 - z_2) \\ x_1 & 0 & (1.3)^2 z_1 & 0 & 0 & 0 \end{bmatrix} \tag{6.45}$$

and the vector $\mathbf{b}$ is given by the 4-vector

$$\mathbf{b}(t) = -\begin{bmatrix} \dot{x}_1^2 + \dot{y}_1^2 + \dot{z}_1^2 \\ \dot{x}_2^2 + \dot{y}_2^2 + \dot{z}_2^2 \\ (\dot{x}_2 - \dot{x}_1)^2 + (\dot{y}_2 - \dot{y}_1)^2 + (\dot{z}_2 - \dot{z}_1)^2 \\ \dot{x}_1^2 + (1.3)^2 \dot{z}_1^2 \end{bmatrix}. \tag{6.46}$$

Example 6.1 (continued)

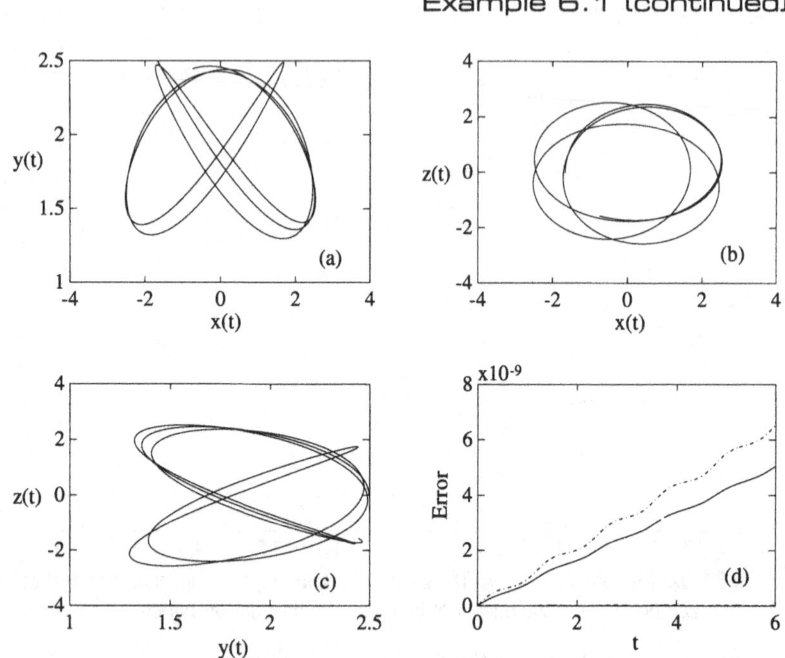

Figure 6.4. Motion of mass m_1 and the error in the satisfaction of the constraints

The motion of the system is described, as usual, by equation (6.37) with these altered quantities A and **b**. The initial position of the two masses is taken to be the same as before; the initial velocities are, as before, $\dot{z}_1 = 10, \dot{z}_2 = 0$.

The motion of the system, obtained by numerically integrating equation (6.37), is shown in Figure 6.5. Comparing Figures 6.4(b) and 6.5(b) we see that the motion of the system now satisfies the equation (6.43) in addition to the three constraints (6.32)–(6.34). The forces that need to be exerted on the system (consistent with Gauss's principle) are given, as usual, by

$$\mathbf{F}^c = M^{1/2}(AM^{-1/2})^+(\mathbf{b} - A\mathbf{a}). \tag{6.47}$$

They are graphed below in Figures 6.6(a) and 6.6(b). Note that these are the forces that are needed to be exerted on the two masses to not only cause the motion of mass m_1 in the XZ-plane to be elliptic but also to maintain the prescribed fixed distances between the two masses, and between them and the point of support O.

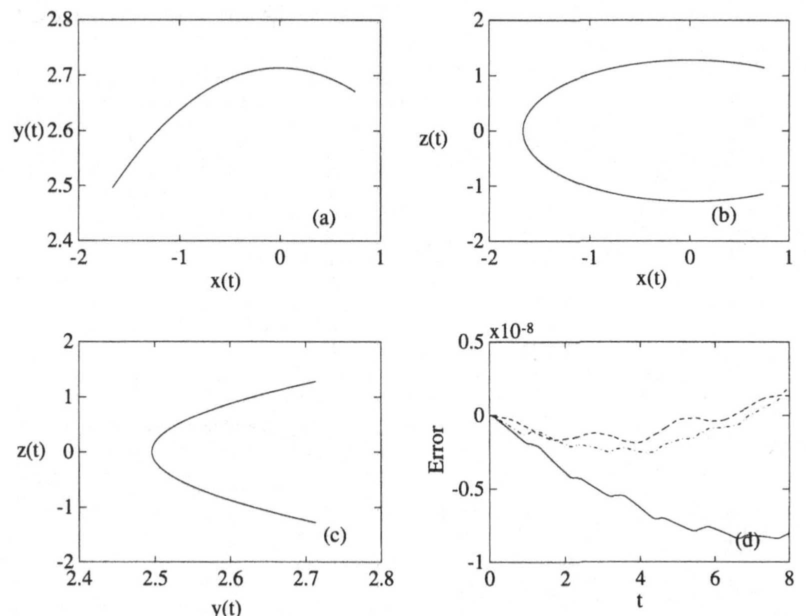

Figure 6.5. Motion of the mass m_1, and the error in the satisfaction of the constraints

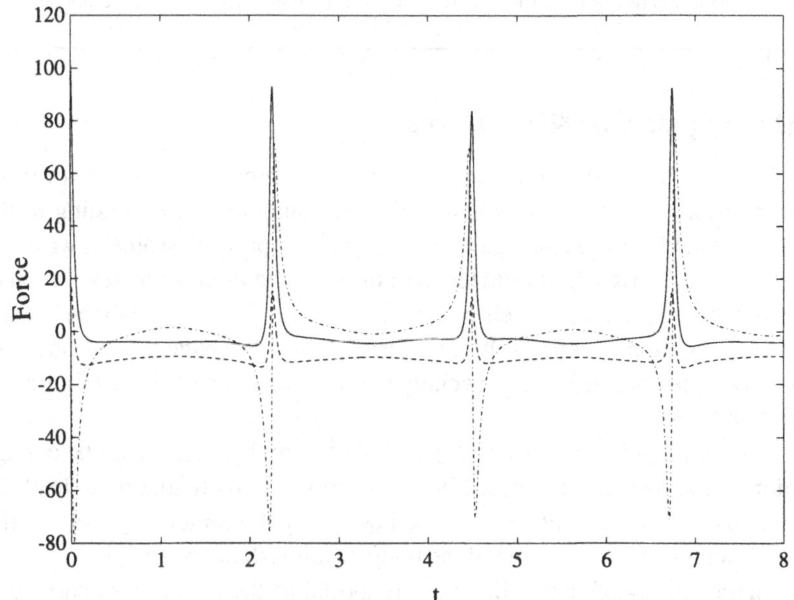

Figure 6.6(a). Components of the force acting on mass m_1. The solid line is the X-component, the dashed line the Y-component and the dash–dot line the Z-component

Example 6.1 (continued)

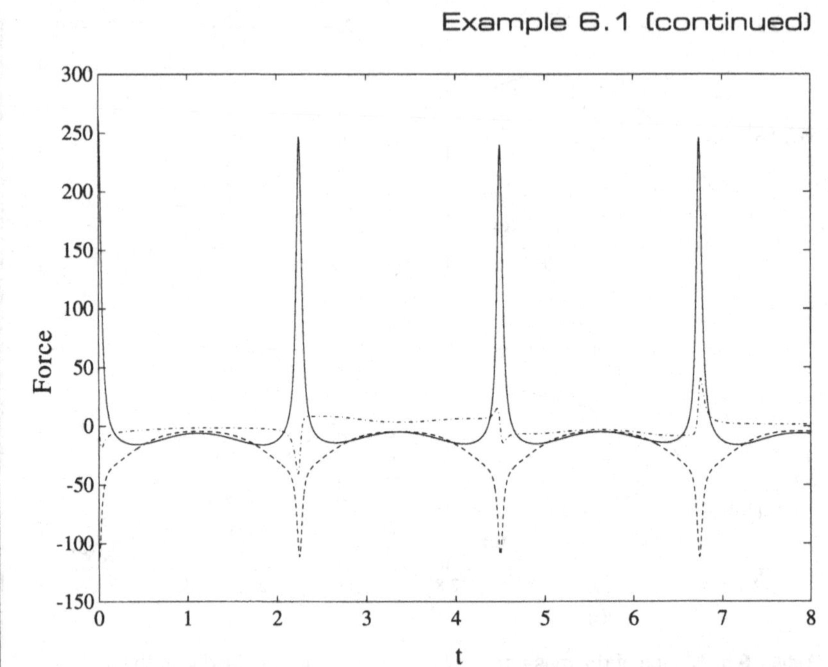

Figure 6.6(b). The components of the force acting on mass m_2. The solid line is the X-component, the dashed line the Y-component and the dash–dot line the Z-component

Description of Rigid Bodies

When we use the fundamental equation to study the motion of one or more rigid bodies, we need to be able to write the equations corresponding to the "unconstrained" motion of the system. For this, we need to be able to write the kinetic energy of rigid bodies in terms of the Lagrangian coordinates chosen to describe their position and their orientation. We are therefore led to the discussion of the motion of rigid bodies. In this section we provide a brief introduction to rigid body dynamics, especially to the description of the configuration of rigid bodies.

Consider a dynamical system consisting of a single rigid body. The configuration of the system is described by the position of a particular particle of the rigid body (i.e., of a point fixed on the body), say the center of gravity of the body G, and by the orientation of the body. Konig's theorem allows us to write the kinetic energy of such a body as composed of the sum of two parts, one dependent on the motion of G, and the other determined by the motion of the body relative to G. The latter motion may be thought of as a change in orientation, with G considered fixed. We prove this result as follows.

Consider an inertial Cartesian coordinate frame of reference. If we denote $S\{*\}$ to be the sum over all the particles of the body of the quantity "$*$," then we can write the kinetic energy as

$$T = \frac{1}{2}S\{m[\dot{x}^2 + \dot{y}^2 + \dot{z}^2]\} \tag{6.48}$$

where $\dot{x}$, $\dot{y}$ and $\dot{z}$ are the components of the velocity of a typical particle of mass m in the rigid body, relative to a fixed inertial frame of reference. Denoting the coordinates of the center of gravity of the body by $\bar{x}, \bar{y}$ and $\bar{z}$, and the position of a typical particle relative to G by α, β and γ, we get

$$T = \frac{1}{2}S\{m[(\dot{\bar{x}} + \dot{\alpha})^2 + (\dot{\bar{y}} + \dot{\beta})^2 + (\dot{\bar{z}} + \dot{\gamma})^2]\} \tag{6.49}$$

But since G is the location of the center of gravity, we get

$$S\{m[\dot{\alpha}]\} = S\{m[\dot{\beta}]\} = S\{m[\dot{\gamma}]\} = 0, \tag{6.50}$$

and we have Konig's theorem, which says that

$$\begin{aligned} T &= \frac{1}{2}S\{m[(\dot{\bar{x}}^2 + \dot{\bar{y}}^2 + \dot{\bar{z}}^2)] + m[\dot{\alpha}^2 + \dot{\beta}^2 + \dot{\gamma}^2]\} \\ &= \frac{1}{2}(\dot{\bar{x}}^2 + \dot{\bar{y}}^2 + \dot{\bar{z}}^2)S\{m\} + \frac{1}{2}S\{m[\dot{\alpha}^2 + \dot{\beta}^2 + \dot{\gamma}^2]\} \\ &= \frac{1}{2}\overline{m}(\dot{\bar{x}}^2 + \dot{\bar{y}}^2 + \dot{\bar{z}}^2) + \frac{1}{2}S\{m[\dot{\alpha}^2 + \dot{\beta}^2 + \dot{\gamma}^2]\}. \end{aligned} \tag{6.51}$$

Here the total mass of the rigid body $\overline{m} = S\{m\}$. The kinetic energy is expressed as a sum of two parts. The first term is obtained by imagining that the entire mass $\overline{m}$ of the body is concentrated at its center of gravity. The second term is obtained by imagining that the body is fixed at G and rotated about G. It is convenient to isolate the second component of the kinetic energy and study it separately. We shall next proceed to do just that.

Description of the Orientation of a Rigid Body

Consider the point O fixed to the center of gravity of a rigid body. Let the tetrahedron formed by the mutually perpendicular axes OA, OB and OC be fixed to the rigid body and move with it. Let the coordinate axes Ox, Oy and Oz be fixed in space. We want to define the orientation of the tetrahedron OABC relative to Oxyz, i.e., the orientation of the rigid body relative to the axes Oxyz which are fixed in space.

Perhaps the easiest way of doing this is through the use of the Euler angles θ, φ, ψ. To get from the orientation Oxyz to the orientation OABC of the rigid body, we do the following in sequence: (1) rotate through the angle φ about the axis Oz; (2) rotate about the carried axis Oy, which now is in the position OS, through the angle θ; and (3) rotate about the carried axis Oz, which now is in the position OC, through the angle ψ. The three steps are shown in Figure 6.7 below.

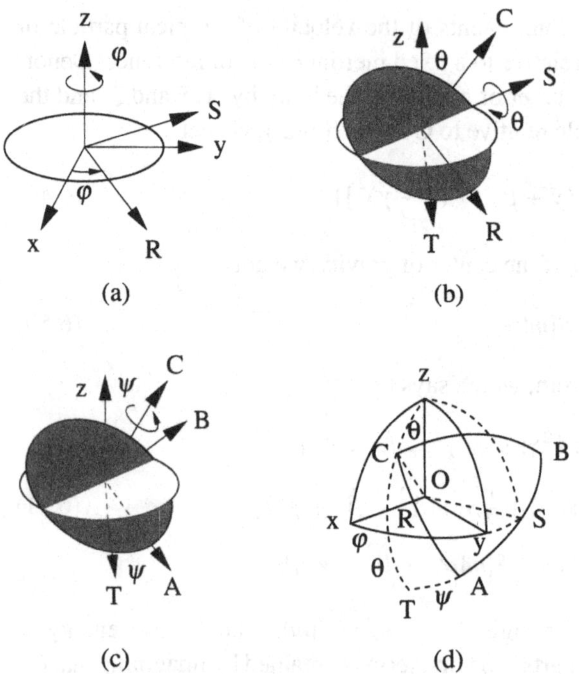

Figure 6.7. Rotations showing the three Euler angles

To describe these rotations in a quantitative manner, we need to know something about the theory of rotations, a topic we take up next.

Theory of Rotations

Consider a rigid body which is fixed at a point O about which it can rotate. Let us say that we have a coordinate frame Oxyz whose origin is at O and whose axes Ox, Oy and Oz are fixed in space. Let us consider another coordinate frame ORSz which is fixed to the body at O and which coincides with Oxyz at some time t_0. In time, the body rotates about the fixed point O, and the coordinate frame ORSz which is fixed to the body moves with it.

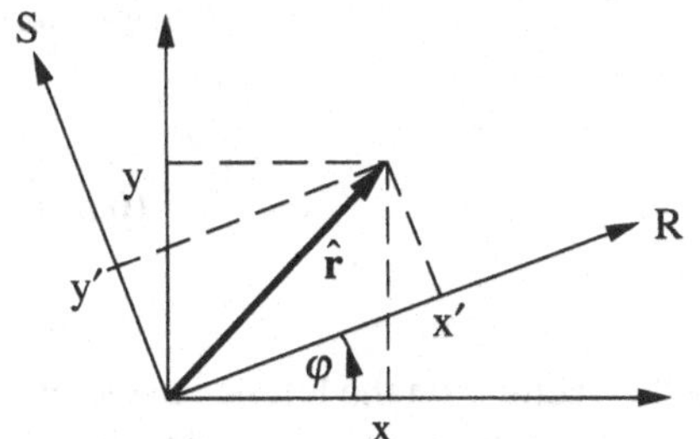

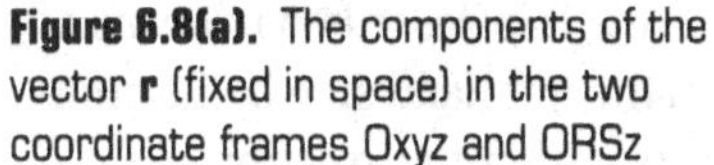
Figure 6.8(a). The components of the vector **r** (fixed in space) in the two coordinate frames Oxyz and ORSz

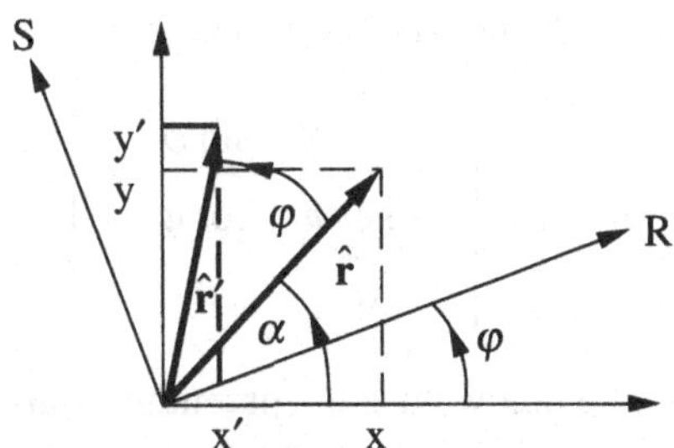

Figure 6.8(b). The components of the vectors **r** and **r′** in the fixed frame Oxyz

We are obviously considering a simple form of rotation of the body about the Oz axis as shown in Figure 6.8(a) by an angle φ so that now at some time t, the coordinate frame which is fixed in the body occupies the position ORSz, as shown.

We now intend to concern ourselves, for the time being, only with these two coordinate frames, Oxyz and ORSz. Consider a vector **r** emanating from the point O which is *fixed* in space. This vector can be resolved into its components in either the Oxyz coordinate frame or the ORSz coordinate frame. Let the components of this vector **r** in the Oxyz frame be denoted by the 3-vector $[x \;\; y \;\; z]^T$ and its components in the ORSz coordinate system be denoted by $[x' \;\; y' \;\; z']^T$. We want to express the primed components of the fixed vector **r** in terms of its unprimed components.

We note that the coordinate frames share the same axes Oz and hence the z-component of the vector **r** will remain unchanged in the two coordinate frames. Hence $z' = z$. Let us project the vector **r** on to the xy (or RS) plane (see Figure 6.8a); denote the projected vector by $\hat{\mathbf{r}}$. We now find that $x' = x\cos\varphi + y\sin\varphi$ and $y' = -x\sin\varphi + y\cos\varphi$. Hence we have

$$\begin{bmatrix} x' \\ y' \\ z' \end{bmatrix} = \begin{bmatrix} \cos\varphi & \sin\varphi & 0 \\ -\sin\varphi & \cos\varphi & 0 \\ 0 & 0 & 1 \end{bmatrix} \begin{bmatrix} x \\ y \\ z \end{bmatrix}. \tag{6.52}$$

Were we to denote the components of the fixed (in space) vector **r** in the coordinate frame ORSz by the 3-vector $(\mathbf{r})'$, and the components of **r** in the coordinate frame Oxyz by **r**, we could rewrite equation (6.52) more compactly as

$$(\mathbf{r})' = R_3(\varphi)\mathbf{r}, \tag{6.53}$$

where the matrix $R_3(\varphi)$ is given by

$$R_3(\varphi) = \begin{bmatrix} \cos\varphi & \sin\varphi & 0 \\ -\sin\varphi & \cos\varphi & 0 \\ 0 & 0 & 1 \end{bmatrix}. \tag{6.54}$$

Notice that we use the parentheses around the vector **r** on the left hand side of equation (6.53) to denote that we are talking about the *same* fixed vector **r** on both sides of equation (6.53) – on the right hand side it is described in terms of its components in the Oxyz coordinate frame, on the left hand side by its components in the ORSz frame. The matrix $R_3(\varphi)$ is called the rotation matrix, denoting a counterclockwise rotation of the frame ORSz with respect to the frame Oxyz about the axis Oz. The reader should observe the structure of the matrix $R_3(\varphi)$; its third row and third column (signifying rotation about the Oz axis) consists of zeros except for the 3-3 element, which is unity.

By direct multiplication we find that

$$R_3(\varphi)R_3^T(\varphi) = R_3^T(\varphi)R_3(\varphi) = I \tag{6.55}$$

so that the matrix $R_3(\varphi)$ is an orthogonal matrix, and therefore

$$[(\mathbf{r})']^T(\mathbf{r})' = \mathbf{r}^T R_3^T(\varphi)R_3(\varphi)\mathbf{r}^T = \mathbf{r}^T\mathbf{r}. \tag{6.56}$$

The last relation may have been directly inferred by observing that we are talking about the same vector **r** in two different coordinate frames and hence, its "length" as determined from its components in either the Oxyz frame or the ORSz frame must be identical. Again, the equation (6.53) provides how the components of a given fixed (in space) vector change when expressed in a different frame of coordinates. This frame of coordinates is described by the rotation matrix $R_3(\varphi)$. We shall call this interpretation of the rotation matrix the passive interpretation.

We now look at what is referred to as the active interpretation of the transformation matrix $R_3(\varphi)$. Let us now consider a vector **r** which, instead of being fixed in space, is fixed in the body. At time t_0 the two coordinate frames Oxyz and ORSz coincide; after a time t because the body rotates about the axis Oz through the angle φ, the vector **r**, which is now fixed in the body, rotates to the new position **r**′, as shown in Figure 6.8(b). Notice that as this vector **r** rotates to **r**′ through the angle φ, the coordinate frame ORSz which was originally coincident with Oxyz rotates through the same angle φ. Let the components of the vector **r**′ (the vector after the body is rotated) in the fixed coordinate frame Oxyz be denoted by the 3-vector $[x' \;\; y' \;\; z']^T$ as shown in Figure 6.8(b). Denote the components of **r** (the vector before rotation) in the same frame Oxyz by the 3-vector $[x \ y \ z]^T$. We shall now find how the primed components are related to the unprimed components.

Since the axis Oz remains unchanged, as before, $z' = z$. Let $\hat{r}$ be the projection of **r** on the xy plane. Now the x and y components of **r** are $\hat{r}\cos\alpha$ and $\hat{r}\sin\alpha$ respectively where $\hat{r}$ denotes the length of the vector $\hat{\mathbf{r}}$. Since we have rotated the body about the Oz axis, the lengths of the vectors **r** and **r′** are the same; the lengths of their projections on the xy plane are likewise identical. The x and y components of **r′** are $x' = \hat{\mathbf{r}}\cos(\alpha + \varphi) = \hat{\mathbf{r}}\cos\alpha\cos\varphi - \hat{\mathbf{r}}\sin\alpha\sin\varphi = x\cos\varphi - y\sin\varphi$ and $y' = \hat{\mathbf{r}}\sin(\alpha + \varphi) = y\cos\varphi + x\sin\varphi$. Hence we have

$$\begin{bmatrix} x' \\ y' \\ z' \end{bmatrix} = \begin{bmatrix} \cos\varphi & -\sin\varphi & 0 \\ \sin\varphi & \cos\varphi & 0 \\ 0 & 0 & 1 \end{bmatrix} \begin{bmatrix} x \\ y \\ z \end{bmatrix} \tag{6.57}$$

which can be written compactly (see equation (6.54)) as

$$\mathbf{r}' = R_3^T(\varphi)\mathbf{r} = S_3(\varphi)\mathbf{r}, \tag{6.58}$$

where we denote the matrix R_3^T by the matrix S_3. Premultiplying equation (6.58) by $R_3(\varphi)$ and noting equation (6.55) we get

$$\mathbf{r} = R_3(\varphi)\mathbf{r}'. \tag{6.59}$$

Notice that though the interpretation of the left hand sides are completely different, the form of equations (6.53) and (6.58) (ignoring the brackets around **r**) are identical; in (6.53) we have the rotation matrix $R_3(\varphi)$ representing a counterclockwise rotation φ, in (6.58) the matrix $S_3(\varphi) = R_3(-\varphi)$ representing a clockwise rotation φ.

Let us now get back to our passive interpretation of the transformation. Having determined the components of the vector **r**, which is fixed in space in the coordinate frame ORSz as given by equations (6.53) and (6.54), let us now further rotate the frame ORSz about the axis OS through the angle θ as shown in Figure 6.7(b). We now arrive at the new frame of reference OTSC. The components of the vector **r** in the two frames ORSz and OTSC are then similarly connected by the rotation matrix

$$R_2(\theta) = \begin{bmatrix} \cos\theta & 0 & -\sin\theta \\ 0 & 1 & 0 \\ \sin\theta & 0 & \cos\theta \end{bmatrix}. \tag{6.60}$$

and

$$(\mathbf{r})'' = R_2(\theta)(\mathbf{r})' = R_2(\theta)R_3(\varphi)\mathbf{r} \tag{6.61}$$

where the 3-vector $(\mathbf{r})'' = [x''\ y''\ z'']^T$, and by x", y'' and z'' we mean the components of the vector **r** (fixed in space) along the coordinates OT, OS and OC respectively. In the last equality in equation (6.61) we have used equation (6.53).

A further rotation of the coordinate frame OTSC about the axis OC through the counterclockwise angle ψ is similarly characterized by the matrix $R_3(\psi)$ given by

$$R_3(\psi) = \begin{bmatrix} \cos\psi & \sin\psi & 0 \\ -\sin\psi & \cos\psi & 0 \\ 0 & 0 & 1 \end{bmatrix}. \tag{6.62}$$

The new coordinate frame engendered by this rotation is OABC (see Figure 6.7(c)). As before,

$$(\mathbf{r})''' = R_3(\psi)(\mathbf{r})'' = R_3(\psi)R_2(\theta)(\mathbf{r})' = R_3(\psi)R_2(\theta)R_3(\varphi)\mathbf{r} \tag{6.63}$$

where the 3-vector $(\mathbf{r})''' = [x'''\ y'''\ z''']^T$ and the quantities x‴, y''' and z''' are now the components of the fixed vector **r** along the directions OA, OB and OC respectively. Denoting the product matrix $R_3(\psi)R_2(\theta)R_3(\varphi)$ by R_E, and using equations (6.54), (6.60) and (6.62) we have

$$R_E(\theta,\varphi,\psi) = \begin{bmatrix} c_1c_2c_3 - s_2s_3 & c_1s_2c_3 + c_2s_3 & -s_1c_3 \\ -c_1c_2s_3 - s_2c_3 & -c_1s_2s_3 + c_2c_3 & s_1s_3 \\ s_1c_2 & s_1s_2 & c_1 \end{bmatrix} \tag{6.64}$$

where the subscripts 1, 2, 3 refer to θ, φ, ψ respectively; c_1 means $\cos\theta$, s_1 means $\sin\theta$ and so on. Notice from equation (6.63) that the total rotation has been obtained by "compounding" the successive rotations φ about Oz, θ about OS and ψ about OC. Such a compounding of rotations manifests itself mathematically through a multiplication of rotation matrices as seen on the right hand side of equation (6.63).

The matrix R_E is made up of the product of orthogonal matrices and therefore, by Example 2.16, it is also an orthogonal matrix. Since any general motion of a rigid body with a fixed point O can be regarded as a sequence of rotations as shown in Figure 6.7, we find that any arbitrary motion of such a body must always be expressed, in general, by a rotation matrix R which is orthogonal. Also, the order of the rotations is important, for the product of two matrices depends on the order in which they are multiplied. Since a given rotation about one axis followed by another given rotation about another axis does not necessarily yield the same final orientation of the body were the order of the rotations reversed (while the magnitudes of the rotation about each axis are kept the same), rotations do *not* qualify to be regarded as vectors.

Having established that the transformation matrix corresponding to every physically realizable rotation of a rigid body about a fixed point is orthogonal, we next ask the converse question, can every 3 by 3 orthogonal matrix R be interpreted as a rigid body rotation? We will show that in order to represent physically meaningful rigid body rotations we need to restrict the class of 3 by 3 orthogonal matrices.

Since R is orthogonal $R^T R = I$. On equating the determinants of the matrices on both sides of this equation we get

$$Det(R^T R) = Det(R^T)Det(R) = [Det(R)]^2 = Det(I) = 1, \tag{6.65}$$

which means that the determinant of R is either +1 or −1. But an orthogonal matrix R whose determinant is −1 cannot represent a rigid body rotation. For, if we considered the simple orthogonal 3 by 3 matrix $R = -I$ whose determinant is −1, then the transformation $(\mathbf{r})''' = R\mathbf{r} = -I\mathbf{r}$ has the effect of changing the sign of each component of the vector $\mathbf{r}$. But this would require that the rotation operation has transformed a right handed coordinate frame Oxyz into a left-handed one OABC, with OA pointed along the direction opposite to Ox, OB pointed opposite to Oy, and OC pointed opposite to Oz. But such an "inversion" of the coordinate frame is impossible through any rotation of a rigid body. Thus an inversion cannot correspond to a physical rotation of a rigid body. What is true of the special matrix $R = -I$ is true of any rotation matrix $\hat{R}$ whose determinant is −1, for such a matrix $\hat{R}$ could always be conceived as a compound rotation expressed as $\hat{R} = -IR$. where the rotation matrix R had determinant +1. Thus the compound rotation represented by $\hat{R}$ would have to include an inversion if its determinant were −1, something impossible to happen in a rigid body rotation. Hence the determinant of an orthogonal matrix R must be +1 for it to represent rigid body rotation; rotations of this kind are called *proper* rotations.

We remind the reader that, as before (see equation (6.58)), an equation such as (6.63), of the form

$$(\mathbf{r})''' = R\mathbf{r} \tag{6.66}$$

where R is a rotation matrix, can be given its active interpretation as

$$\mathbf{r} = R\mathbf{r}', \text{ or, } \mathbf{r}' = R^T\mathbf{r} = S\mathbf{r}, \tag{6.67}$$

where now we think of the vector **r** as *fixed in the rotating body* and the vector **r′** as its new position after the sequence of rotations described by R. Note that the matrix $S = R^T$ is an orthogonal matrix since R is orthogonal. The vectors **r** and **r′** in equation (6.67) each have three components; these components of the two vectors are expressed with respect to the fixed (in space) coordinate axes Oxyz.

In fact, were there to exist a vector **r** fixed in the rotating body which remains unmoved after the sequence of proper rotations described by the orthogonal matrix R, this vector would obviously then be the axis of rotation. Hence the axis of rotation is defined by the relation

$$\mathbf{r}' = \mathbf{r} \tag{6.68}$$

where the components of these vectors are in the Oxyz coordinate frame which is fixed in space. By equation (6.67) the axis of rotation then is a vector that satisfies the equation

$$R\mathbf{r} = \mathbf{r}. \tag{6.69}$$

Hence the axis of rotation is simply the eigenvector of the matrix R corresponding to the eigenvalue unity. That unity is always an eigenvalue of R is easy to show. Since R is orthogonal, $(R - I)R^T = I - R^T$. On taking the determinant of each side of this equation and noting that $Det(R) = Det(R^T) = +1$, we get

$$Det\{(R - I)R^T\} = Det(R - I) = Det(I - R^T) = Det(I - R) \tag{6.70}$$

so that

$$Det(R - I) = Det(I - R). \tag{6.71}$$

But since R is a 3 by 3 matrix, $Det(I - R) = (-1)^3 Det(R - I)$, and so by equation (6.71), $Det(R - I) = 0$, which in turn implies that unity is an eigenvalue of the matrix R. The eigenvector corresponding to this eigenvalue of unity then yields the axis of rotation **r** which remains unmoved by the displacement of the body. We have hence shown that any motion of a rigid body with one point fixed can always be construed as a rotation about some axis passing through that fixed point, a result first discovered by Euler in 1776.

From the rotation matrix R we have adduced the axis of rotation; we next need to find the angle through which the rigid body has rotated about this axis, given the rotation matrix R.

To do this we need to reconsider the two alternative interpretations of equations (6.66) and (6.67), both of which use the matrix operator R. Consider first the vector **r** fixed in a rigid body which is subjected to a proper rotation relative to a fixed coordinate frame Oxyx. We represent the new vector $\mathbf{r}_1$ obtained from rotating **r** by

$$\mathbf{r}_1 = S\mathbf{r} \tag{6.72}$$

where the components of vectors **r** and $\mathbf{r}_1$ are expressed in the frame Oxyz which is fixed in space. The matrix S is orthogonal with determinant +1. Now if the coordinate system Oxyz is transformed to a new coordinate system by a proper rotation represented by the matrix T, the components of the vector $\mathbf{r}_1$ in the new space-fixed coordinate system will be related to those in the frame Oxyz by the relation

$$(\mathbf{r}_1)' = T\mathbf{r}_1 = TS\mathbf{r} \tag{6.73}$$

where the 3-vector $(\mathbf{r}_1)'$ as usual denotes the components of the vector $\mathbf{r}_1$ in the new coordinate system. We can rewrite equation (6.73) as

$$(\mathbf{r}_1)' = \{TST^{-1}\}T\mathbf{r} = TST^{-1}(\mathbf{r})' \tag{6.74}$$

where we have denoted by $(\mathbf{r})' = T\mathbf{r}$ the components of the vector **r** in the new space-fixed coordinate frame. Hence the transformation S which physically rotated the vector **r** to $\mathbf{r}_1$, where the components of each of these vectors are expressed in the space-fixed coordinate frame Oxyz can be represented in the new space-fixed frame of reference through equation (6.74) – here both $(\mathbf{r})'$ and $(\mathbf{r}_1)'$ are *physically the same* as the vectors **r** and $\mathbf{r}_1$, but their components are expressed in the new space-fixed frame of reference. Thus in the new frame of reference the operator S which signifies the actual rotation of the vector **r** to $\mathbf{r}_1$ is represented by the operator $S' = TST^{-1}$. The matrices S and S' are called similar matrices and they have the same eigenvalues (see Example 2.15).

We are now ready to determine the angle of rotation (about the axis of rotation) corresponding to a given transformation matrix R, such as the matrix R_E of equation (6.64). By equation (6.72), using the active interpretation, we have

$$\mathbf{r}_1 = S\mathbf{r}, \tag{6.75}$$

where S is the transpose of the matrix R. The vectors **r** and $\mathbf{r}_1$ have components in the Oxyz frame of reference which is fixed in space. Suppose we knew the axis of rotation to be the vector **a** (recall, **a** is the eigenvector of R corresponding to the eigenvalue unity) whose components are also described in the Oxyz coordinate frame. We could make a transformation of coordinates so that the fixed frame Oxyz is now so oriented that its axis Ox lies along the vector **a**. Let the rotation matrix which corresponds to such a coordinate change be *T*. Now in this new space-fixed coordinate frame, whose axis Ox lies along the axis of rotation of the body, the rotation of the rigid body is described simply by the angle of rotation, say κ, about this new Ox direction (which is now along the vector **a**, the axis of rotation). Hence in this new space-fixed frame of reference we must have by equation (6.67)

$$(\mathbf{r}_1)' = \begin{bmatrix} 1 & 0 & 0 \\ 0 & \cos\kappa & -\sin\kappa \\ 0 & \sin\kappa & \cos\kappa \end{bmatrix} (\mathbf{r}) = S_a(\mathbf{r}), \tag{6.76}$$

where we have specifically indicated that both the vectors $\mathbf{r}$ and $\mathbf{r}_1$ are expressed in terms of components in the new fixed coordinate system whose axis Ox coincides with **a**. But, as we saw earlier, such a change of axes also causes equation (6.75) to transform to (6.74). Hence we must have

$$S_a = TST^{-1}. \tag{6.77}$$

In particular, the trace of the matrix on the left must equal the trace of the matrix on the right. But the trace of the matrix on the right equals the sum of its eigenvalues. But the eigenvalues of *S* and TST^{-1} are identical. Hence the trace of the matrix on the right equals the sum of the eigenvalues of *S*, which in turn equals the trace of the matrix *S*. Hence we have, from equation (6.77),

$$Trace\{S_a\} = 1 + 2\cos\kappa = Trace\{S\}. \tag{6.78}$$

For the transformation R_E given by equation (6.64) corresponding to the Euler angles, we have $S = S_E = R_E^T$. Then using equation (6.64) to determine the $Trace\{S\}$, we can thus obtain the rotation angle in terms of the Euler angles by equation (6.78). After some algebra, we get

$$\cos(\kappa/2) = \cos\left(\frac{\varphi + \psi}{2}\right)\cos(\theta/2). \tag{6.79}$$

It should be noted that there still remains some ambiguity related to the direction of the rotation axis and the sense of rotation because if **r** is an eigenvector of (6.69) then so is –**r**. We can reduce this ambiguity by fixing the sense of the axis of rotation by using the right hand screw rule.

Orientation of a Rigid Body Using Angles φ_1, φ_2 and φ_3

The use of Euler angles requires two rotations about the z-like coordinate and one about the y-like coordinate. A more symmetrical approach to determining the orientation of a rigid body would be as follows. Suppose we start with the triad OABC which is fixed to the body coincident with the triad Oxyz which is fixed in space, and then we arrive at its final position by the following sequence of rotations: a rotation φ_1 about the axis OA (which is actually the axis Ox), followed by a rotation φ_2 about the carried position of OB, followed by a rotation φ_3 about the carried position of OC. Denoting

$$R_1(\varphi) = \begin{bmatrix} 1 & 0 & 0 \\ 0 & \cos\varphi & \sin\varphi \\ 0 & -\sin\varphi & \cos\varphi \end{bmatrix}, \tag{6.80}$$

we can obtain the matrix R, which yields the coordinate transformation as we did before as

$$(\mathbf{r})''' = R_3(\varphi_3)R_2(\varphi_2)R_1(\varphi_1)\mathbf{r} = R(\varphi_1, \varphi_2, \varphi_3)\mathbf{r} \tag{6.81}$$

where the matrices $R_2(\varphi_2)$ and $R_3(\varphi_3)$ are defined in equations (6.60) and (6.54). In equation (6.81) the vector $\mathbf{r}$ is fixed in space and its components are expressed in the fixed Oxyz coordinate frame. The vector $(\mathbf{r})'''$ denotes the 3-vector whose components are the components of the vector $\mathbf{r}$ in the final configuration of the frame OABC. Carrying out the matrix multiplication we get

$$R(\varphi_1, \varphi_2, \varphi_3) = \begin{bmatrix} c_2c_3 & c_1s_3 + s_1s_2c_3 & s_1s_3 - c_1s_2c_3 \\ -c_2s_3 & c_1c_3 - s_1s_2s_3 & s_1c_3 + c_1s_2s_3 \\ s_2 & -s_1c_2 & c_1c_2 \end{bmatrix} \tag{6.82}$$

where c_i and s_i stand for $\cos\varphi_i$ and $\sin\varphi_i$, respectively.

We could be more general in allowing the body-fixed triad OABC to be initially *noncoincident* with the space-fixed triad Oxyz. In fact consider a rotation matrix $R_0(\varphi_1^0, \varphi_2^0, \varphi_3^0)$ to be given, thereby describing the frame OABC, or equivalently, the relationship between the components of any fixed (in space) vector in the fixed (in space) frame Oxyz, and its components in the body-fixed frame OABC. Now let us say that *further* rotations described by the same sequence of operations as before are made, namely, a rotation φ_1 about the axis OA, followed by a rotation φ_2 about the carried position of OB, followed by a rotation φ_3 about the carried position of OC. Then the new transformation matrix would simply be

$$R = R_3(\varphi_3)R_2(\varphi_2)R_1(\varphi_1)R_0. \tag{6.83}$$

This transformation matrix would transform the components in the frame Oxyz of any fixed vector (in space) to its components along the final positions of OA,

OB and OC. Since the matrices do not commute in general, the order of the rotations is important.

Infinitesimal Rotations and Angular Velocity

We have already seen that the resultant orientation of a rigid body fixed at a point O after two rotations about two different axes is dependent, in general, on the order of the rotations. Hence rotations do not qualify as vectors. However, infinitesimal rotations do.

Consider, as before, a body fixed at O and the usual two coordinate frames of reference: the space-fixed frame Oxyz, and the body-fixed frame OABC. Let us begin by assuming that the two frames are initially coincident. We provide infinitesimal rotations in the following sequence: a rotation ε_1 about the axis OA, followed by a rotation ε_2 about the carried position of OB, followed by a rotation ε_3 about the carried position of OC. Here the quantities ε_i, $i = 1, 2, 3$ are each infinitesimal so that we will neglect squares and products of these numbers. Then the rotation matrix representing this transformation will be obtained as

$$R = R_3(\varepsilon_3)R_2(\varepsilon_2)R_1(\varepsilon_1) \tag{6.84}$$

where, as before, the matrices $R_3(\varepsilon_3)$, $R_2(\varepsilon_2)$ and $R_1(\varepsilon_1)$ are given in equations (6.54), (6.60) and (6.80). Equation (6.82) gives this product in explicit form. Since the ε_i are infinitesimal, $\sin\varepsilon_i \approx \varepsilon_i$ and $\cos\varepsilon_i \approx 1$. Ignoring the squares and products of the ε_i's, the matrix R describing the sequence of infinitesimal rotations, by equation (6.82), becomes

$$R = \begin{bmatrix} 1 & \varepsilon_3 & -\varepsilon_2 \\ -\varepsilon_3 & 1 & \varepsilon_1 \\ \varepsilon_2 & -\varepsilon_1 & 1 \end{bmatrix} = I + R_\varepsilon. \tag{6.85}$$

In fact, we would obtain the same matrix R independent of the order in which the rotations ε_1 about OA, ε_2 about OB and ε_3 about OC were performed. Hence R would be the rotation matrix independent of the sequence in which the infinitesimal rotations were executed. This property of infinitesimal rotations qualifies them for being considered as vectors. Notice that we have expressed R as a sum of the identity matrix and the matrix R_ε which is skew-symmetric, i.e., $R_\varepsilon^T = -R_\varepsilon$.

Let us now think for a moment about the active interpretation of this transformation matrix R. Then we have the vector **r** fixed to the body rotated into a new position **r′**. Let the components of the 3-vector **r** in the space-fixed Oxyz frame be x, y and z and those of $\mathbf{r}'$ in the same space-fixed frame be x', y' and z'. Then the 3-vectors $\mathbf{r}$ and $\mathbf{r}'$ satisfy the relation (see equation (6.67))

$$\mathbf{r}' = R^T\mathbf{r} = (I + R_\varepsilon)^T\mathbf{r} = (I - R_\varepsilon)\mathbf{r} = \mathbf{r} - R_\varepsilon\mathbf{r} \tag{6.86}$$

where we have made use of the skew-symmetric property of the matrix R_ε. If the components of the vector $\mathbf{r}$ in Oxyz are x, y and z, we then get from equation (6.86)

$$\mathbf{r}' - \mathbf{r} = -R_\varepsilon\mathbf{r} = \begin{bmatrix} 0 & -\varepsilon_3 & \varepsilon_2 \\ \varepsilon_3 & 0 & -\varepsilon_1 \\ -\varepsilon_2 & \varepsilon_1 & 0 \end{bmatrix}\mathbf{r} = R_\varepsilon^T\mathbf{r} = \begin{bmatrix} z\varepsilon_2 - y\varepsilon_3 \\ x\varepsilon_3 - z\varepsilon_1 \\ y\varepsilon_1 - x\varepsilon_2 \end{bmatrix}. \tag{6.87}$$

Since the rotation is infinitesimal, the rotated vector $\mathbf{r}'$ differs from $\mathbf{r}$ only slightly. Let us denote $x' - x = dx$, $y' - y = dy$ and $z' - z = dz$. Furthermore if we denote the vector $\boldsymbol{\varepsilon} = [\varepsilon_1\ \varepsilon_2\ \varepsilon_3]^T$, and $d\mathbf{r} = [dx\ dy\ dz]^T$ equation (6.87) can be rewritten as

$$d\mathbf{r} = \boldsymbol{\varepsilon} \times \mathbf{r} \tag{6.88}$$

where the right hand side denotes the usual vector cross-product of the two vectors $\boldsymbol{\varepsilon}$ and $\mathbf{r}$.

If this motion $d\mathbf{r}$ is assumed to have taken place in the interval of time dt, then we can write

$$\frac{d\mathbf{r}}{dt} = \boldsymbol{\omega} \times \mathbf{r}, \tag{6.89}$$

where we define the angular velocity $\boldsymbol{\omega}$ by the relation

$$\boldsymbol{\varepsilon} = \boldsymbol{\omega} dt. \tag{6.90}$$

The three components ω_1, ω_2 and ω_3 of the vector $\boldsymbol{\omega}$ are the components of the angular velocity along OA, OB and OC.

We recall that we started this discussion by assuming that the body-fixed frame was initially coincident with the space-fixed frame OABC. More generally, at some instant of time t let these frames be noncoincident. The body-fixed coordinate frame OABC at time t may be described by the transformation R_0. The sequence of infinitesimal rotations that occurs in the succeeding duration of time dt could be defined again, as before, by: a rotation ε_1 about the axis OA, followed by a rotation ε_2 about the carried position of OB, followed by a rotation ε_3 about the carried position of OC. (Note that since the rotations are infinitesimal, their actual sequence does not really matter. We chose a sequence only for definiteness.) This would yield the transformation matrix (see equation (6.83))

$$R = R_3(\varepsilon_3)R_2(\varepsilon_2)R_1(\varepsilon_1)R_0 \tag{6.91}$$

which would, as before (see equation (6.85)), yield

$$R = (I + R_\varepsilon)R_0. \tag{6.92}$$

Hence we get

$$\Delta R = R - R_0 = R_\varepsilon R_0 = -(-R_\varepsilon)R_0 = -R_\varepsilon^T R_0. \tag{6.93}$$

Since the infinitesimal rotations about OA, OB and OC occur in the interval of time dt, we can write,

$$R_\varepsilon^T = \Omega dt = \begin{bmatrix} 0 & -\omega_3 & \omega_2 \\ \omega_3 & 0 & -\omega_1 \\ -\omega_2 & \omega_1 & 0 \end{bmatrix} dt \tag{6.94}$$

where ω_1, ω_2 and ω_3 are now the components of the angular velocity vector $\boldsymbol{\omega}$ in the directions OA, OB and OC respectively. The reader must have realized that equation (6.94) is simply another way of writing equation (6.90)! Then dividing by dt and taking limits of equation (6.93) we get

$$\frac{dR_0}{dt} = -\Omega R_0. \tag{6.95}$$

Postmultiplying both sides of this equation by R_0^T and noting that R_0 is orthogonal, we get

$$\Omega = -\frac{dR_0}{dt} R_0^T. \tag{6.96}$$

This equation determines the angular velocity components along OA, OB and OC at time t in terms of the transformation matrices R and $\frac{dR}{dt}$ at that time.

Components of the Angular Velocity

We can now determine the components of the angular velocity vector $\boldsymbol{\omega}$ of a rigid body in the directions OA, OB, OC, that is, along the axes fixed to the body. If the orientation of the body is described by the Euler angles as described in Figure 6.7, then the angular velocity is the vector sum of $\dot{\varphi}$ about Oz, $\dot{\theta}$ about OS and $\dot{\psi}$ about OC. The components along OA, OB and OC of the vector $\mathbf{w}_1 = [0\ \ 0\ \ \dot{\varphi}]^T$ whose components are in the Oxyz frame, are simply given by $R_E\mathbf{w}_1$, where the matrix R_E is defined in (6.64). The components along OA, OB and OC of the vector $\mathbf{w}_2 = [0\ \ \dot{\theta}\ \ 0]^T$ whose components are expressed in the frame OTSC, are given by $R_3(\psi)\mathbf{w}_2$, where the matrix $R_3(\psi)$ is given in equation (6.62). Lastly the components along OA, OB and OC of the vector $\mathbf{w}_3 = [0\ \ 0\ \ \dot{\psi}]^T$ whose components are expressed in the frame OABC are given by $\mathbf{w}_3$ itself. Hence we have the components of the angular velocity in the frame OABC given by

$$\begin{bmatrix} \omega_1 \\ \omega_2 \\ \omega_3 \end{bmatrix} = R_E\mathbf{w}_1 + R_3(\psi)\mathbf{w}_2 + \mathbf{w}_3 \tag{6.97}$$

which yield the following three relations:

$$\omega_1 = \dot{\theta}\sin\psi - \dot{\varphi}\sin\theta\cos\psi; \tag{6.98}$$

$$\omega_2 = \dot{\theta}\cos\psi + \dot{\varphi}\sin\theta\sin\psi; \tag{6.99}$$

$$\omega_3 = \dot{\varphi}\cos\theta + \dot{\psi}. \tag{6.100}$$

Curiously enough, because of the asymmetrical way in which the Euler angles are measured, when the tetrahedron OABC coincides with Oxyz, we have $\theta = \varphi = \psi = 0$, and the first equation above gives $\omega_1 = 0$ no matter what finite values $\dot{\theta}$, $\dot{\psi}$ and $\dot{\varphi}$ have. Thus Euler angles are ill-suited to a situation in which at some instant of time OABC coincides with Oxyz, unless indeed at that instant the angular velocity vector lies in the plane Oyz and ω_1 is in fact zero. More generally, when $\theta = 0$, the formulae (6.98)–(6.100) above would imply that $\omega_1 = \omega_2 \tan\psi$, which is not true in general.

When we use the angles φ_1, φ_2 and φ_3 to describe the orientation of the rigid body as in equations (6.81) and (6.82), the expressions for the angular velocity components in the directions OA, OB and OC are given by

$$\omega_1 = c_2c_3\dot{\varphi}_1 + s_3\dot{\varphi}_2, \tag{6.101}$$

$$\omega_2 = -c_2s_3\dot{\varphi}_1 + c_3\dot{\varphi}_2, \tag{6.102}$$

$$\omega_3 = s_2\dot{\varphi}_1 + \dot{\varphi}_3, \tag{6.103}$$

where c_i and s_i stand for $\cos\varphi_i$ and $\sin\varphi_i$ respectively. These components obviously reduce to $\dot{\varphi}_1, \dot{\varphi}_2$ and $\dot{\varphi}_3$ when $\varphi_1 = \varphi_2 = \varphi_3 = 0$. We leave it as an exercise for the reader to verify these relations.

Kinetic Energy of a Rigid Body

Consider first a body that has one point O fixed in it. Let OA, OB and OC be the principal axes of inertia at O, and the components of the angular velocity in the directions OA, OB and OC are ω_1, ω_2 and ω_3. The kinetic energy can be written as

$$T = \frac{1}{2} S\{m[(\omega_2 c - \omega_3 b)^2 + (\omega_3 a - \omega_1 c)^2 + (\omega_1 b - \omega_2 a)^2]\} \tag{6.104}$$

where m is the mass of a typical particle with coordinates (a, b, c) referred to the axes OA, OB and OC. By $S\{*\}$ we mean, as before, the summation of the quantity "$*$" over the entire rigid body. Since the axes OA, OB and OC are the principal axes, we must have

$$S\{mbc\} = S\{mca\} = S\{mab\} = 0 \tag{6.105}$$

so that equation (6.104) becomes

$$\begin{aligned} T &= \frac{1}{2} S\left\{m[\omega_1^2(b^2 + c^2) + \omega_2^2(c^2 + a^2) + \omega_3^2(a^2 + b^2)]\right\} \\ &= \frac{1}{2}\omega_1^2 S\left\{m(b^2 + c^2)\right\} + \frac{1}{2}\omega_2^2 S\left\{m(c^2 + a^2)\right\} + \frac{1}{2}\omega_3^2 S\left\{m(a^2 + b^2)\right\} \\ &= \frac{1}{2}\omega_1^2 I_1 + \frac{1}{2}\omega_2^2 I_2 + \frac{1}{2}\omega_3^2 I_3. \end{aligned} \tag{6.106}$$

The quantities $I_1 = S\{m(b^2 + c^2)\}$, $I_2 = S\{m(c^2 + a^2)\}$ and $I_3 = S\{m(a^2 + b^2)\}$ are the moments of inertia of the body about OA, OB and OC. They are the principal moments of inertia at O.

Suppose next that the body moves in space. Let the coordinates of the center of gravity of the body G when referred to a fixed set of coordinates Oxyz in space, be $\bar{x}, \bar{y}$ and $\bar{z}$, and let the principal axes of inertia at the point G be GA, GB and GC. If ω_1, ω_2 and ω_3 are the components of the angular velocity in the directions GA, GB and GC then, by Konig's theorem, we have (see Problem 6.2)

$$T = \frac{1}{2} M(\dot{\bar{x}}^2 + \dot{\bar{y}}^2 + \dot{\bar{z}}^2) + \frac{1}{2}(\omega_1^2 I_1 + \omega_2^2 I_2 + \omega_3^2 I_3). \tag{6.107}$$

We need to express ω_1, ω_2 and ω_3 in terms of the Lagrangian coordinates. In terms of the Euler angles, for example, we need to use relations (6.98)–(6.100) and thus express the kinetic energy which is given by equation (6.107) in terms of these angles.

Example 6.2

Consider a rigid body with axial symmetry, like a top. One point O along the axis of the body is fixed. The body is allowed to move about this fixed point. Let us write the equations of motion of this body when the only forces acting on it are those due to gravity.

Consider the axes OABC fixed in the top, with OC along the axis of the top. Then OA, OB and OC are the principal axes of inertia at O and the moments of inertia about OA and OB are equal. Let the principal moments of inertia at O be denoted by I_A, I_A and I_C.

Now consider the coordinate system Oxyz which is fixed in space with Oz vertically upwards. Using Euler angles to define the orientation of the tetrahedron OABC, we have

$$T = \tfrac{1}{2}\omega_1^2 I_A + \tfrac{1}{2}\omega_2^2 I_A + \tfrac{1}{2}\omega_3^2 I_C, \tag{6.108}$$

where ω_1, ω_2 and ω_3 are the angular velocities about OA, OB and OC. Using equations (6.98)–(6.100) we get,

$$\begin{aligned} T &= \tfrac{1}{2} I_A(\dot{\theta}\sin\psi - \dot{\varphi}\sin\theta\cos\psi)^2 + \tfrac{1}{2} I_A(\dot{\theta}\cos\psi + \dot{\varphi}\sin\theta\sin\psi)^2 \\ &\quad + \tfrac{1}{2} I_C(\dot{\varphi}\cos\theta + \dot{\psi})^2 \\ &= \tfrac{1}{2} I_A(\dot{\theta}^2 + \dot{\varphi}^2\sin^2\theta) + \tfrac{1}{2} I_C(\dot{\varphi}\cos\theta + \dot{\psi})^2. \end{aligned} \tag{6.109}$$

If l is the distance from O along the axis OC at which the center of gravity G of the top is located, its vertical distance from the horizontal plane is $l\cos\theta$. The work done by the force of gravity under a virtual displacement is

$$-Mg\delta(l\cos\theta) = Mgl\sin\theta\delta\theta. \tag{6.110}$$

The generalized force related to the θ coordinate is then $Mgl\sin\theta$, and the corresponding Lagrange equation of motion becomes

$$I_A(\ddot{\theta} - \dot{\varphi}^2\sin\theta\cos\theta) + I_C(\dot{\varphi}\cos\theta + \dot{\psi})\dot{\varphi}\sin\theta = Mgl\sin\theta. \tag{6.111}$$

We leave the determination of the other two Lagrange equations of motion as an exercise for the reader.

Example 6.3

Consider a penny of mass m and radius a rolling on a horizontal surface without slipping. We want to write the equations of motion describing the penny as it rolls on the surface with gravity acting downwards. We could choose the fixed coordinate frame of reference Oxyz with O lying in the plane on which the penny rolls and Oz vertically upwards. We can use the Euler angles θ, φ and ψ to describe the orientation of the penny, and the coordinates of the center of gravity G of the penny to be given by $(\bar{x}, \bar{y}, a \sin\theta)$. Figure 6.9 shows how we use the Euler angles

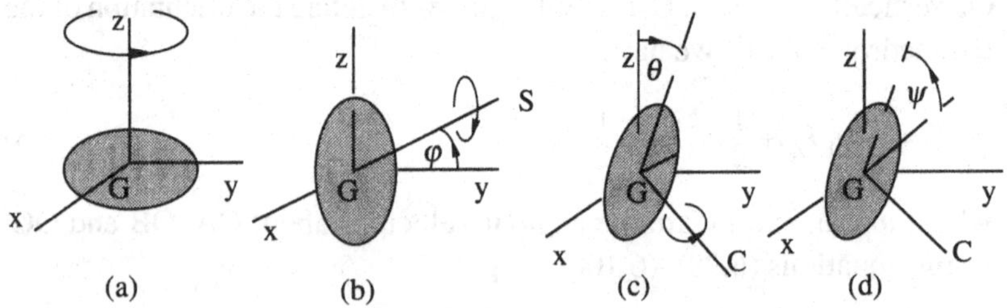

Figure 6.9. The Euler angles for the rolling penny. Note the orientation of the two perpendicular lines in the plane of the penny

We thus have five Lagrangian coordinates to describe the configuration of the penny – the three Euler angles, $\bar{x}$ and $\bar{y}$. But the following two non-holonomic conditions apply because the penny rolls on the horizontal surface without slipping:

$$\dot{\bar{x}} \cos\varphi + \dot{\bar{y}} \sin\varphi - a\dot{\theta} \sin\theta = 0, \tag{6.112}$$

and

$$-\dot{\bar{x}} \sin\varphi + \dot{\bar{y}} \cos\varphi + a\dot{\psi} + a\dot{\varphi}\cos\theta = 0. \tag{6.113}$$

We leave it as an exercise to the reader to show that these conditions describe the constraints between the five coordinates that describe the configuration of the rolling penny.

The kinetic energy of the penny is then simply

$$\begin{aligned} T = \frac{1}{2} m(\dot{\bar{x}}^2 + \dot{\bar{y}}^2 + a^2\dot{\theta}^2 \cos^2\theta) + \frac{1}{2} I_A(\dot{\theta}^2 + \dot{\varphi}^2 \sin^2\theta) \\ + \frac{1}{2} I_C(\dot{\psi} + \dot{\varphi}\cos\theta)^2 \end{aligned} \tag{6.114}$$

where the principal moments of inertia at G of the penny are I_A, I_A and I_C with $I_C = 2I_A$.

The only force impressed on the penny is the force of gravity. The virtual work done by this force of gravity is $-mg\delta(a\sin\theta) = -mga\cos\theta\delta\theta$.

Lagrange's equations for the "unconstrained" system, where we consider the five coordinates as independent, can now be obtained. These equations corresponding to the coordinates $\bar{x}$, $\bar{y}$, θ, φ and ψ are respectively

$$m\ddot{\bar{x}} = 0, \tag{6.115}$$

$$m\ddot{\bar{y}} = 0, \tag{6.116}$$

$$\begin{aligned}\frac{d}{dt}(ma^2\dot{\theta}\cos^2\theta + I_A\dot{\theta}) = -ma^2\dot{\theta}^2\cos\theta\sin\theta + I_A\dot{\varphi}^2\cos\theta\sin\theta \\ -I_C(\dot{\psi} + \dot{\varphi}\cos\theta)\dot{\varphi}\sin\theta - mga\cos\theta,\end{aligned} \tag{6.117}$$

$$\frac{d}{dt}\{I_A\dot{\varphi}\sin^2\theta + I_C(\dot{\psi} + \dot{\varphi}\cos\theta)\cos\theta\} = 0 \tag{6.118}$$

and

$$I_C\frac{d}{dt}(\dot{\psi} + \dot{\varphi}\cos\theta) = 0. \tag{6.119}$$

Along with these equations that correspond to the "unconstrained" system we must add the differentiated constraint equation given by the two rolling constraints (6.112) and (6.113). We then obtain the constraint equation

$$\begin{bmatrix} \cos\varphi & \sin\varphi & -a\sin\theta & 0 & 0 \\ -\sin\varphi & \cos\varphi & 0 & a\cos\theta & a \end{bmatrix} \begin{bmatrix} \ddot{\bar{x}} \\ \ddot{\bar{y}} \\ \ddot{\theta} \\ \ddot{\varphi} \\ \ddot{\psi} \end{bmatrix} \tag{6.120}$$

$$= \begin{bmatrix} \dot{\bar{x}}\dot{\varphi}\sin\varphi - \dot{\bar{y}}\dot{\varphi}\cos\varphi + a\dot{\theta}^2\cos\theta \\ \dot{\bar{x}}\dot{\varphi}\cos\varphi + \dot{\bar{y}}\dot{\varphi}\sin\varphi + a\dot{\theta}\dot{\varphi}\sin\theta \end{bmatrix}$$

which is now in our standard form.

Example 6.3 (continued)

Equations (6.115)–(6.120) now provide a complete description of the rolling penny.

If, in addition, we want the point G of the penny to traverse a specific trajectory, say, given by

$$f(\bar{x}, \bar{y}) = h(t), \tag{6.121}$$

where f is a given function, and $h(t)$ is a known function of time t, we simply append the twice differentiated equation (6.121) to the set (6.120) to obtain the set of augmented constraint equations.

Example 6.4

Consider a knife edge EF of length L and mass m which moves in the XY-plane. The center of mass G of the knife edge is at a distance r from its point of contact, C, with the plane XY. The velocity of the knife edge at its point of contact C must always be along its length. We want to obtain the equations of motion for the knife edge assuming that the given impressed forces acting at C are $F_x(t)$ in the X-direction, and $F_y(t)$ in the Y-direction. The moment of inertia of the knife edge about an axis through point C which is normal to the XY-plane is I_C. See Figure 6.10.

Denote the coordinates of the point of contact C of the knife edge with the plane by (x, y) and those of its center of mass G by $(\bar{x}, \bar{y})$. We shall describe the configuration of the knife edge by the coordinates x, y and θ. The knife edge has a velocity at its point of contact directed along its length. Hence we have the constraint equation

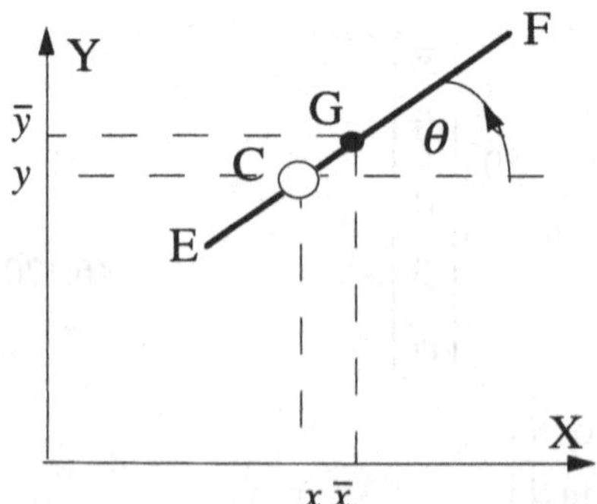

Figure 6.10. A knife edge whose point of contact C is a distance r from its center of mass G

$$\dot{y} = \dot{x} \tan \theta. \tag{6.122}$$

The components of the velocity of the center of mass of the knife edge can be written as

$$\dot{\bar{x}} = \dot{x} - r\dot{\theta} \sin \theta, \tag{6.123}$$

and

$$\dot{\bar{y}} = \dot{y} + r\dot{\theta} \cos \theta. \tag{6.124}$$

Also, the moment of inertia about the center of mass G is given by

$$I_G = I_C - mr^2. \tag{6.125}$$

The kinetic energy of the knife edge, taking the three coordinates that describe its configuration as though they are independent of each other, may now be written as

$$T = \frac{1}{2} m(\dot{\bar{x}}^2 + \dot{\bar{y}}^2) + \frac{1}{2} I_G \dot{\theta}^2, \tag{6.126}$$

which, on using equations (6.123)–(6.125), reduces to

$$T = \frac{1}{2} m(\dot{x}^2 + \dot{y}^2) + mr\dot{\theta}(\dot{y} \cos \theta - \dot{x} \sin \theta) + \frac{1}{2} I_C \dot{\theta}^2. \tag{6.127}$$

The work done under virtual displacements by the impressed forces can be written as

$$\delta W = F_x \delta x + F_y \delta y. \tag{6.128}$$

Using Lagrange's equations we now obtain, for the unconstrained system, the equations

$$\begin{bmatrix} m & 0 & -mr \sin \theta \\ 0 & m & mr \cos \theta \\ -mr \sin \theta & mr \cos \theta & I_C \end{bmatrix} \begin{bmatrix} \ddot{x} \\ \ddot{y} \\ \ddot{\theta} \end{bmatrix} = \begin{bmatrix} F_x + mr\dot{\theta}^2 \cos \theta \\ F_y + mr\dot{\theta}^2 \sin \theta \\ 0 \end{bmatrix}. \tag{6.129}$$

Example 6.4 (continued)

Differentiating (6.122) once with respect to time we get the constraint equation

$$[\tan\theta \quad -1 \quad 0]\begin{bmatrix}\ddot{x}\\ \ddot{y}\\ \ddot{\theta}\end{bmatrix} = -\dot{x}\dot{\theta}\sec^2\theta. \tag{6.130}$$

Equation (6.129), which describes the unconstrained system together with the constraint equation (6.130), now constitutes a complete description of the system. The matrices M and A and the vectors **a** and **b** are directly obtained from these equations.

All we need do now is simply to use equation (6.24) to write down the equations of motion, and equation (6.26) to obtain the forces of constraint.

It should be observed that when generalized coordinates are used, it is advantageous to use the fundamental equation in a form which does not require the determination of $M^{-1/2}$, such as,

$$\mathrm{M}\ddot{\mathbf{q}}(t) = \mathrm{F}(t) + \mathrm{A}^T(\mathrm{AM}^{-1}\mathrm{A}^T)^+(\mathbf{b} - \mathrm{AM}^{-1}\mathrm{F}) \tag{6.131}$$

PROBLEMS

6.1 Prove the expression (6.104) given in the text.

6.2 Show that the expression (6.107) is correct as follows. Let the direction cosines (with respect to Oxyz) of GA be (l_1,m_1,n_1), those of GB be (l_2,m_2,n_2) and those of GC be (l_3,m_3,n_3). A particle at location (a,b,c) in the coordinate frame GABC has its x-coordinates $x = \bar{x} + al_1 + bl_2 + cl_3$, etc. Hence $\dot{x} = \dot{\bar{x}} + a\dot{l}_1 + b\dot{l}_2 + c\dot{l}_3$.

Since $S\{ma\} = S\{mb\} = S\{mc\} = S\{mab\} = S\{bc\} = S\{ca\} = 0$, show that the kinetic energy $T = \frac{1}{2}S\{m(\dot{x}^2 + \dot{y}^2 + \dot{z}^2)\}$ can be expressed as

$$2T = M(\dot{\bar{x}}^2 + \dot{\bar{y}}^2 + \dot{\bar{z}}^2) + S\{ma^2(\dot{l}_1^2 + \dot{m}_1^2 + \dot{n}_1^2)\} + \cdots$$

Now note that the direction cosines of Ox in the GABC frame are (l_1,l_2,l_3). Since this is a unit vector in a fixed direction, show that $\dot{l}_1 = l_2\omega_3 - l_3\omega_2$. Use this in the expression for $2T$, and hence prove (6.107).

6.3 Obtain the other two equations of motion for the spinning top described in Example 6.2.

6.4 Verify the relations (6.52)–(6.54) and relation (6.60).

6.5 Use Euler angles to reformulate the description of the system shown in Figure 6.1. Obtain the equations of motion in terms of these angles. Integrate these equations and verify the results in Figures 6.2–6.6.

6.6 A dumbbell consisting of two masses m_1 and m_2 connected by a weightless rod of length L is constrained to move in the X-direction. Find the forces of constraint on the two masses if: (1) they move with a constant velocity, (2) they move with a constant acceleration, (3) they move with a time-varying acceleration.

6.7 The dumbbell described in Problem 6.6 is pinned at the midpoint P of the rod and constrained to rotate in the XY-plane about this pivot P. See Figure 6.11. Its angular acceleration is given by $\ddot{\theta}(t) = A\sin(\omega t^2)$, where A and ω are constants. Find the forces of constraint on the two masses.

6.8 What would be the forces of constraint on the masses in the above problem if in addition to the rotation, the pivot P moves in the Z-direction (normal to the XY-plane) so that its motion is given by $z_P = A_1 \sin(bt)$ where A_1 and b are constants.

6.9 Prove the two relations (6.112) and (6.113) which pertain to the constraints on the motion of the penny described in Example 6.3.

6.10 Prove the relations (6.101)–(6.103) using equation (6.96).

6.11 Verify that equation (6.126) is correct.

6.12 In Figure 6.12 the irregularly shaped area of mass m, with its center of gravity at G, is suspended from the point O. The area swings about O in the XY-plane. If the distance OG is L, and the moment of inertia of the area about an axis normal to the area through G is I_G, then write the equation of motion of the swinging pendulum so formed. Gravity is acting in the downward direction.

6.13 Obtain the equations of motion for the small vibrations in the vertical plane of a uniform straight bar of length L which is suspended by a weightless string of length s from one of its ends.

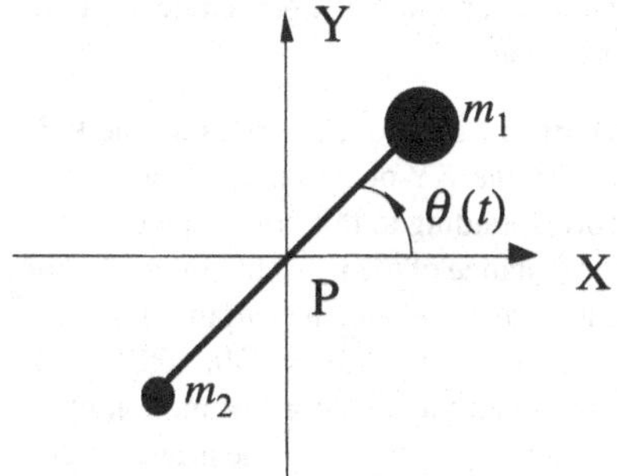

Figure 6.11. A rotating dumbbell

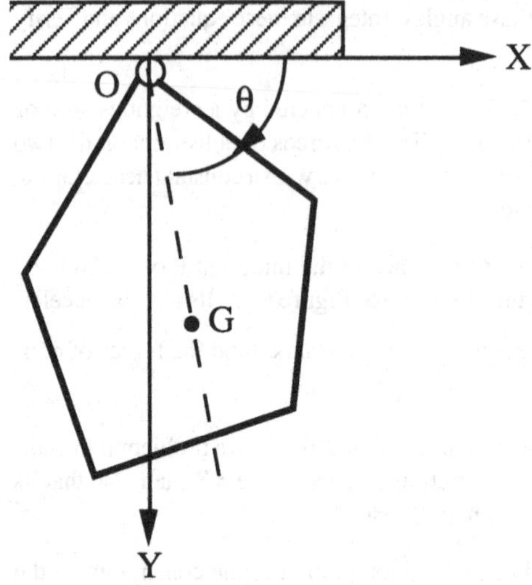

Figure 6.12. A planar area suspended from O

6.14 Show that equation (6.85) follows from (6.84). Show that the same equation (6.85) would be obtained independent of the order of the infinitesimal rotations.

6.15 Find the vectors **a** and **b** and the matrices M and A for the penny of Example 6.3 so that equation (6.24) can be directly used to obtain its equations of motion. The center of gravity of the penny, in addition, is constrained to move along a circle of radius r.

6.16 Consider Example 4.2 again, but this time include the mass of the two rods. Let the mass of the rod of length L_1 be m_1, and the mass of the rod of length L_2 be m_2. Discuss the motion of the system. Compare this with that obtained in equation (4.42).

6.17 A uniform ladder of length L and mass m has its two ends on the floor and against a wall respectively. If both the floor and the wall are smooth, describe the motion of the ladder until it hits the floor. The ladder starts from rest in a vertical plane perpendicular to the wall and makes an angle of 45 degrees with the horizontal.

6.18 In Example 6.4 assume that $F_x(t) = F_y(t) = 0$. In addition, assume that the knife edge is constrained so that its point of contact C with the XY-plane moves in a circle of radius 50 units. Deduce the equations of motion corresponding to this constrained system. Take the total mass of the knife edge as 10 units, the distance of the point of contact C from the end E (along the knife edge) as 2 units, the distance r as 1 unit and the length L as 6 units. See Figure 6.10. Use the initial conditions $x(0) = 0$, $y(0) = 50$, $\theta(0) = 0$, $\dot{x}(0) = 10$, $\dot{y}(0) = 0$ and $\dot{\theta}(0) = 0$. Numerically integrate the equations of motion of the constrained system and show numerically that the trajectory in the XY-plane is indeed a circle of radius 50 units. Plot the forces of constraint needed to execute such a motion as a function of the distance along the trajectory in the XY-plane.

6.19 In Example 6.4, if the forces F_x and F_y were impressed on the knife edge at its center of mass G instead of at its point of contact with the XY-plane, C, how would equation (6.129) change?

6.20 Ten particles, each of equal mass, are connected each to the other by rigid weightless rods. How many total constraints does the system have? How many of these constraints are independent ?

6.21 Consider two particles of masses m_1 and m_2 respectively at the hubs of two uniform massless rims, each of radius r (see Figure 6.13). The masses are connected by a massless inextensible rod of length L as shown below. The rims roll without slipping on an inclined plane, inclined to the horizontal at the angle θ. How many degrees of freedom does this system have? Write the constraint equations and then the equations of motion for the system.

6.22 Consider the same problem above (see Figure 6.13) except that the masses of the rims are now M_1 and M_2 respectively, and the mass of the rod connecting the masses m_1 and m_2 is m_L.

6.23 Consider the three-bar linkage under the action of the impressed forces shown in Figure 6.14. The lengths of the three bars (numbered 1, 2 and 3) are L_1, L_2 and L_3, respectively, and their masses are m_1, m_2 and m_3. Two of the joints are subjected to impressed forces as shown. There are elastic springs between the links which have spring constants k_1, k_2 and k_3. Their potential function is $V = (1/2)(k_1\phi_{10}^2 + k_2\phi_{21}^2 + k_3\phi_{32}^2)$, where ϕ_{ij} is the angle between link i and link j (link 0 is the horizontal ground). Also, there are dissipative elements between the links which have damping values c_1, c_2 and c_3. Their dissipation function is $R_D = (1/2)(c_1\dot{\phi}_{10}^2 + c_2\dot{\phi}_{21}^2 + c_3\dot{\phi}_{32}^2)$. Gravity acts in the negative Y-direction.

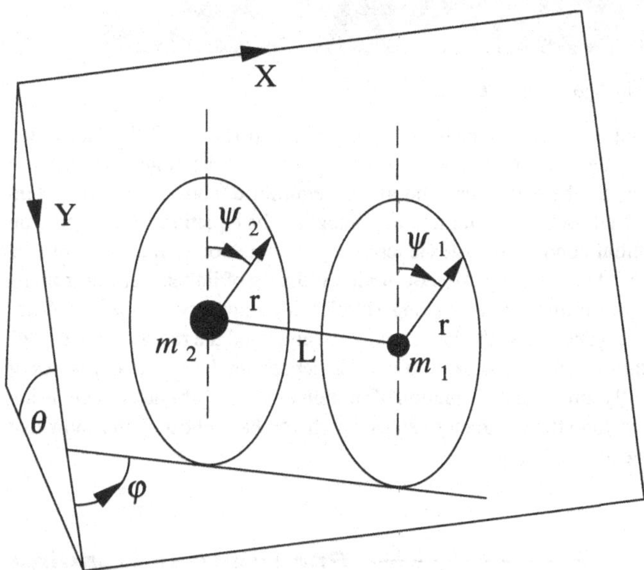

Figure 6.13. Two weightless rims rolling down a plane

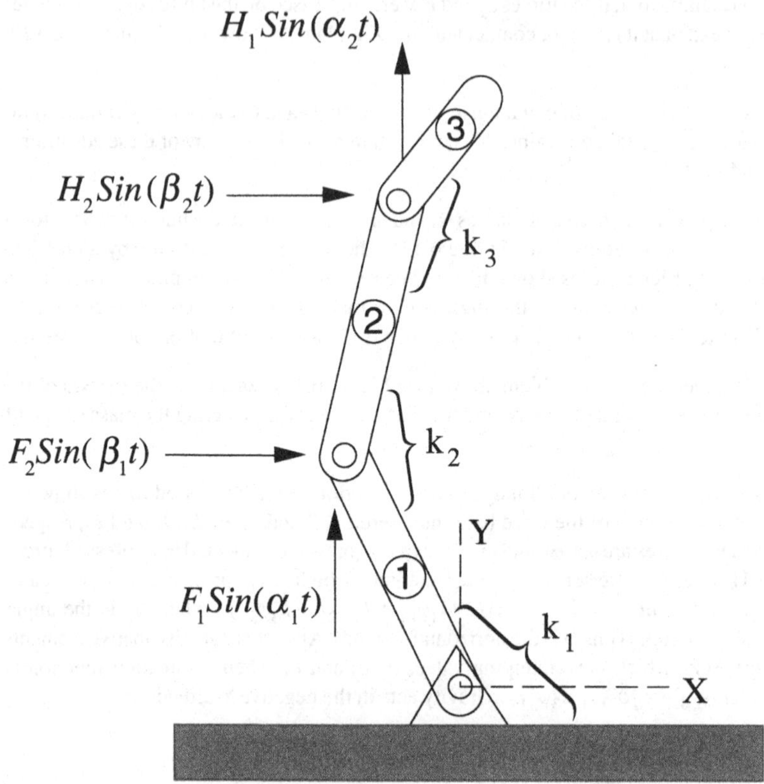

Figure 6.14. Three-bar linkage under gravity

Choose the three angles that the three bars make with the horizontal X-direction as the three generalized coordinates. (a) Obtain the equation of motion of the system. (b) Choose suitable numerical values for the parameters and find the equilibrium position of the system in the absence of impressed forces. (c) Numerically integrate the equations of motion you obtain (with suitable initial conditions). (d) Generalize your equation to the problem of n links instead of three. (e) Obtain the equation of motion if the tip of link 3, starting from its equilibrium position, is constrained to move along a circle with frequency f in the XY-plane. (f) How does your solution depend on the location of the center and the radius of the circle? How do the initial conditions which you need to provide depend on the center and radius of the circle? (g) Numerically integrate the equation of motion of the constrained system using suitable initial conditions (and the parameter values which you have chosen) and show that indeed the tip does move along a circle.

For Further Reading

1. We suggest that the reader go back to Chapter 2 of this book and get more familiar with all material in it, especially the four results that we have used in arriving at the fundamental equation in generalized coordinates.

2. P1 The description of the configuration of rigid bodies, their kinetic energies and rotations are well discussed in Chapter 7 of this book. Example 6.2 about the rolling penny has been taken from page 120 of this book.

3. R1 Chapters 6 and 7 of this book treat the kinematics and the kinetics of rigid bodies in some detail. The reader will find some interesting examples in each of these chapters as well as in Chapters 15 and 18. Example 6.4 is a generalization of a problem on page 335 of this book.

4.G5 This book gives an account of the Cayley–Klein parameters, another way of describing the orientation of a rigid body. This form of representation is useful mainly in areas like quantum mechanics and the interested reader may want to refer to this approach.

5.K1 Problem 6.21 has been taken from a paper by Kane where he introduces a method of considering nonholonomically constrained systems. Problem 6.22 is a variant of an example from the same paper.

7 Gauss's Principle Revisited

The reader will recall that we began our study of constrained motion in Chapter 3 by invoking Gauss's principle. There, we simply stated the principle as a basic principle of mechanics. We obtained the fundamental equation which described the motion of systems constrained by holonomic and nonholonomic constraints using this principle. It behooves us to understand Gauss's principle in greater depth now, moving full circle, as it were, by coming around to where we started from. This will be our agenda for this chapter.

We will begin with a simple proof of Gauss's principle in Cartesian coordinates, based on equation (5.18), the *basic equation of analytical mechanics* which we introduced in Chapter 5. We will next interpret this equation physically to demonstrate its aesthetic beauty, and then move on to prove the principle in terms of generalized coordinates. Using this principle, we will then provide an alternative proof for the fundamental equation in generalized coordinates.

Statement of Gauss's Principle in Cartesian Coordinates

Consider a system of n particles. The inertial Cartesian coordinates of the n particles can be described by the $3n$-vector $\mathbf{x} = [x_1\, x_2\, x_3 \cdots x_{3n-1}\, x_{3n}]^T$ in which the first three elements of the vector $\mathbf{x}$ correspond to the X-, Y- and Z- components of the position of the first particle, the next three elements correspond to the X-, Y- and Z-components of the position of the second particle, etc. We can also assemble the diagonal matrix $M = Diag\{m_1, m_2, m_3, m_4, \ldots, m_{3n-2}, m_{3n-1}, m_{3n}\}$. Here the successive diagonal elements are in sets of three, so that $m_{3r-2} = m_{3r-1} = m_{3r}$, each of these being the mass of the rth particle. Similarly we can assemble the given impressed forces acing on the n particles into a $3n$-vector $\mathbf{F} = [F_1\ F_2\ F_3 \cdots F_{3n-1}\ F_{3n}]^T$, where the components of the forces in the x, y and z directions on the rth particle are F_{3r-2}, F_{3r-1} and F_{3r} respectively. In the absence of any constraints, the equations of motion of the system of n particles would then be simply written as

$$m_i \ddot{x}_i = F_i, \quad i = 1, 2, \ldots, 3n, \tag{7.1}$$

or, more compactly, as

$$M\ddot{\mathbf{x}} = \mathbf{F}. \tag{7.2}$$

Thus, in the absence of any constraints, the acceleration of the particles of the system can be obtained as

$$\mathbf{a}(t) = M^{-1}\mathbf{F} \tag{7.3}$$

where we have denoted the $3n$-vector of accelerations by $\mathbf{a}(t)$. Equation (7.3), in component form, would simply read

$$a_i = \frac{F_i}{m_i}, \qquad i = 1, 2, \ldots, 3n. \tag{7.4}$$

We now assume that the system of n particles is constrained. We suppose that the configuration $\mathbf{x}(t)$ and the velocity $\dot{\mathbf{x}}(t)$ of this constrained system at time t are prescribed. We consider the function

$$\mathrm{G}(\ddot{\mathbf{x}}) = (\ddot{\mathbf{x}} - \mathbf{a})^T M(\ddot{\mathbf{x}} - \mathbf{a}) = \sum_{i=1}^{3n} m_i\left(\ddot{x}_i - \frac{F_i}{m_i}\right)^2 \tag{7.5}$$

as a function of the $3n$ quantities $\ddot{x}_1, \ddot{x}_2, \ldots, \ddot{x}_{3n-1}, \ddot{x}_{3n}$. The values of these $3n$ accelerations (or the $3n$-vector $\ddot{\mathbf{x}}$) which are considered in expression (7.5) are all those which are *possible* for the system to possess with the given configuration and the given velocity at the time t. By *possible* accelerations we mean that these accelerations must satisfy whatever constraints are imposed on the system at that time.

Gauss's principle now states that, from this class of values of $\ddot{\mathbf{x}}$, Nature appears to "choose" the value of the *actual* acceleration of the constrained system at time t to be such as to minimize G. In other words, the *actual* acceleration vector of the system when plugged into the right hand side of equation (7.5) causes G to be less than for any other *possible* acceleration vector.

To fix our ideas, let us consider the types of constraints that we have been dealing with all along, namely constraints that can be expressed (perhaps after a suitable number of differentiations) in our standard form as

$$A(\mathbf{x}, \dot{\mathbf{x}}, t)\ddot{\mathbf{x}}(t) = \mathbf{b}(\mathbf{x}, \dot{\mathbf{x}}, t) \tag{7.6}$$

where the matrix A is given (and has dimensions m by $3n$), and the vector $\mathbf{b}$ is given.

It should be noticed that these equations may be thought of as resulting from equations of constraint which take the general form

$$\phi_i(\mathbf{x}, \dot{\mathbf{x}}, t) = 0, \; i = 1, 2, \ldots, m. \tag{7.7}$$

These m equations, which do not need to be independent, when differentiated, will yield the system of constraint equations described by equation (7.6). Thus our results, as mentioned before, are applicable to a much wider class of constraints than those which fall within the usual rubric of Lagrangian mechanics. The reader should compare equation (7.6) with (3.28) and equation (7.7) with (3.26). Note that the matrix A in equation (7.6) is a function of $\mathbf{x}$ and t as well as of $\dot{\mathbf{x}}$; the constraints (7.7) could be nonlinear functions of the $\dot{x}_j$'s.

In the presence of the constraint given by equation (7.6), the equation of motion describing the constrained system, becomes

$$M\ddot{\mathbf{x}} = \mathbf{F} + \mathbf{F}^c \tag{7.8}$$

where $\mathbf{F}^c$ is the force of constraint that is brought into play so that the system, while under the influence of the given impressed force $\mathbf{F}$, satisfies the constraint (7.6) *and* Gauss's priniciple.

We assume that at time t the position and velocity of the particles (i.e., the vectors $\mathbf{x}(t)$ and $\dot{\mathbf{x}}(t)$) of the system are known and vary continuously with time. Our object is to determine the acceleration $\ddot{\mathbf{x}}(t)$ of the system at time t.

If the *actual* acceleration of the constrained system at time t is $\ddot{\mathbf{x}}$, then it must clearly satisfy the constraint equation (7.6). (Note that we assume that $\mathbf{x}$ and $\dot{\mathbf{x}}$ are prescribed and are continuous at time t.) Now consider any *possible* acceleration vector $\ddot{\mathbf{x}} + \Delta\ddot{\mathbf{x}}$ at the same time, i.e., any acceleration vector at time t which satisfies the constraints, and differs from the actual acceleration $\ddot{\mathbf{x}}$ by $\Delta\ddot{\mathbf{x}}$. Since this *possible* acceleration vector must be such that

$$A(\mathbf{x}, \dot{\mathbf{x}}, t)(\ddot{\mathbf{x}} + \Delta\ddot{\mathbf{x}}) = \mathbf{b}(\mathbf{x}, \dot{\mathbf{x}}, t), \tag{7.9}$$

by subtracting equation (7.6) from equation (7.9) we therefore deduce that

$$A(\mathbf{x}, \dot{\mathbf{x}}, t)\Delta\ddot{\mathbf{x}} = \mathbf{0}. \tag{7.10}$$

Equation (7.10) is valid for finite vectors $\Delta\ddot{\mathbf{x}}$. We note two things about this equation: (1) $\Delta\ddot{\mathbf{x}}$ is the difference between any *possible* acceleration $(\ddot{\mathbf{x}} + \Delta\ddot{\mathbf{x}})$ and the *actual* acceleration $\ddot{\mathbf{x}}$ of the system, and (2) if $\Delta\ddot{\mathbf{x}}$ satisfies equation (7.10) then for any constant β, $\beta\Delta\ddot{\mathbf{x}}$ also satisfies equation (7.10). The latter result is simply a consequence of the fact that equation (7.10) is homogeneous.

We shall now show that the vector $\Delta\ddot{\mathbf{x}}$ can be thought of as a virtual displacement vector. Hence, we show, as promised in Chapter 6, that for constraint equations which can be described in the standard form of equation (7.6), any (nonzero) vector $\mathbf{v}$ which satisfies

$$A(\mathbf{x}, \dot{\mathbf{x}}, t)\mathbf{v} = \mathbf{0} \tag{7.11}$$

constitutes a (generalized) virtual displacement vector at time t. To accomplish this, we need a deeper understanding of the concept of virtual displacements, a concept which we had introduced in Chapter 5 in detail, although in a somewhat elementary manner.

What we need to take along with us from this section is simply that the vector $\Delta\ddot{\mathbf{x}}$ which is the difference between any *possible* acceleration vector at time t and the *actual* acceleration vector at that time, always satisfies equation (7.10).

Virtual Displacements Revisited

Consider a typical particle of the constrained system which has mass m, as shown in Figure 7.1. Let a be the position of this particle at time t. Let us say that the Cartesian coordinates of a are $(x(t),y(t),z(t))$, and that at time t, the Cartesian components of the particle's velocity are $\dot{x}$, $\dot{y}$ and $\dot{z}$. We will assume that the particle is not subjected to impulses at any time, and so the velocity components of the particle at time t, are continuous functions of time. Let the components of the given impressed force on the particle at time t be F_x, F_y, F_z.

The representative particle's position and velocity at time t, as well as the forces impressed on it at that time, are thus assumed to be known. The position

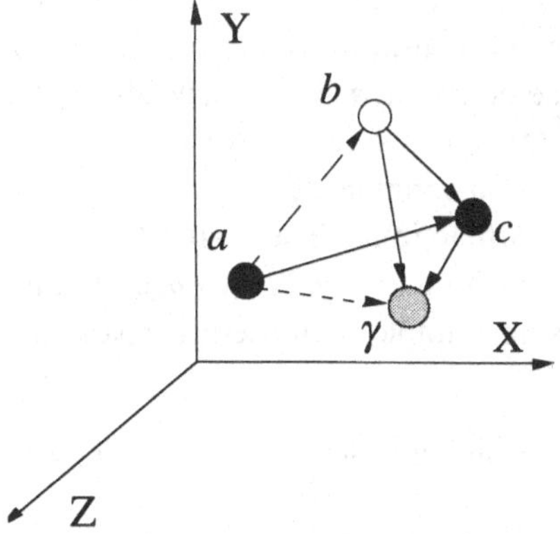

Figure 7.1. A representative particle of the constrained system at time t is located at a. After a time dt it moves to c. Had there been no constraints it would have moved to b. The position γ is a *possible* position of the particle after time dt

and velocity of the particle at time t are therefore no longer the subject of hypothesis or question. It is the future positions of the particle that constitute the focus of our inquiry. As we will see below, this amounts to determining the acceleration components of each and every particle, more particularly, of our representative particle, at time t.

Let us denote the time *immediately after* the time t by $t + dt$. We are thus interested in the position of the representative particle in the immediate future; i.e., the position after a vanishingly small (infinitesimal) increment dt of time succeeding the instant at time t. At time $t + dt$ this particle, under the influence of the impressed forces and the constraints will, in general, be located at, say, position c. Thus c represents the position of the particle which is a consequence of the *actual* motions that ensue in the constrained system. The representative particle moves from a to c in the infinitesimal time interval dt. However, had there been no constraints on the system and had the forces acting on the particle been only the known impressed forces, then at time $t + dt$ in the immediate future, this particle would have occupied, in general, a different position, say b.

The X-component of the actual acceleration of the particle at time t which enables it to reach location c at the time $t + dt$ in the immediate future is $\ddot{x}$. In fact, we know that for this representative particle,

$$m\ddot{x} = F_x + F_x^c \tag{7.12}$$

where F_x is the X-component of the impressed force, and F_x^c is the X-component of the force of constraint induced by the presence of the constraints, which causes the particle to be at c instead of at b in the immediate future.

In addition to these positions, let us consider a *possible* position of our representative particle at time $t + dt$ – a position which the particle is capable of reaching from a while satisfying the constraints at time $t + dt$. Let us denote this position by γ. The X-component of the acceleration at a, at time t, necessary to obtain this position will be denoted by $\hat{\ddot{x}}(t) = \ddot{x}(t) + \Delta\ddot{x}(t)$. We call such an acceleration component a *possible* acceleration component.

Since the position c is acquired by the particle immediately after time t, at time $t + dt$, clearly the X-component of this position is simply $x(t + dt)$. Assuming sufficient smoothness, we then have, using a simple Taylor series expansion,

$$x(t + dt) = x(t) + (dt)\dot{x}(t) + \frac{(dt)^2}{2}\ddot{x}(t) + \text{higher order terms.} \tag{7.13}$$

But the X-component of a is $x(t)$, so from equation (7.13), the X-component of c, denoted, c_x, is approximately

$$c_x \approx a_x + (dt)\dot{x}(t) + \frac{(dt)^2}{2}\ddot{x}(t). \tag{7.14}$$

In the above expression we have denoted the X-component of the position a by a_x. In the absence of any constraints, this particle would have at time $t + dt$ occupied the position b so that

$$b_x \approx a_x + (dt)\dot{x} + \frac{(dt)^2}{2}\frac{F_x}{m}. \tag{7.15}$$

Note that we obtain equation (7.15) in a manner similar to (7.14) by realizing that the acceleration of the particle at time t, when it is only under the influence of the impressed force acting on it, is simply $\frac{F_x}{m}$. Similarly, the X-component of the position γ is obtained as

$$\gamma_x \approx a_x + (dt)\dot{x} + \frac{(dt)^2}{2}\hat{\ddot{x}}(t), \tag{7.16}$$

realizing that the X-component of the acceleration at time t needed to get to γ from a is $\hat{\ddot{x}}$. Expressions similar to equations (7.14)–(7.16) can also be obtained for the Y- and Z-components of the points c, b and γ. The X-component of the directed segment $\overline{c\gamma}$ can be expressed as

$$\overline{c\gamma}_x \approx \frac{(dt)^2}{2}[\hat{\ddot{x}}(t) - \ddot{x}(t)] = \frac{(dt)^2}{2}\Delta\ddot{x}(t). \tag{7.17}$$

Likewise, the Y- and Z-components of $\overline{c\gamma}$ are $\frac{(dt)^2}{2}\Delta\ddot{y}$ and $\frac{(dt)^2}{2}\Delta\ddot{z}(t)$ respectively.

We now observe that the directed segment $\overline{c\gamma}$ constitutes, by definition, a virtual displacement which we shall denote by $\delta x(t + dt)$; because it represents a displacement at the instantaneous time $t + dt$ from an actual position of the representative particle (namely, from c at time $t + dt$), to a possible position (namely, to γ at time $t + dt$) consistent with the constraints. Thus the components of the virtual displacement $\overline{c\gamma}$, which occurs immediately after the instant at time t, namely at time $t + dt$, are the quantities $\frac{(dt)^2}{2}\Delta\ddot{x}(t)$, $\frac{(dt)^2}{2}\Delta\ddot{y}(t)$ and $\frac{(dt)^2}{2}\Delta\ddot{z}(t)$. It is important to realize that the latter three acceleration quantities are evaluated at time t. Thus the vector $\frac{(dt)^2}{2}[\Delta\ddot{x}(t)\ \ \Delta\ddot{y}(t)\ \ \Delta\ddot{z}(t)]^T$

constitutes a virtual displacement vector at time $t + dt$ for the representative particle under study.

Let us now go back to our original system consisting of n particles. We then see from the last paragraph that the $3n$-vector $\tilde{\mathbf{v}} = \frac{(dt)^2}{2} \Delta\ddot{\mathbf{x}}(t)$ constitutes a virtual displacement vector for our system of n particles. But we have seen from the previous section that the $3n$-vector $\Delta\ddot{\mathbf{x}}(t)$ satisfies equation (7.10), i.e., the equation $A\,\Delta\ddot{\mathbf{x}}(t) = \mathbf{0}$. Hence the vector $\tilde{\mathbf{v}}$ must also satisfy equation (7.10). We have thus shown that every virtual displacement vector $\tilde{\mathbf{v}}$ must satisfy equation (7.10). Remember that $\tilde{\mathbf{v}}$ serves as a virtual displacement $\delta\mathbf{x}$ at a time immediately following time t, namely at time $t + dt$. Thus the *basic equation of analytical mechanics* may be written at time $t + dt$, to yield

$$\delta\mathbf{x}(t + dt)^T\, \mathbf{F}^c(t + dt) = 0 = \Delta\ddot{\mathbf{x}}(t)^T [M\ddot{\mathbf{x}} - \mathbf{F}]_{t+dt}. \tag{7.18}$$

In the last expression above, we have cancelled out the $\frac{(dt)^2}{2}$ term since it is nonzero. Note that the term in square brackets in this expression needs to be evaluated at time $t + dt$.

Now the last term on the right of equation (7.18) is zero. This shows that one can indeed think of the vector $\Delta\ddot{\mathbf{x}}(t)$ itself as a virtual displacement vector at time $t + dt$! Furthermore, every such virtual displacement vector $\Delta\ddot{\mathbf{x}}(t)$ must satisfy equation (7.10). This should, of course, come as no surprise because $\Delta\ddot{\mathbf{x}}(t)$ and $\tilde{\mathbf{v}}$ differ by only a constant multiple. (See the remark after equation (7.10)).

Taking the limit as $dt \to 0$ of the last term in equation (7.18), and assuming that $\mathbf{F}^c(t)$ is continuous at time t, we get

$$\lim_{dt\to 0}\left[\Delta\ddot{\mathbf{x}}(t)^T [M\ddot{\mathbf{x}} - \mathbf{F}]_{t+dt}\right] = \Delta\ddot{\mathbf{x}}(t)^T \{M\ddot{\mathbf{x}}(t) - \mathbf{F}\} = \Delta\ddot{\mathbf{x}}(t)^T \{\mathbf{F}^c\} = 0 \tag{7.19}$$

where the quantities in the curly brackets are now evaluated at time t.

We note that the last equation above can also be written in terms of our representative particle as

$$S\{\overline{c\gamma} \cdot \mathrm{F}^c\} = S\{\overline{c\gamma}_x\, F_x^c + \overline{c\gamma}_y\, F_y^c + \overline{c\gamma}_y\, F_y^c\} = 0, \tag{7.20}$$

where F^c is the constraint force on the representative particle at time t, and the notation $\mathrm{S}\{*\}$ simply implies the sum over all such particles in the system of the quantity "$*$." For "one-sided" constraints we can replace the last equality by an equal-to-or-greater-than sign.

The limiting process performed in equation (7.19) can be understood physically as follows. Equation (7.18) is applicable at time $t + dt$, immediately after the instant at time t. The value of the vector $\ddot{\mathbf{x}}$, so determined, is of course the very quantity we wish to obtain, since the acceleration of the system at a given instant does not denote anything different from its acceleration immediately after that instant, assuming that this acceleration is a continuous function at time t. Alternatively put, the virtual displacement $\overline{c\gamma}$ which occurs at the instant immediately following time t is nothing but the virtual displacement at time t itself.

We have thus established the following result. The *basic equation of analytical mechanics* demands (compare with equation (5.18)) that at every time t

$$\Delta\ddot{\mathbf{x}}^T(M\ddot{\mathbf{x}} - \mathbf{F}) = 0 \tag{7.21}$$

for all *possible* acceleration vectors at time t that deviate from the *actual* acceleration of the constrained system at time t by $\Delta\ddot{\mathbf{x}}(t)$, i.e., for all vectors $\Delta\ddot{\mathbf{x}}(t)$ that satisfy equation (7.10) at time t. (For "one-sided" constraints the equality in equation (7.21) can be replaced by a greater-than-or-equal-to sign. However we shall not labor this point further.)

Alternately stated, at each instant of time t, for all vectors $\mathbf{v}$ such that $A\mathbf{v} = \mathbf{0}$, we must have $\mathbf{v}^T\mathbf{F}^c(t) = 0$. This is then a succinct statement of the *basic principle of analytical mechanics*. Recall that this statement of the basic principle requires the assumption that $\mathbf{F}^c(t)$ is a continuous function of time at time t.

Any such vector $\mathbf{v}$ which satisfies the equation $A\mathbf{v} = \mathbf{0}$ can be thought of as being the difference between a *possible* and an *actual* acceleration, and hence, as a (generalized) virtual displacement vector.

Gauss's Principle Again

It is now an easy matter to prove Gauss's principle using the basic equation of analytical dynamics. For the value of G, for any possible acceleration vector $\ddot{\mathbf{x}} + \Delta\ddot{\mathbf{x}}$ at time t, where $\ddot{\mathbf{x}}$ is the *actual* acceleration of the constrained system at time t, is given by

$$\begin{aligned} \mathrm{G}(\ddot{\mathbf{x}} + \Delta\ddot{\mathbf{x}}) &= (\ddot{\mathbf{x}} + \Delta\ddot{\mathbf{x}} - \mathbf{a})^T M(\ddot{\mathbf{x}} + \Delta\ddot{\mathbf{x}} - \mathbf{a}) \\ &= (\ddot{\mathbf{x}} - \mathbf{a})^T M(\ddot{\mathbf{x}} - \mathbf{a}) + 2\Delta\ddot{\mathbf{x}}^T M(\ddot{\mathbf{x}} - \mathbf{a}) + \Delta\ddot{\mathbf{x}}^T M\Delta\ddot{\mathbf{x}}. \end{aligned} \tag{7.22}$$

But the first term on the right hand side in the last expression is just $\mathrm{G}(\ddot{\mathbf{x}})$, the second term on the right hand side simplifies to zero because

$$2\Delta\ddot{\mathbf{x}}^T M(\ddot{\mathbf{x}} - \mathbf{a}) = 2\Delta\ddot{\mathbf{x}}^T(M\ddot{\mathbf{x}} - M\mathbf{a}) = 2\Delta\ddot{\mathbf{x}}^T(M\ddot{\mathbf{x}} - \mathbf{F}) = 0, \tag{7.23}$$

where we have made use of equations (7.3) and (7.21), and the third term is always positive as long as the vector $\Delta\ddot{\mathbf{x}} \neq \mathbf{0}$ because the masses of the particles are all positive. Hence we get

$$G(\ddot{\mathbf{x}} + \Delta\ddot{\mathbf{x}}) = G(\ddot{\mathbf{x}}) + \Delta\ddot{\mathbf{x}}^T M \Delta\ddot{\mathbf{x}} > G(\ddot{\mathbf{x}}) \tag{7.24}$$

unless $\Delta\ddot{\mathbf{x}} = \mathbf{0}$. We have proved Gauss's principle: Of all the *possible* acceleration vectors pertinent to the constrained system, the acceleration vector which *actually materializes* is the one for which G is a minimum – a truly aesthetic result!

Let us now obtain a deeper physical feel for this result. We begin by noting from equations (7.14) and (7.15) that for the representative particle, the X-component of $\overline{bc}$ is

$$\overline{bc}_x \approx \frac{1}{2}\left(\ddot{x} - \frac{F_x}{m}\right)(dt)^2 = \frac{1}{2}\frac{F_x^c}{m}(dt)^2, \tag{7.25}$$

where in the last equality we have used equation (7.12). Similar expressions can be found for the Y- and Z-components of $\overline{bc}$. Since (see Figure 7.1)

$$\overline{b\gamma} = \overline{bc} + \overline{c\gamma}, \tag{7.26}$$

taking the dot product on both sides we get

$$\overline{b\gamma} \cdot \overline{b\gamma} = \left(\overline{bc} + \overline{c\gamma}\right) \cdot \left(\overline{bc} + \overline{c\gamma}\right), \tag{7.27}$$

which implies that

$$(b\gamma)^2 = (bc)^2 + (c\gamma)^2 + 2\overline{c\gamma} \cdot \overline{bc}. \tag{7.28}$$

Multiplying equation (7.28) by m on both sides and summing over all particles, as we have done before in Chapter 6, we get

$$S\{m(b\gamma)^2\} = S\{m(bc)^2\} + S\{m(c\gamma)^2\} + 2S\{m\overline{c\gamma} \cdot \overline{bc}\} \tag{7.29}$$

where by S{∗} we mean, as usual, the sum of "∗" over all the particles of the system. But noting equation (7.25) we get

$$2S\{m\overline{c\gamma} \cdot \overline{bc}\} = (dt)^2 S\{\overline{c\gamma}_x F_x^c + \overline{c\gamma}_y F_y^c + \overline{c\gamma}_z F_z^c\} = (dt)^2 S\{\overline{c\gamma} \cdot \mathbf{F}^c\} \tag{7.30}$$

where F^c is the constraint force acting on our representative particle. But the last term on the right hand side of equation (7.30) is simply the sum total of the work done by the forces of constraint over the virtual displacements (for, as stated earlier, $\overline{c\gamma}$ is a representative virtual displacement). As determined from equation (7.20), by the basic principle of analytical dynamics, the sum total of this work must be zero (or, more generally for one-sided constraints, nonnegative). Using this in equation (7.29) we get the inequality

$$S\{m(b\gamma)^2\} = S\{m(bc)^2\} + S\{m(c\gamma)^2\} + 2S\{m\overline{c\gamma}\cdot\overline{bc}\} \geq S\{m(bc)^2\}, \tag{7.31}$$

with the equality occurring when $\overline{c\gamma} = 0$, i.e., when the possible position of each particle coincides with its actual position. Thus the position c that our representative particle occupies in the actual motion is such that the sum $S\{m(b\gamma)^2\}$ is minimized over all *possible* positions γ of the representative particle. This is Gauss's principle.

Using (7.14)–(7.16) the distance

$$(bc)^2 \approx \frac{1}{4}\left[\left\{\ddot{x} - \frac{F_x}{m}\right\}^2 + \left\{\ddot{y} - \frac{F_y}{m}\right\}^2 + \left\{\ddot{z} - \frac{F_z}{m}\right\}^2\right](dt)^4, \tag{7.32}$$

and the distance

$$(b\gamma)^2 \approx \frac{1}{4}\left[\left\{\hat{\ddot{x}} - \frac{F_x}{m}\right\}^2 + \left\{\hat{\ddot{y}} - \frac{F_y}{m}\right\}^2 + \left\{\hat{\ddot{z}} - \frac{F_z}{m}\right\}^2\right](dt)^4. \tag{7.33}$$

Thus by (7.31), the actual accelerations, i.e., the $\ddot{x}$'s, $\ddot{y}$'s and $\ddot{z}$'s, at each instant of time, are such that

$$\begin{aligned} G(\hat{\ddot{\mathbf{x}}}) &= S\left\{m\left(\hat{\ddot{x}} - \frac{F_x}{m}\right)^2 + m\left(\hat{\ddot{y}} - \frac{F_y}{m}\right)^2 + m\left(\hat{\ddot{z}} - \frac{F_z}{m}\right)^2\right\} \\ &\geq S\left\{m\left(\ddot{x} - \frac{F_x}{m}\right)^2 + m\left(\ddot{y} - \frac{F_y}{m}\right)^2 + m\left(\ddot{z} - \frac{F_z}{m}\right)^2\right\} = G(\ddot{\mathbf{x}}) \end{aligned} \tag{7.34}$$

for any set of *possible* accelerations, i.e. $\hat{\ddot{x}}$'s, $\hat{\ddot{y}}$'s and $\hat{\ddot{z}}$'s, which are compatible with the constraints. The minimum of G over all *possible* accelerations occurs for those accelerations which are the *actual* accelerations in the system.

At each instant of time, the actual accelerations of the particles of the constrained system may be thought of as constituting *one* set among the many sets of *possible* accelerations. What distinguishes this one specific set from among the others is that for this specific set of *possible* accelerations, the left hand side of the relation (7.31) is a minimum. Hence the minimization over all *possible* accelerations of the function G (taken as a function of the accelerations) at each instant of time will yield the *actual* accelerations of the constrained system at that instant. This is an alternative statement of Gauss's principle.

If we thought of the quantity $S\{m(bc)^2\}$ as a measure of the departure of the constrained system from the motion it would have had were there no constraints, then the fact that G is a minimum over all *possible* positions of the system can be interpreted as follows. Gauss's principle states that at each instant of time, the motion of the constrained system departs as little as possible from the unconstrained motion of the system, while still remaining compatible with the constraints.

Besides its deep aesthetic beauty, Gauss's principle enjoys comprehensive generality in a certain regard. The reader will recall that in arriving at equation (7.34) in our latter discussion we did *not* need to invoke any specific form of the constraints (such as equation (7.6)). Thus Gauss's principle is general enough to be applicable to constraints whose explicit form does not need specification – indeed, a truly remarkable result.

Gauss's Principle in Generalized Coordinates

In this section we ask what form Gauss's principle would take were we to use generalized coordinates. Let us assume that we have a system of n particles and that we use s generalized coordinates to describe them. We describe the system, as usual, in two steps. First we develop the "unconstrained" equations of motion, pretending that the coordinates used to describe the configuration of the system are all independent of one another. Then we provide the equations of constraint which this unconstrained system must satisfy in addition. As we saw in Chapter 5, the number of these constraint equations (and their nature) will depend on the number (and nature) of coordinates which are chosen as being independent while specifying the unconstrained system.

More specifically, we shall obtain the "unconstrained" equations of motion, as we did before, using Lagrange equations by treating the s coordinates as though they were all independent, i.e., as though virtual displacements could be prescribed independently for each of the s coordinates. Thus the "unconstrained" equations of motion would then, as in equation (5.149), be

$$\mathrm{M}(\mathbf{q},t)\ddot{\mathbf{q}} = \mathrm{F}(\mathbf{q},\dot{\mathbf{q}},t) \tag{7.35}$$

where the matrix $\mathrm{M}(\mathbf{q},t)$, as before, is an s by s positive definite matrix. The generalized force $\mathrm{F}(\mathbf{q}, \dot{\mathbf{q}},t)$, is the given impressed force acting on the system;

by given, we mean that F is a known function of $\mathbf{q}$, $\dot{\mathbf{q}}$ and t. The generalized acceleration of the "unconstrained" system, denoted by $\mathbf{a}(t)$, can then be obtained directly from equation (7.35) as

$$\mathbf{a}(t) = M^{-1}(\mathbf{q}, t)\mathrm{F}(\mathbf{q}, \dot{\mathbf{q}}, \mathrm{t}). \tag{7.36}$$

Now, when additional constraints described by the equation

$$\mathrm{A}(\mathbf{q}, \dot{\mathbf{q}}, t)\ddot{\mathbf{q}}(t) = \mathbf{b}(\mathbf{q}, \dot{\mathbf{q}}, t) \tag{7.37}$$

are imposed on this unconstrained system, additional forces of constraint $\mathrm{F}^c(\mathbf{q}, \dot{\mathbf{q}}, t)$ are brought into play so that these forces of constraint along with the given impressed forces F cause the constrained system to satisfy the constraints (7.37) at each instant of time t. The matrix A in equation (7.37) is m by s, signifying the presence of m (not necessarily independent) constraints.

As before, we note that equation (7.37) may be thought of as obtained from the set of m constraints that have the general form

$$\varphi_i(\mathbf{q}, \dot{\mathbf{q}}, t) = 0, \; i = 1, 2, \ldots, m. \tag{7.38}$$

Differentiating the set of equations (7.38) appropriately, will yield the constraint equation in the form (7.37).

Thus the motion of the constrained system is described by the equation

$$\mathrm{M}(\mathbf{q}, t)\ddot{\mathbf{q}} = \mathrm{F}(\mathbf{q}, \dot{\mathbf{q}}, t) + \mathrm{F}^c(\mathbf{q}, \dot{\mathbf{q}}, t). \tag{7.39}$$

Assuming that at time t, the configuration $\mathbf{q}(t)$ and the velocities $\dot{\mathbf{q}}(t)$ of the *constrained system* are known, our aim, as before, is to obtain the generalized accelerations of the constrained system. We assume that both $\mathbf{q}(t)$ and $\dot{\mathbf{q}}(t)$ are continuous at time t.

We now assert that of all the *possible* generalized acceleration vectors $\ddot{\mathbf{q}}(t)$ which satisfy the constraint equation (7.38) at time t, the one that actually materializes (i.e., the actual acceleration of the constrained system), is the one that minimizes the function

$$\mathrm{G}(\ddot{\mathbf{q}}) = [\ddot{\mathbf{q}} - \mathbf{a}(t)]^T \mathrm{M}(\mathbf{q}, t)[\ddot{\mathbf{q}} - \mathbf{a}(t)], \tag{7.40}$$

where the vector $\mathbf{a}(t)$ corresponds to the acceleration at time t of the unconstrained system as explicitly provided by equation (7.36). This is Gauss's principle in terms of generalized coordinates. We now go on to prove this.

Let $\ddot{\mathbf{q}}(t)$ be the actual acceleration vector of the constrained system which materializes at time t. Consider the vector $\ddot{\mathbf{q}}(t) + \Delta\ddot{\mathbf{q}}(t)$ which deviates from the actual acceleration vector by the deviation $\Delta\ddot{\mathbf{q}}(t)$ and also satisfies the constraint equation (7.37). Then since both vectors $\ddot{\mathbf{q}}(t)$and $\ddot{\mathbf{q}}(t) + \Delta\ddot{\mathbf{q}}(t)$ satisfy equation (7.37), we have

$$\mathrm{A}(\mathbf{q}, \dot{\mathbf{q}}, t)\ddot{\mathbf{q}}(t) = \mathbf{b}(\mathbf{q}, \dot{\mathbf{q}}, t) = \mathrm{A}(\mathbf{q}, \dot{\mathbf{q}}, t)\{\ddot{\mathbf{q}}(t) + \Delta\ddot{\mathbf{q}}(t)\} \tag{7.41}$$

from which it follows that

$$\mathrm{A}(\mathbf{q}, \dot{\mathbf{q}}, t)\Delta\ddot{\mathbf{q}}(t) = 0. \tag{7.42}$$

As before, we can consider $\Delta\ddot{\mathbf{q}}(t)$ as a generalized virtual displacement vector $\delta\mathbf{q}(t+dt)$ at time $t+dt$, where $t+dt$ is the instant immediately following the instant of time t. The basic equation of analytical mechanics then becomes (similar to equations (7.18) and (7.19), after applying the limiting argument as $dt \to 0$, and assuming that F^c is continuous at time t)

$$\Delta\ddot{\mathbf{q}}(t)^T\{\mathrm{M}(\mathbf{q}, t)\ddot{\mathbf{q}} - \mathrm{F}(\mathbf{q}, \dot{\mathbf{q}}, t)\} = \Delta\ddot{\mathbf{q}}(t)^T\mathrm{M}(\mathbf{q}, t)\{\ddot{\mathbf{q}} - \mathbf{a}(t)\} = 0. \tag{7.43}$$

Again, the vector $\Delta\ddot{\mathbf{q}}(t)$ – the difference between a *possible* generalized acceleration vector and the *actual* generalized acceleration vector – has the two properties stated in equations (7.42) and (7.43). Hence, as before, any vector $\mathbf{v}$ at time t which satisfies the equation

$$\mathrm{A}(\mathbf{q}, \dot{\mathbf{q}}, t)\mathbf{v} = 0 \tag{7.44}$$

can be thought of as a virtual displacement vector, and for every such vector we must have $\mathbf{v}^T\mathrm{F}^c(t) = 0$. This then is the statement of the basic principle of analytical mechanics in Lagrangian coordinates. Note that in contrast to equation (6.8), in the approach followed here the matrix A is a function not only of $\mathbf{q}(t)$ and t, but also of $\dot{\mathbf{q}}(t)$.

Evaluating $\mathrm{G}[\ddot{\mathbf{q}}(t) + \Delta\ddot{\mathbf{q}}(t)]$, we then get (for convenience we drop the arguments of the various quantities),

$$\begin{aligned}\mathrm{G}(\ddot{\mathbf{q}} + \Delta\ddot{\mathbf{q}}) &= (\ddot{\mathbf{q}} + \Delta\ddot{\mathbf{q}} - \mathbf{a})^T\mathrm{M}(\ddot{\mathbf{q}} + \Delta\ddot{\mathbf{q}} - \mathbf{a}) \\ &= (\ddot{\mathbf{q}} - \mathbf{a})^T\mathrm{M}(\ddot{\mathbf{q}} - \mathbf{a}) + 2\Delta\ddot{\mathbf{q}}^T\mathrm{M}(\ddot{\mathbf{q}} - \mathbf{a}) + \Delta\ddot{\mathbf{q}}^T\mathrm{M}\Delta\ddot{\mathbf{q}} \\ &= \mathrm{G}(\ddot{\mathbf{q}}) + \Delta\ddot{\mathbf{q}}^T\mathrm{M}\Delta\ddot{\mathbf{q}} > \mathrm{G}(\ddot{\mathbf{q}})\end{aligned} \tag{7.45}$$

unless $\Delta\ddot{\mathbf{q}}(t) = 0$. In carrying out the simplifications above, we have made use of equation (7.43) and the fact that M is a positive definite matrix. We have thus proved Gauss's principle in generalized coordinates. Notice that we proved that the minimum of $\mathrm{G}(\ddot{\mathbf{q}})$, which occurs when the acceleration vector $\ddot{\mathbf{q}}$ is the actual acceleration of the constrained system, is a *global* minimum, for no restrictions on the magnitude of the deviation vector $\Delta\ddot{\mathbf{q}}$ are imposed.

The determination of the vector $\ddot{\mathbf{q}}(t)$ which:

(1) minimizes the function $G(\ddot{\mathbf{q}}(t))$, as provided by (7.40), and
(2) belongs to the class of vectors which satisfy the constraint equation (7. 37),

thus yields the *actual* acceleration of the constrained system at time t.

Our task is now to explicitly find this vector. But we have already done this before in Chapter 3! All we need do is replace the $3n$-vector $\mathbf{x}$ by the s-vector $\mathbf{q}$, the $3n$ by $3n$ constant diagonal matrix M by the s by s positive definite matrix $M(\mathbf{q},t)$, the m by $3n$ matrix A by the m by s matrix $A(\mathbf{q},\dot{\mathbf{q}},t)$. We then obtain the result, mutatis mutandis, as

$$\ddot{\mathbf{q}}(t) = \mathbf{a}(t) + M^{-1/2}(AM^{-1/2})^{+}(\mathbf{b} - A\mathbf{a}) \tag{7.46}$$

or, through the use of equation (2.88), as

$$\ddot{\mathbf{q}}(t) = \mathbf{a}(t) + M^{-1}A^{T}(AM^{-1}A^{T})^{+}(\mathbf{b} - A\mathbf{a}), \tag{7.47}$$

which by now the reader is familiar with. The expression for the force of constraint is similarly given by

$$F^{c}(\mathbf{q}, \dot{\mathbf{q}}, t) = M^{1/2}(AM^{-1/2})^{+}(\mathbf{b} - A\mathbf{a}) \tag{7.48}$$

or, again, by

$$F^{c}(\mathbf{q}, \dot{\mathbf{q}}, t) = A^{T}(AM^{-1}A^{T})^{+}(\mathbf{b} - A\mathbf{a}). \tag{7.49}$$

We are back to the fundamental equation which we first adduced in Chapter 5, this time, however, starting from Gauss's principle!

The equation of motion of the constrained system can also be expressed by premultiplying both sides of equations (7.46) and (7.47) by the matrix $M(\mathbf{q},t)$. We then obtain, using equations (7.46) and (7.36), the equation of motion of the constrained system as,

$$M\ddot{\mathbf{q}}(t) = F(t) + M^{1/2}(AM^{-1/2})^{+}(\mathbf{b} - AM^{-1}F). \tag{7.50}$$

Furthermore were we to use the Lagrange multiplier m-vector $\boldsymbol{\lambda}$, we could have expressed the constrained equation of motion as

$$M\ddot{\mathbf{q}} = F + A^{T}\boldsymbol{\lambda}. \tag{7.51}$$

As we showed in Chapter 4, such an m-vector $\boldsymbol{\lambda}$ must exist because the column spaces of the matrices $(\mathrm{AM}^{-1/2})^{+}$ and $(\mathrm{AM}^{-1/2})^{T}$ are identical. Hence, at each instant of time, the equation

$$(\mathrm{AM}^{-1/2})^{T}\boldsymbol{\lambda} = (\mathrm{AM}^{-1/2})^{+}(\mathbf{b} - \mathrm{A}\mathbf{a}) \tag{7.52}$$

must be consistent for some m-vector $\boldsymbol{\lambda}$. Equation (7.52) then yields the relation

$$\mathrm{A}^{T}\boldsymbol{\lambda} = \mathrm{M}^{1/2}(\mathrm{AM}^{-1/2})^{+}(\mathbf{b} - \mathrm{A}\mathbf{a}) = \mathrm{A}^{T}(\mathrm{AM}^{-1}\mathrm{A}^{T})^{+}(\mathbf{b} - \mathrm{A}\mathbf{a}) = \mathrm{F}^{c}(\mathbf{q}, \dot{\mathbf{q}}, t). \tag{7.53}$$

Equation (7.53) is a linear equation in $\boldsymbol{\lambda}$; solving it enables us to directly determine the Lagrange multiplier vector for the constrained system. The closed form solution for $\boldsymbol{\lambda}$ is directly given by (see equation (2.107))

$$\begin{aligned}\boldsymbol{\lambda} &= (\mathrm{A}^{T})^{+}\mathrm{F}^{c} + [I - (\mathrm{A}^{T})^{+}\mathrm{A}^{T}]\mathbf{h} \\ &= (\mathrm{AA}^{+})(\mathrm{AM}^{-1}\mathrm{A}^{T})^{+}(\mathbf{b} - \mathrm{A}\mathbf{a}) + [I - (\mathrm{AA}^{+})]\mathbf{h},\end{aligned} \tag{7.54}$$

where $\mathbf{h}$ is an arbitrary m-vector. (We have made use of the fact that AA^{+} is a symmetric matrix.) Notice that though the Lagrange multiplier vector $\boldsymbol{\lambda}$ may be nonunique, the equations of motion governing the constrained system certainly are, as are the forces of constraint given by the vector F^{c}. Furthermore, when the rank of the matrix A equals the number of its rows m, then $I = (\mathrm{A}^{T})^{+}\mathrm{A}^{T}$, as pointed out in equation (2.90), and the second term on the right hand side of equation (7.54) vanishes, making the Lagrange multipliers unique. Equations (7.47) and (7.49) then simplify considerably because $(\mathrm{AM}^{-1}\mathrm{A}^{T})^{+} = (\mathrm{AM}^{-1}\mathrm{A}^{T})^{-1}$.

We are back to the fundamental equation which we adduced in Chapter 5, this time, however, starting from Gauss's principle and including constraints of the general form given by equations (7.7). We hope that as we have traced this full circle, the reader has acquired a deeper appreciation of some of the central concepts of mechanics.

The interested reader may want to go back to Chapter 6 and now re-derive the fundamental formula using D'Alembert's principle and the principle of virtual work, exactly as we did there, this time taking the definition of a virtual displacement vector $\mathbf{v}$ as any vector satisfying equation (7.44) instead of (6.8). Thereby the fundamental equation for the more general standard constraint of the form given by equation (7.37) instead of equation (6.6) can be proved.

The power of Gauss's principle, and the elegant results obtained in equations (7.46)–(7.49) which it engenders through its easy application, bear testimony to its deep aesthetic value, its simplicity and, above all, to the genius of Gauss.

PROBLEMS

7.1 In equation (7.54) note that the extent to which the acceleration corresponding to the unconstrained system does not satisfy the constraint equation (7.37) is given by the vector $\mathbf{e} = \mathbf{b} - \mathrm{A}\mathbf{a}$. Using equation (7.52), show that the Lagrange multiplier vector $\boldsymbol{\lambda}$ is such that $\partial\lambda_i/\partial e_j = \partial\lambda_j/\partial e_i$. Interpret this "reciprocity" relation physically.

7.2 When the rank of A is m, show that $\boldsymbol{\lambda} = (\mathrm{AM}^{-1}\mathrm{A}^T)^{-1}(\mathbf{b} - \mathrm{A}\mathbf{a})$.

For Further Reading

1. D1 The first two lectures of this four lecture compendium by Dirac deal with conservative systems which have constraints of the form, $\varphi_i(\mathbf{p},\mathbf{q}) = 0$, $i = 1, 2, \ldots, m$, where $\mathbf{p}$ denotes the generalized momentum. Dirac provides a general step by step recursive approach (algorithm) to arrive at a linear set of equations involving the Lagrange multipliers, the solution to which provides the multipliers. The method, however, is difficult to use when there are several tens of constraints.

2. G3 The original short paper of Gauss published in 1829 is well worth reading, except that it is written in German. In this chapter we have essentially used and amplified the arguments presented by Gauss in his original masterpiece.

3. G4 Gibbs's paper, published in 1879, is another masterpiece. It has an excellent treatment of the concept of a virtual displacement, and develops a minimum principle more general than Gauss's principle, from which, of course, Gauss's principle can be deduced. The last section of the paper introduces the concept of "quasi-coordinates" and develops what are today called the Gibbs–Appell equations. We shall look at these equations briefly in the next chapter.

8 Connections among Different Approaches

In the last seven chapters we have developed some of the basic concepts of analytical mechanics and have moved along various fundamental threads of thought, among them Gauss's principle and the basic equation (principle) of analytical mechanics. In the previous chapter we came full circle, as it were, by deriving Gauss's principle from the basic principle of analytical mechanics. We then used it to obtain, in a straightforward manner, the fundamental equation of analytical mechanics in Lagrangian coordinates.

In this, our last chapter, we shall attempt to further deepen our understanding by looking at alternative formulations of the problem of constrained motion and by developing the various interconnections between them, thereby weaving the various threads that we had previously laid out into the fabric that is analytical dynamics. This, we hope, will provide the reader with a more holistic picture of the subject.

We shall begin with some of the subtler ideas related to the fundamental equation, and provide the general form of the fundamental equation. Then we will move on to the Lagrange multiplier method and show its close connection to the fundamental equation and to the basic principle of analytical mechanics. The general form of the Lagrange multiplier vector will be obtained. We shall also develop the equations of motion through the direct use of the basic principle of analytical mechanics. This approach re-emphasizes the importance of linear algebra even when considering highly nonlinear mechanical systems. We shall then focus our attention on methods that rely on elimination to reduce the number of independent coordinates that we need to deal with in the presence of constraints. Within this context, we will broach the Gibbs–Appell equations and then show the close connection between these equations and Gauss's principle, thus having pushed our thinking back to where this book started from, namely from Gauss's principle.

Forms of the Fundamental Equation

In this chapter we shall introduce the main currents of thought in analytical dynamics in terms of Cartesian coordinates rather than generalized coordinates. This is so that the major concepts can be first highlighted and the interconnections understood before we embark on the use of generalized coordinates. As we shall see, it is, in general, a relatively simple matter to extend the basic ideas and concepts, once established in Cartesian coordinates, when using Lagrangian coordinates. We hope the reader has already noticed this in our earlier discussions, especially those in Chapters 5 and 7.

As usual we consider an inertial Cartesian coordinate frame and consider n particles of mass m_i, $i = 1, 2, \ldots, n$. The ith particle has coordinates x_i, y_i, z_i, and the equations of motion of the unconstrained system are given by

$$M\ddot{\mathbf{x}} = \mathbf{F}(\mathbf{x}, \dot{\mathbf{x}}, t), \tag{8.1}$$

where as usual, the $3n$ by $3n$ diagonal matrix M contains the masses of the n particles, in sets of three, down its main diagonal, and the $3n$-vector $\mathbf{F}(\mathbf{x}, \dot{\mathbf{x}}, t)$ of impressed forces is a known function of its arguments. The matrix M is positive definite, and the acceleration $\mathbf{a}(t)$ of the unconstrained system is obtained from (8.1) as

$$\mathbf{a}(t) = M^{-1}\mathbf{F}(\mathbf{x}, \dot{\mathbf{x}}, t). \tag{8.2}$$

The system is now further constrained by the m equations $\varphi_i(\mathbf{x}(t), \dot{\mathbf{x}}(t), t) = 0, i = 1, 2, \ldots, m$, which we assume to be sufficiently smooth. Differentiating these m equations with respect to time t we get our constraint in standard form as

$$A(\mathbf{x}, \dot{\mathbf{x}}, t)\ddot{\mathbf{x}} = \mathbf{b}(\mathbf{x}, \dot{\mathbf{x}}, t). \tag{8.3}$$

Equation (8.1) describes the unconstrained system, equation (8.3) describes the constraints. Then these two sets of equations along with a set of $6n$ initial conditions $\mathbf{x}(t)$, $\dot{\mathbf{x}}(t)$ which are compatible with the m equations $\varphi_i = 0$, $i = 1, 2, \ldots, m$, describe the problem of constrained motion of mechanical systems.

Our main job is the determination of: (1) the $3n$-vector $\ddot{\mathbf{x}}(t)$ of acceleration of the constrained system, and (2) the $3n$-vector $\mathbf{F}^c$ of the force of constraint at time t. The vector $\mathbf{F}^c$ is brought into play by the presence of the m constraints $\varphi_i = 0$, so that we have,

$$M\ddot{\mathbf{x}}(t) = \mathbf{F}(\mathbf{x}, \dot{\mathbf{x}}, t) + \mathbf{F}^c. \tag{8.4}$$

We have seen the answer to this problem many times by now. It is simply

$$\ddot{\mathbf{x}}(t) = \mathbf{a}(\mathbf{x}, \dot{\mathbf{x}}, t) + M^{-1/2}(AM^{-1/2})^{+}(\mathbf{b} - A\mathbf{a}). \tag{8.5}$$

But here we shall provide an additional nuance which was missing earlier in our understanding of the basic problem which will help us answer the somewhat perplexing questions we had raised at the end of Chapter 3. So we start from equation (8.5) as our point of departure here, only to return back to it later on in the section, as we shall see.

To begin with, we note that since the $6n$ quantities $\mathbf{x}(t)$ and $\dot{\mathbf{x}}(t)$ are known at time t, the impressed force $\mathbf{F}(\mathbf{x}, \dot{\mathbf{x}}, t)$ is also known at time t, and hence by equation (8.2) the vector $\mathbf{a}(t)$ is known (and fixed) at time t. For the same reason, the elements of the matrix A and the vector $\mathbf{b}$ in equation (8.3) are known (and fixed) at time t.

Gauss's principle asserts that of all the possible acceleration vectors $\hat{\ddot{\mathbf{x}}}$ which satisfy the constraints at time t, the one that actually materializes in the constrained system at time t minimizes the Gauss function

$$\mathrm{G}(\hat{\ddot{\mathbf{x}}}) = \left[\hat{\ddot{\mathbf{x}}} - \mathbf{a}(t)\right]^T M\left[\hat{\ddot{\mathbf{x}}} - \mathbf{a}(t)\right]. \tag{8.6}$$

Hence we need to find $\hat{\ddot{\mathbf{x}}}$ which minimizes G subject to the constraint (8.3). Thus we require any candidate $\hat{\ddot{\mathbf{x}}}$ to satisfy the equation $A(\mathbf{x}, \dot{\mathbf{x}}, t)\hat{\ddot{\mathbf{x}}} = \mathbf{b}(\mathbf{x}, \dot{\mathbf{x}}, t)$. Now we can express (8.6) as

$$\mathrm{G}(\mathbf{y}) = \mathbf{y}^T\mathbf{y}, \tag{8.7}$$

where,

$$\mathbf{y} = M^{1/2}(\hat{\ddot{\mathbf{x}}} - \mathbf{a}). \tag{8.8}$$

Also, if we use equation (8.8), equation (8.3) becomes

$$B\mathbf{y} = (\mathbf{b} - A\mathbf{a}), \tag{8.9}$$

where the constraint matrix $B = AM^{-1/2}$. The right hand side of equation (8.9) is known at time t, as is the matrix B. Gauss's principle thus demands that $\mathbf{y}$ be found such that G($\mathbf{y}$) is minimum and $\mathbf{y}$ satisfies equation (8.9). Thus, to determine the acceleration $\ddot{\mathbf{x}}$ of the constrained system at time t, we need to find the $3n$-vector $\mathbf{y}$ such that: (1) its length is minimum, and, (2) it satisfies equation (8.9). But we know (see page 66) that the minimum norm vector $\mathbf{y}$ which satisfies equation (8.9) is simply given by

$$\mathbf{y} = B^{\{1,4\}}(\mathbf{b} - A\mathbf{a}), \tag{8.10}$$

where $B^{\{1,4\}}$ is *any* {1,4}-inverse of the matrix B (see Chapter 2). Using (8.8) to express $\mathbf{y}$ in terms of $\ddot{\mathbf{x}}$, now leads to

$$\boxed{\ddot{\mathbf{x}} = \mathbf{a} + M^{-1/2}(AM^{-1/2})^{\{1,4\}}(\mathbf{b} - A\mathbf{a})}. \tag{8.11}$$

We note that equation (8.11) looks deceptively similar to equation (8.5), yet there is a big difference. Equation (8.5) requires the MP-inverse of the constraint matrix $AM^{-1/2}$, while equation (8.11) requires any {1,4}-inverse of $AM^{-1/2}$. We note (see Example 2.36) that since the vector $\mathbf{b} - A\mathbf{a}$ belongs to the column space of $AM^{-1/2}$, the right hand side of equation (8.11) is uniquely determined no matter which particular {1,4}-inverse of the constraint matrix is used. Hence the acceleration of the constrained system is uniquely determined.

The reader may recall that in Chapter 3 we had argued that perhaps it might be possible to replace the MP-inverse in the fundamental equation (8.5) by a less restrictive one. We have shown, by equation (8.11), that we have now arrived at the least restrictive generalized inverse which when used instead of the MP-inverse will yield the proper acceleration of the constrained mechanical system in concordance with Gauss's principle.

But we know still some more. From Example 2.35 we know that for any matrix B, one specific $B^{\{1,4\}}$ is $B^{\{1,4\}} = B^T(BB^T)^{\{1\}}$. Using this relation in equation (8.11), we obtain an alternative form of the fundamental equation

$$\boxed{\ddot{\mathbf{x}} = \mathbf{a} + M^{-1}A^T(AM^{-1}A^T)^{\{1\}}(\mathbf{b} - A\mathbf{a})}, \tag{8.12}$$

where we have got rid of $M^{-1/2}$ from the equation. Premultiplying by M we obtain

$$\boxed{M\ddot{\mathbf{x}} = \mathbf{F} + A^T(AM^{-1}A^T)^{\{1\}}(\mathbf{b} - A\mathbf{a})}. \tag{8.13}$$

The forms (8.11)–(8.13) of the fundamental equations appear to be the most parsimonious forms in the sense that they employ the least restrictive generalized inverse.

Since {1,4}-inverses, {1,2,4}-inverses, {1,3,4}-inverses and the {1,2,3,4}-inverse form subsets of the set of {1}-inverses of any matrix, we can express equations (8.12) and (8.13) in the alternative forms

$$\boxed{\ddot{\mathbf{x}} = \mathbf{a} + M^{-1}A^T(AM^{-1}A^T)^*(\mathbf{b} - A\mathbf{a})}, \tag{8.14}$$

and,

$$\boxed{M\ddot{\mathbf{x}} = \mathbf{F} + A^T(AM^{-1}A^T)^*(\mathbf{b} - A\mathbf{a})} \tag{8.15}$$

where the superscript "*" stands for *any* {1}-inverse or *any* subset of the set of {1}-inverses. Equations (8.14) and (8.15) are then the most general fundamental forms *regardless of the rank of A*.

In Chapter 3 the reader may recall that we had identified three different locations in our verification of the fundamental equation where we needed the properties of generalized inverses. While the first two locations did not specifically require the use of the MP-inverse (only the {1}-inverse was needed), we identified the third location as requiring at least the {1,2,4}-inverse. Recall that we wanted the third (and second) member on the right hand side of equation (3.43) to vanish. Our choice of the {1,2,4}-inverse was dictated by our thinking that we wanted the term $(M^{1/2}\mathbf{v})^T(AM^{-1/2})^*$ to equal zero whenever $AM^{-1/2}(M^{1/2}\mathbf{v}) = \mathbf{0}$, where the "*" indicates a suitable generalized inverse of the Constraint Matrix.

How could this square with the fact that we seem to require, by equation (8.11), only the {1,4}-inverse? In other words, how could we show that the third term on the right hand side of equation (3.43) vanishes even when the MP-inverse is replaced by the {1,4} inverse? The perceptive reader will realize that we cannot show that $(M^{1/2}\mathbf{v})^T(AM^{-1/2})^*$ equals zero whenever $AM^{-1/2}(M^{-1/2}\mathbf{v}) = \mathbf{0}$ where the superscript * signifies the {1,4}-inverse! But, what we can show is that when * signifies any {1,4}-inverse, and $\mathbf{z}$ is any vector which lies in the range space of $AM^{-1/2}$, then $(M^{1/2}\mathbf{v})^T(AM^{-1/2})^*\mathbf{z} = \mathbf{0}$ whenever $AM^{-1/2}(M^{1/2}\mathbf{v}) = \mathbf{0}$. Since the vector $(\mathbf{b} - A\mathbf{a})$ lies in the range space of $AM^{-1/2}$, we can then show that the entire third (and second) member on the right hand side of equation (3.43) is zero! We leave the detailed proof of this result as an exercise for the reader (see Problem 8.2). Thus the proof of our validation (along the lines shown in Chapter 3) will go through then even when the MP-inverse is replaced by the {1,4}-inverse in equation (8.5).

Notice that the vector $A\mathbf{a}$ always belongs to the range space of A, and therefore to the range space of $AM^{-1/2}$. Hence for $(\mathbf{b} - A\mathbf{a})$ to belong to the range space of $AM^{-1/2}$, the vector $\mathbf{b}$ must belong to the range space of $AM^{-1/2}$. What if the constraint equation is imprecisely specified? What if the vector $\mathbf{b}$ does not belong to the range space of the matrix A? In that case, perhaps all we can hope for is the satisfaction of the constraint equation (8.3) or, alternatively of equation (8.9), in the least squares sense. Hence we would need to determine the vector $\mathbf{y}$ such that it has minimum length and satisfies equation (8.9) in the least squares sense. But in Chapter 2 we have shown (see page 67) that the minimum length least squares solution of the equation $B\mathbf{y} = (\mathbf{b} - A\mathbf{a})$ is simply $\mathbf{y} = B^+(\mathbf{b} - A\mathbf{a})$. Remembering from equation (8.8) that $\ddot{\mathbf{x}} = \mathbf{a} + M^{-1/2}\mathbf{y}$, we are led back to equation (8.5), our point of departure in this section! (See reference F1.)

We shall now move on to alternate ways of formulating, and arriving at, the equation of motion for constrained systems. But before we do that, we point out that all the forms that we have dealt with in this section yield the equation of motion in terms of the original coordinates used to describe the motion of the unconstrained system; no elimination of coordinates in the face of the connections created between them by the presence of constraints is done. The equations are obtained directly through a knowledge of the matrices M and A, and the vectors $\mathbf{F}$ and $\mathbf{b}$, these being the only four quantities involved in the fundamental forms.

We have not dealt with generalized coordinates here, but the reader will have realized by now that the equivalent fundamental forms in generalized coordinates can be directly obtained by replacing $\ddot{\mathbf{x}}$ by $\ddot{\mathbf{q}}$, and the diagonal matrix M by the symmetric matrix $\mathrm{M}(\mathbf{q},\mathrm{t})$. The elements of the matrix A and those of the vectors F and $\mathbf{b}$ are now functions of $\mathbf{q}$, $\dot{\mathbf{q}}$, and time. The unconstrained equation of motion $\mathrm{M}(\mathbf{q},\mathrm{t})\ddot{\mathbf{q}} = \mathrm{F}(\mathbf{q},\dot{\mathbf{q}},\mathrm{t})$ is now of course a consequence of Lagrange's equation.

The Fundamental Form and the Lagrange Multiplier

In Chapter 4 we had established that the constrained equation of motion can always be expressed as (see equation 4.19)

$$M\ddot{\mathbf{x}} = \mathbf{F} + A^T\boldsymbol{\lambda} \tag{8.16}$$

for a suitable m-vector $\boldsymbol{\lambda}$, which is called the Lagrange multiplier vector. This equation, along with the constraint equation (8.3), provides a system of equations for the determination of the $3n$-vector $\ddot{\mathbf{x}}$ and the m-vector $\boldsymbol{\lambda}$.

In Chapter 4, we established equation (8.16) through the observation that the range spaces of the matrices $(\mathrm{AM}^{-1/2})^T$ and $(\mathrm{AM}^{-1/2})^+$ are identical. Though this was a rigorous enough proof of equation (8.16), it also behooves us now to see how this result follows directly from the basic principle of analytical mechanics.

Accordingly we shall show the equivalence of the following two propositions.

P1: For all vectors $\mathbf{v}$ such that $A\mathbf{v} = \mathbf{0}$, $\mathbf{v}^T\mathbf{F}^c = 0$.
P2: There must exist an m-vector $\boldsymbol{\lambda}$ such that $\mathrm{F}^c = A^T\boldsymbol{\lambda}$.

We first show that P2 implies P1. If P2 is true, then $\mathbf{v}^T\mathbf{F}^c = \mathbf{v}^T A^T\boldsymbol{\lambda} = (A\mathbf{v})^T\boldsymbol{\lambda}$, which is zero whenever $A\mathbf{v} = \mathbf{0}$. Now we show that P1 implies P2. If $A\mathbf{v} = \mathbf{0}$, then by equation (2.107), $\mathbf{v} = (I - A^+A)\mathbf{h}$, and P1 says that $\mathbf{h}^T(I - A^+A)^T\mathbf{F}^c = 0$, for all vectors $\mathbf{h}$. Since the matrix $(I - A^+A)$ is symmetric, this implies that $\mathbf{F}^c = A^+A\mathbf{F}^c$. But the column space of A^+A is the same as the column space of A^T so that there must exist a $\boldsymbol{\lambda}$ such that $A^+A\mathbf{F}^c = A^T\boldsymbol{\lambda}$. Hence $\mathbf{F}^c = A^T\boldsymbol{\lambda}$. Notice that we did *not* require that the m by n matrix A have rank m; i.e., the constraint equations (8.3) need not be independent.

Using P2 in equation (8.4) yields equation (8.16). We have thus shown that the basic principle of analytical mechanics implies equation (8.16)! Using equations (8.15) and (8.16), we then obtain the explicit equation for the Lagrange multiplier m-vector $\boldsymbol{\lambda}$ (see equation (2.107)) as

$$\boldsymbol{\lambda} = C(AM^{-1}A^T)^*(\mathbf{b} - A\mathbf{a}) + (I - C)\mathbf{h}, \tag{8.17}$$

where $C = C^T = AA^+$, and the vector $\mathbf{h}$ is an arbitrary m-vector. The "*" as usual denotes *any* {1}-inverse. This then becomes the most general form of the Lagrange multiplier vector. In general, because of the arbitrariness of $\mathbf{h}$, the Lagrange multiplier m-vector is nonunique. However, when A has rank m, $C = I$, and the second member on the right of equation (8.17) vanishes, the Lagrange multiplier vector then becomes unique.

The general form of the Lagrange multiplier in Lagrangian coordinates can be written down directly from equation (8.17) by making the usual substitutions.

Direct Use of the Basic Principle of Analytical Mechanics

We have obtained the fundamental forms through the use of Gauss's principle and the Lagrange multiplier vector by first establishing equation (8.16) and then comparing its form with the fundamental form. We now show how we can obtain an equivalent explicit formulation of the equations of motion for constrained systems by the direct use of the basic principle of analytical mechanics.

For convenience, let us premultiply equation (8.4) by $M^{-1/2}$ and rewrite it as

$$\ddot{\mathbf{x}}_s - \mathbf{F}_s^c = \mathbf{a}_s, \tag{8.18}$$

where we denote by $M^{1/2}\ddot{\mathbf{x}}$ the "scaled" acceleration of the constrained system $\ddot{\mathbf{x}}_s$, by $M^{1/2}\mathbf{a}$ the "scaled" acceleration of the unconstrained system $\mathbf{a}_s$ and by $M^{-1/2}\mathbf{F}^c$ the "scaled" force of constraint $\mathbf{F}_s^c$. Since the matrix M is known, determination of the acceleration of the constrained system $\ddot{\mathbf{x}}$ and the corresponding constraint force $\mathbf{F}^c$ at time t is tantamount to determining the "scaled" acceleration and the "scaled" constraint force. Our constraint equation at time t can likewise be expressed as

$$B\ddot{\mathbf{x}}_s = \mathbf{b}, \tag{8.19}$$

where $B = AM^{-1/2}$ is the Constraint Matrix. Now the basic principle of analytical mechanics stated in proposition P1 earlier says that for all vectors $\mathbf{v}$ such that $A\mathbf{v} = (AM^{-1/2})(M^{1/2}\mathbf{v}) = \mathbf{0}$, we must have $\mathbf{v}^T\mathbf{F}^c = (M^{1/2}\mathbf{v})^T M^{-1/2}\mathbf{F}^c = 0$. Denoting $\mathbf{u} = M^{1/2}\mathbf{v}$, thus this principle then alternatively states the following:

$$\text{For all vectors } \mathbf{u} \text{ such that } B\mathbf{u} = 0, \text{ we must have } \mathbf{u}^T\mathbf{F}_s^c = 0. \tag{8.20}$$

For how many independent vectors $\mathbf{u}$ would the above relation be true? If the rank of the m by $3n$ matrix B is r ($r < m$ and $r < 3n$), then clearly the null space of B will have dimension $(3n - r)$. Hence there will be $(3n - r)$ such linearly independent vectors such that $B\mathbf{u}_i = \mathbf{0}$, $i = 1, 2, \ldots, (3n - r)$. Let us create a matrix U whose columns are these vectors $\mathbf{u}_i$ so that $U = [\mathbf{u}_1\ \mathbf{u}_2\ \mathbf{u}_3 \ldots \mathbf{u}_{3n-r}]$. The basic principle of analytical mechanics can then be equivalently stated as follows: For the matrix U defined as we have

$$U^T\mathbf{F}_s^c = \mathbf{0}. \tag{8.21}$$

Equations (8.18), (8.19) and (8.21) form a system of equations for the determination of the $3n$ components of the vector $\ddot{\mathbf{x}}$ and the $3n$ components of the vector $\mathbf{F}_s^c$. They may be written in a more compact form, $\mathbf{Lr} = \mathbf{s}$, as

$$\mathbf{Lr} = \begin{bmatrix} I_{3n} & -I_{3n} \\ B_{m \text{ by } 3n} & \mathbf{0} \\ \mathbf{0} & U^T_{(3n-r) \text{ by } 3n} \end{bmatrix} \begin{bmatrix} \ddot{\mathbf{x}}_s \\ \mathbf{F}_s^c \end{bmatrix} = \begin{bmatrix} \mathbf{a}_s \\ \mathbf{b} \\ \mathbf{0} \end{bmatrix} = \mathbf{s}, \tag{8.22}$$

where the vector $\mathbf{r}^T = \begin{bmatrix} \ddot{\mathbf{x}}_s^T & \mathbf{F}_s^{cT} \end{bmatrix}^T$. It is not difficult to show that the rank of the matrix $\mathbf{L}$ is $6n$ (see Problem 8.4). Premultiplying each side of the equation $\mathbf{Lr} = \mathbf{s}$ by $\mathbf{L}^T$ we can now solve for the vector $\mathbf{r}$ to yield

$$\mathbf{r} = (\mathbf{L}^T\mathbf{L})^{-1}\mathbf{L}^T\mathbf{s}. \tag{8.23}$$

We note that since the rank of the matrix $\mathbf{L}$ equals the number of its columns (i.e., $6n$), the $6n$ by $6n$ matrix $(\mathbf{L}^T\mathbf{L})$ has rank $6n$, and is hence invertible. Writing out equation (8.23) in its entirety, we get

$$\begin{bmatrix} \ddot{\mathbf{x}}_s \\ \mathbf{F}_s^c \end{bmatrix} = \begin{bmatrix} I + K & -I_{3n} \\ -I_{3n} & I + J \end{bmatrix}^{-1} \begin{bmatrix} \mathbf{a}_s + B^T\mathbf{b} \\ -\mathbf{a}_s \end{bmatrix}, \tag{8.24}$$

where the matrix $K = B^T B$, and the matrix $\mathrm{J} = UU^T$.

We have thus obtained explicit relations for the scaled acceleration of the constrained system $\ddot{\mathbf{x}}_s$ and the scaled constraint force $\mathbf{F}_s^c$ at the time t! Of course, the acceleration $\ddot{\mathbf{x}}$ given by (8.24) (as also the constraint force $\mathbf{F}^c$) must be identical to that given by the fundamental form, say, equation (8.5). That this is indeed so is left for the reader to show.

Using equation (8.24), in Lagrangian coordinates the corresponding equation for constrained motion can be written down directly, as before, by simply making the appropriate substitutions. It should be remembered, though, that now the matrix M is a function of **q** and time.

We point out that the explicit equation of motion obtained in equation (8.24) differs in form quite dramatically from the fundamental equation (8.5). Not only do we need to find the matrix U at each instant of time, but we lose the physical insights developed in Chapters 4 and 6 on how constrained motion in a mechanical system evolves. Yet, equation (8.24) appears to be perhaps the most direct and basic way of obtaining the equation of motion for constrained systems. It does not need notions such as the generalized inverse of a matrix or the Gibbs function, which we describe below, or the extended Poisson brackets which Dirac used.

Equations of Motion through Methods of Elimination

We will consider two approaches to obtaining the equations of motion through the concept of elimination of coordinates. The first uses the basic principle of analytical mechanics, the second was invented by Gibbs and Appell, each working independently.

Elimination Using the Basic Principle of Analytical Mechanics

The unconstrained system, as we have discussed it in the previous section, is fully described by $3n$ Cartesian coordinates. These $3n$ coordinates are independent, i.e., the components of the virtual displacement $3n$-vector can be independently chosen. The constraints described by equation (8.3) bring about relations between these independent coordinates. We have already shown that a generalized virtual displacement vector at time t is any vector $\mathbf{v}$ that satisfies the equation $A\mathbf{v} = \mathbf{0}$. The $3n$ components of the vector $\mathbf{v}$ can obviously not be independently chosen, precisely because they must satisfy the equation $A\mathbf{v} = \mathbf{0}$. The basic equation of analytical mechanics states that,

$$\mathbf{v}^T\mathbf{F}^c = \mathbf{v}^T(M\ddot{\mathbf{x}} - \mathbf{F}) = 0 \text{ for all vectors } \mathbf{v} \text{ which satisfy the relation } A\mathbf{v}=\mathbf{0}. \tag{8.25}$$

In the absence of constraints, the components of **v** can be independently chosen and we get back, from the above relation, the unconstrained equation of motion $M\ddot{\mathbf{x}} - \mathbf{F} = 0$.

Let us assume for convenience that the m by $3n$ matrix A has rank m and further that the $3n$ components in the vector $\ddot{\mathbf{x}}$ which describes the configuration of the constrained system are so numbered that the left hand m by m submatrix of A has an ordinary inverse. We can then partition the column vector **v**, and write it as $\mathbf{v} = [\mathbf{v}_e^T \quad \mathbf{v}_I^T]^T$, where the subvectors $\mathbf{v}_e$ and $\mathbf{v}_I$ are column vectors of dimension m and $(3n - m)$ respectively. The relation $A\mathbf{v} = \mathbf{0}$ can now be expressed as

$$A\mathbf{v} = [A_e \quad A_I]\begin{bmatrix} \mathbf{v}_e \\ \mathbf{v}_I \end{bmatrix} = A_e\mathbf{v}_e + A_I\mathbf{v}_I = 0, \tag{8.26}$$

where, by stipulation, the m by m matrix A_e has an ordinary inverse. From this equation we obtain the relation

$$\mathbf{v}_e = -A_e^{-1}A_I\mathbf{v}_I \tag{8.27}$$

which clearly shows that the subvector $\mathbf{v}_e$ can be expressed in terms of the subvector $\mathbf{v}_I$; the components of $\mathbf{v}_I$ can then be chosen independently.

Also, we can partition the $3n$-vector $M\ddot{\mathbf{x}} - \mathbf{F}$ into the m-vector $M_e\ddot{\mathbf{x}}_e - \mathbf{F}_e$ and the $(3n - m)$ vector $M_I\ddot{\mathbf{x}}_I - \mathbf{F}_I$. Note here that since M is diagonal, M_e and M_I are m by m and $(3n - m)$ by $(3n - m)$ diagonal matrices respectively. Relation (8.25) can now be expressed in terms of these subvectors as

$$[\mathbf{v}_e^T \quad \mathbf{v}_I^T]^T\begin{bmatrix} M_e\ddot{\mathbf{x}}_e - \mathbf{F}_e \\ M_I\ddot{\mathbf{x}}_I - \mathbf{F}_I \end{bmatrix} = \left[-\mathbf{v}_I^T(A_e^{-1}A_I)^T \quad \mathbf{v}_I^T\right]^T\begin{bmatrix} M_e\ddot{\mathbf{x}}_e - \mathbf{F}_e \\ M_I\ddot{\mathbf{x}}_I - \mathbf{F}_I \end{bmatrix} = 0, \tag{8.28}$$

or equivalently,

$$\mathbf{v}_I^T\left[-(A_e^{-1}A_I)^T(M_e\ddot{\mathbf{x}}_e - \mathbf{F}_e) + M_I\ddot{\mathbf{x}}_I - \mathbf{F}_I\right] = 0. \tag{8.29}$$

Since each component of the $(3n - m)$ vector $\mathbf{v}_I$ can be chosen independently, equation (8.29) implies that

$$-(A_e^{-1}A_I)^T(M_e\ddot{\mathbf{x}}_e - \mathbf{F}_e) + M_I\ddot{\mathbf{x}}_I - \mathbf{F}_I = \mathbf{0}. \tag{8.30}$$

To express this equation in terms of the $(3n - m)$-vector $\ddot{\mathbf{x}}_I$, all we need do is express the constraint equation (8.3) in terms of partitioned matrices as

$$\left[A_e \quad A_I\right]\begin{bmatrix} \ddot{\mathbf{x}}_e \\ \ddot{\mathbf{x}}_I \end{bmatrix} = A_e\ddot{\mathbf{x}}_e + A_I\ddot{\mathbf{x}}_I = \mathbf{b}. \tag{8.31}$$

From the last equality it follows that

$$\ddot{\mathbf{x}}_e = -A_e^{-1}A_I\ddot{\mathbf{x}}_I + A_e^{-1}\mathbf{b}. \tag{8.32}$$

Using this expression for $\ddot{\mathbf{x}}_e$ in equation (8.30) we obtain the equation of motion for the constrained system in terms of the acceleration vector $\ddot{\mathbf{x}}_I$ as

$$\left[R^T M_e R + M_I\right]\ddot{\mathbf{x}}_I = \mathbf{F}_I + R^T\left[M_e A_e^{-1}\mathbf{b} - \mathbf{F}_e\right], \tag{8.33}$$

where the matrix R denotes

$$R = A_e^{-1}A_I. \tag{8.34}$$

We have thus eliminated the vector $\ddot{\mathbf{x}}_e$ and obtained the equations of motion in terms of the independent acceleration vector $\ddot{\mathbf{x}}_I$!

Example 8.1

Consider a particle of mass m moving in 3-dimensional space subjected to the constraint $\dot{y} = z\dot{x}$. We shall determine the equations of motion of this particle. There are no impressed forces acting.

We have looked at this problem before in Chapter 3 (see Example 3.9), where we obtained the equations of motion using the fundamental equation. We now compare our previous method with the method of elimination, which we have outlined above. Differentiating the constraint equation we obtain

$$\begin{bmatrix}1 & -z & 0\end{bmatrix}\begin{bmatrix}\ddot{y}\\ \ddot{x}\\ \ddot{z}\end{bmatrix} = \dot{z}\dot{x}, \tag{8.35}$$

which is in our standard constraint form. Taking the submatrices $A_e = 1$ and $A_I = [-z \quad 0]$ (recall that A_e must be invertible), we obtain $R = A_e^{-1}A_I = [-z \quad 0]$. The vector $\ddot{\mathbf{x}} = [\ddot{y} \quad \ddot{x} \quad \ddot{z}]^T$ can be split into two subvectors and written as $\ddot{\mathbf{x}} = [\ddot{\mathbf{x}}_e^T \quad \ddot{\mathbf{x}}_I^T]^T$, where the subvector $\ddot{\mathbf{x}}_e = \ddot{y}$ and $\ddot{\mathbf{x}}_I = [\ddot{x} \quad \ddot{z}]^T$. The unconstrained equations of motion are

$$M\ddot{\mathbf{x}} = \begin{bmatrix}m & 0 & 0\\ 0 & m & 0\\ 0 & 0 & m\end{bmatrix}\begin{bmatrix}\ddot{y}\\ \ddot{x}\\ \ddot{z}\end{bmatrix} = \begin{bmatrix}M_e & \mathbf{0}\\ \mathbf{0} & M_I\end{bmatrix}\begin{bmatrix}\ddot{\mathbf{x}}_e\\ \ddot{\mathbf{x}}_I\end{bmatrix} = \mathbf{0} = \mathbf{F}, \tag{8.36}$$

where the diagonal matrix M_e is the scalar m, and the 2 by 2 matrix M_I is mI. Since the impressed force is zero, the subvectors $\mathbf{F}_e$ and $\mathbf{F}_I$ are both zero. Then by equation (8.33) we obtain

$$\left[\begin{bmatrix}-z\\0\end{bmatrix}m[-z\quad 0]+m\begin{bmatrix}1&0\\0&1\end{bmatrix}\right]\begin{bmatrix}\ddot{x}\\\ddot{z}\end{bmatrix}=+m\begin{bmatrix}-z\\0\end{bmatrix}\dot{z}\dot{x}, \tag{8.37}$$

which simplifies to

$$\begin{bmatrix}\ddot{x}\\\ddot{z}\end{bmatrix}=\frac{\dot{z}\dot{x}}{(1+z^2)}\begin{bmatrix}-z\\0\end{bmatrix}, \tag{8.38}$$

a result which we had already got in equation (3.100). Notice that the equation for the $\ddot{y}$ component of motion must be obtained using the set (8.38) and equation (8.35). The reader must wonder if this sort of elimination would always work.

To illustrate that the equation set (8.33) may not, by itself, be enough to determine the acceleration of the constrained system, let us modify our problem and say that the particle is subjected to a given impressed force $mu(y)$ in the X-direction and the impressed force $mw(y)$ in the Z-direction. Then the vector $\mathbf{F}=m[0\ \ u(y)\ \ w(y)]^T$ and the subvectors $\mathbf{F}_e=0$ and $\mathbf{F}_I=m[u(y)\ \ w(y)]^T$. Using (8.33) we see that all we need do is add to the right hand side of equation (8.37) the vector $\mathbf{F}_I$, yielding the equation

$$\begin{bmatrix}\ddot{x}\\\ddot{z}\end{bmatrix}=\frac{\dot{z}\dot{x}}{(1+z^2)}\begin{bmatrix}-z\\0\end{bmatrix}+\begin{bmatrix}\dfrac{u(y)}{(1+z^2)}\\w(y)\end{bmatrix}. \tag{8.39}$$

This seemingly innocuous change to our original problem, however, has created a drastic change in things because now equation (8.39) is an equation involving not just x and z and its derivatives, but also y! One would thus need to eliminate y from these equations by expressing it in terms of x and z, and their derivatives. But this cannot be done because the constraint equation is nonholonomic and therefore nonintegrable!

Thus we see that to complete the system of equations provided by equation (8.39) we would need to add the constraint equation $\dot{y}=z\dot{x}$. Alternatively, we could have added on the once differentiated constraint equation given by (8.35) to yield the system of equations

$$\begin{bmatrix}1&-z&0\\0&(1+z^2)&0\\0&0&1\end{bmatrix}\begin{bmatrix}\ddot{y}\\\ddot{x}\\\ddot{z}\end{bmatrix}=\begin{bmatrix}\dot{z}\dot{x}\\-z\dot{z}\dot{x}\\0\end{bmatrix}+\begin{bmatrix}0\\u(y)\\w(y)\end{bmatrix}. \tag{8.40}$$

We thus see that, in general, the equations of motion are then obtained by adding to the set (8.33) the equations of constraint $A\ddot{\mathbf{x}} = \mathbf{b}$. The equation of motion then becomes

$$\begin{bmatrix} A_e & A_I \\ 0 & R^T M_e R + M_I \end{bmatrix} \begin{bmatrix} \ddot{\mathbf{x}}_e \\ \ddot{\mathbf{x}}_I \end{bmatrix} = \begin{bmatrix} \mathbf{b} \\ \mathbf{F}_I + R^T (M_e A_e^{-1} \mathbf{b} - \mathbf{F}_e) \end{bmatrix}, \tag{8.41}$$

where, as before, $R = A_e^{-1} A_I$.

Alternately, we realize that for nonholonomic constraints if any quantity (i.e., elements of A, $\mathbf{F}$ or $\mathbf{b}$) in equation (8.33) is a function of any of the components of the vector $\mathbf{x}_e$ it is unnecessary to eliminate $\ddot{\mathbf{x}}_e$ in equation (8.30); for, in any case then we would need to add the set $A\ddot{\mathbf{x}} = \mathbf{b}$. Then we could use equation (8.30) directly in conjunction with the constraint equation (8.31), to yield

$$\begin{bmatrix} A_e & A_I \\ -R^T M_e & M_I \end{bmatrix} \begin{bmatrix} \ddot{\mathbf{x}}_e \\ \ddot{\mathbf{x}}_I \end{bmatrix} = \begin{bmatrix} \mathbf{b} \\ -R^T \mathbf{F}_e + \mathbf{F}_I \end{bmatrix}. \tag{8.42}$$

Equations (8.41) and (8.42) are two forms of the explicit equations of motion of the constrained mechanical system in Cartesian coordinates.

How would all this change were we to be working in generalized coordinates? To begin with we would have to replace equation (8.1) by Lagrange's equation to describe the unconstrained motion of the system, i.e., with

$$\mathrm{M}(\mathbf{q}, t)\ddot{\mathbf{q}} = \mathrm{F}(\mathbf{q}, \dot{\mathbf{q}}, t), \tag{8.43}$$

where the symmetric s by s matrix M is no longer a diagonal matrix, and F is an s-vector. Also the w constraint equations would be expressed as

$$\mathrm{A}(\mathbf{q}, \dot{\mathbf{q}}, t)\ddot{\mathbf{q}} = \mathbf{b}(\mathbf{q}, \dot{\mathbf{q}}, t), \tag{8.44}$$

where the matrix A is w by s.

Let us again assume that the rank of A is w. Then we may, as before, arrange the components of the s-vector $\mathbf{q}$, such that

$$\mathrm{A}(\mathbf{q}, \dot{\mathbf{q}}, t)\ddot{\mathbf{q}} = \begin{bmatrix} \mathrm{A}_e(\mathbf{q}, \dot{\mathbf{q}}, t) & \mathrm{A}_I(\mathbf{q}, \dot{\mathbf{q}}, t) \end{bmatrix} \begin{bmatrix} \ddot{\mathbf{q}}_e \\ \ddot{\mathbf{q}}_I \end{bmatrix} = \mathbf{b}(\mathbf{q}, \dot{\mathbf{q}}, t), \tag{8.45}$$

where the w by w submatrix $A_e(\mathbf{q}, \dot{\mathbf{q}}, t)$ is invertible, and the vector $\ddot{\mathbf{q}}$ is partitioned such that the subvector $\ddot{\mathbf{q}}_e$ is a w-vector. We can similarly partition the unconstrained equation of motion to read

$$\begin{bmatrix} \mathbf{M}_{ee}(\mathbf{q},t) & \mathbf{M}_{eI}(\mathbf{q},t) \\ \mathbf{M}_{Ie}(\mathbf{q},t) & \mathbf{M}_{II}(\mathbf{q},t) \end{bmatrix} \begin{bmatrix} \ddot{\mathbf{q}}_e \\ \ddot{\mathbf{q}}_I \end{bmatrix} = \begin{bmatrix} \mathbf{F}_e \\ \mathbf{F}_I \end{bmatrix}, \tag{8.46}$$

where the matrix $\mathbf{M}_{ee}(\mathbf{q},t)$ is w by w. The generalized virtual displacement $\mathbf{v}$ must satisfy, as before, the relation

$$\begin{bmatrix} \mathbf{A}_e & \mathbf{A}_I \end{bmatrix} \begin{bmatrix} \mathbf{v}_e \\ \mathbf{v}_I \end{bmatrix} = \mathbf{0}, \tag{8.47}$$

from which we obtain

$$\mathbf{v}_e = -\mathbf{A}_e^{-1}\mathbf{A}_I\mathbf{v}_I = -\mathbf{R}\mathbf{v}_I. \tag{8.48}$$

The basic principle of analytical mechanics now demands that

$$\begin{bmatrix} \mathbf{v}_e^T & \mathbf{v}_I^T \end{bmatrix}^T \left\{ \begin{bmatrix} \mathbf{M}_{ee}(\mathbf{q},t) & \mathbf{M}_{eI}(\mathbf{q},t) \\ \mathbf{M}_{Ie}(\mathbf{q},t) & \mathbf{M}_{II}(\mathbf{q},t) \end{bmatrix} \begin{bmatrix} \ddot{\mathbf{q}}_e \\ \ddot{\mathbf{q}}_I \end{bmatrix} - \begin{bmatrix} \mathbf{F}_e \\ \mathbf{F}_I \end{bmatrix} \right\} = 0. \tag{8.49}$$

Noting (8.48) and the fact that the components of $\mathbf{v}_I$ can be chosen independently yields, much in the same manner as for the Cartesian coordinates discussed earlier,

$$(-\mathbf{R}^T\mathbf{M}_{ee} + \mathbf{M}_{Ie})\ddot{\mathbf{q}}_e + (-\mathbf{R}^T\mathbf{M}_{eI} + \mathbf{M}_{II})\ddot{\mathbf{q}}_I = -\mathbf{R}^T\mathbf{F}_e + \mathbf{F}_I. \tag{8.50}$$

We could use equation (8.45), in a manner similar to what we did before, to eliminate the vector $\ddot{\mathbf{q}}_e$ from equation (8.50) and express it in terms of $\ddot{\mathbf{q}}_I$, as $\ddot{\mathbf{q}}_e = -\mathbf{R}\ddot{\mathbf{q}}_I + \mathbf{A}_e^{-1}\mathbf{b}$. But we know that, in general, we will need equation (8.45) to be added on if the constraints are nonholonomic and any of the entities in equation (8.50) are functions of the components of the vector $\mathbf{q}_e$. Hence we could, as we did in equation (8.42), simply keep the subvector $\ddot{\mathbf{q}}_e$, and append to equation (8.50) the equation of constraint (8.45). This will then yield the complete set of equations of motion given by

$$\begin{bmatrix} \mathbf{A}_e & \mathbf{A}_I \\ (-\mathbf{R}^T\mathbf{M}_{ee} + \mathbf{M}_{Ie}) & (-\mathbf{R}^T\mathbf{M}_{eI} + \mathbf{M}_{II}) \end{bmatrix} \begin{bmatrix} \ddot{\mathbf{q}}_e \\ \ddot{\mathbf{q}}_I \end{bmatrix} = \begin{bmatrix} \mathbf{b} \\ -\mathbf{R}^T\mathbf{F}_e + \mathbf{F}_I \end{bmatrix}. \tag{8.51}$$

These then are the general equations of motion pertinent to the system in Lagrangian coordinates. We note that, in general, the elements of all the matrices $\mathbf{M}_{ee}$, $\mathbf{M}_{Ie}$, etc., the matrices $\mathbf{A}_e$ and $\mathbf{A}_I$ and the vectors $\mathbf{F}_e$, $\mathbf{F}_I$ and $\mathbf{b}$ are functions of

$\mathbf{q}$, $\dot{\mathbf{q}}$ and t. The only loss of generality regarding equation (8.51) arises in our assumption that the number of rows of the matrix A equals its rank.

Now, what would happen if the rank r ($r < w$ and $r < s$) of A is less than w? Clearly, then we would partition the matrix $A = [A_e \quad A_I]$, in a manner such that the matrix A_e, which is of rank r, is w by r and the matrix A_I is w by $(s - r)$. Then solving for the r-vector $\mathbf{v}_e$ from equation (8.47) we would get (see equation (2.107)),

$$\mathbf{v}_e = -A_e^+ A_I \mathbf{v}_I + (I - A_e^+ A_e)\mathbf{h} = -A_e^+ A_I \mathbf{v}_I = -\tilde{R}\mathbf{v}_I \tag{8.52}$$

where we have used the fact that $A_e^+ A_e = I$, because the rank of A_e is r. The reader will realize that our entire previous argument will follow through (now $\ddot{\mathbf{q}}_e$ is an r-vector) with the only change that now the matrix $R = A_e^{-1} A_I$ is replaced by the matrix $\tilde{R} = A_e^+ A_I$. With this replacement of R by $\tilde{R}$, equation (8.51) (and similarly (8.42)) would again be valid! Notice that when the rank of A equals w, $A_e^+ = A_e^{-1}$, and $\tilde{R} = R$. Clearly, equation (8.51) is equivalent to the fundamental forms (expressed in generalized coordinates) that we have discussed earlier. We leave this as an exercise for the reader to prove.

The Gibbs–Appell Approach

We once again revert now to Cartesian coordinates to get the gist of this method. In this approach we consider the Gibbs function S, which is defined as

$$S = \frac{1}{2}\ddot{\mathbf{x}}^T M \ddot{\mathbf{x}}, \tag{8.53}$$

where the $3n$-vector $\ddot{\mathbf{x}}$ represents the acceleration of the constrained system at time t. We know that the acceleration components in this vector are not all independent, because of the constraint equation (8.3), $A(\mathbf{x}, \dot{\mathbf{x}}, t)\ddot{\mathbf{x}} = \mathbf{b}(\mathbf{x}, \dot{\mathbf{x}}, t)$. Let us again assume that the rank of the m by $3n$ matrix A is m. Then, we can partition the vector $\ddot{\mathbf{x}}$ into the m-vector $\ddot{\mathbf{x}}_e$ and the $(3n - m)$-vector $\ddot{\mathbf{x}}_I$ and express the subvector $\ddot{\mathbf{x}}_e$ in terms of the subvector $\ddot{\mathbf{x}}_I$, as we did in equation (8.32). Thus the Gibbs function at time t would then only be a function of the vector $\ddot{\mathbf{x}}_I(t)$. In fact, we have

$$\begin{aligned} 2S = \ddot{\mathbf{x}}^T M \ddot{\mathbf{x}} &= \begin{bmatrix} \ddot{\mathbf{x}}_e^T & \ddot{\mathbf{x}}_I^T \end{bmatrix}^T \begin{bmatrix} M_e & 0 \\ 0 & M_I \end{bmatrix} \begin{bmatrix} \ddot{\mathbf{x}}_e \\ \ddot{\mathbf{x}}_I \end{bmatrix} = \ddot{\mathbf{x}}_e^T M_e \ddot{\mathbf{x}}_e + \ddot{\mathbf{x}}_I^T M_I \ddot{\mathbf{x}}_I \\ &= (-A_e^{-1} A_I \ddot{\mathbf{x}}_I + A_e^{-1}\mathbf{b})^T M_e (-A_e^{-1} A_I \ddot{\mathbf{x}}_I + A_e^{-1}\mathbf{b}) + \ddot{\mathbf{x}}_I^T M_I \ddot{\mathbf{x}}_I \\ &= (-R\ddot{\mathbf{x}}_I + A_e^{-1}\mathbf{b})^T M_e (-R\ddot{\mathbf{x}}_I + A_e^{-1}\mathbf{b}) + \ddot{\mathbf{x}}_I^T M_I \ddot{\mathbf{x}}_I, \end{aligned} \tag{8.54}$$

where, in the second line of the above expression, we have used equation (8.32) to express $\ddot{\mathbf{x}}_e$ in terms of the subvector $\ddot{\mathbf{x}}_I$. We have denoted again $R = A_e^{-1}A_I$.

We next consider the work done at time t under virtual displacements by the impressed force $\mathbf{F}(\mathbf{x}, \dot{\mathbf{x}}, t)$. If $\mathbf{v}$ is a virtual displacement vector at time t, this is simply $\mathbf{v}^T\mathbf{F}(\mathbf{x}, \dot{\mathbf{x}}, t)$. Again we know that because of the presence of the constraints the components of any such virtual displacement vector are not all independent of each other, because they must satisfy the equation $A(\mathbf{x}, \dot{\mathbf{x}}, t)\mathbf{v} = \mathbf{0}$. We can partition the vector $\mathbf{v}$ into the m and $(3n - m)$ component subvectors $\mathbf{v}_e$ and $\mathbf{v}_I$ respectively, as before, to obtain

$$\begin{aligned} \mathbf{v}^T\mathrm{F}(\mathbf{x}, \dot{\mathbf{x}}, t) &= \begin{bmatrix} \mathbf{v}_e^T & \mathbf{v}_I^T \end{bmatrix}^T \begin{bmatrix} \mathbf{F}_e \\ \mathbf{F}_I \end{bmatrix} = \mathbf{v}_e^T\mathbf{F}_e + \mathbf{v}_I^T\mathbf{F}_I \\ &= \mathbf{v}_I^T\left[(-A_e^{-1}A_I)^T\mathbf{F}_e + \mathbf{F}_I\right] = \mathbf{v}_I^T\left[-R^T\mathbf{F}_e + \mathbf{F}_I\right]. \end{aligned} \tag{8.55}$$

In the second line of the expression above we have used equation (8.27). The components of the vector $\mathbf{v}_I$ can now be chosen independently. Let us denote the $(3n - m)$-vector which occurs on the right hand side of equation (8.55) as

$$\mathbf{P} = -R^T\mathbf{F}_e + \mathbf{F}_I. \tag{8.56}$$

Then the Gibbs–Appell equation for the constrained system is

$$\boxed{\frac{\partial S}{\partial \ddot{\mathbf{x}}_I} = \mathbf{P}.} \tag{8.57}$$

Here the scalar S is considered to be a function of the vector $\ddot{\mathbf{x}}_I$ as portrayed in the last expression on the right hand side of equation (8.54). Notice that the Gibbs–Appell equations constitute a set of $(3n - m)$ equations. To show that this is indeed the case, using equation (8.54), the left hand side of equation (8.57) becomes

$$\frac{\partial S}{\partial \ddot{\mathbf{x}}_I} = (R^TM_eR + M_I)\ddot{\mathbf{x}}_I - R^TM_eA_e^{-1}\mathbf{b}. \tag{8.58}$$

Using this and the definition of the vector $\mathbf{P}$ from (8.56), in equation (8.57) we get

$$(R^TM_eR + M_I)\ddot{\mathbf{x}}_I - R^TM_eA_e^{-1}\mathbf{b} = -R^T\mathbf{F}_e + \mathbf{F}_I, \tag{8.59}$$

which is precisely equation (8.33)! As before, in general, we would need to append the constraint equation $A(\mathbf{x}, \dot{\mathbf{x}}, t)\ddot{\mathbf{x}} = \mathbf{b}(\mathbf{x}, \dot{\mathbf{x}}, t)$ to the $(3n - m)$ equations

(8.59) and thereby complete the system of equations. But this, of course, is exactly the equation set (8.41) which we had derived earlier! Hence we have shown that the Gibbs–Appell method yields the equations of motion for constrained mechanical systems. Again, notice that the $(3n - m)$ equations in (8.57) must, in general, be augmented by the m constraint equations in (8.3) to complete the system of equations which describes the motion of the constrained system.

Example 8.2

Consider a particle of unit mass moving in 3-dimensional space and subject to the constraint $\dot{y} = z^2\dot{x}$. Let the impressed force $\mathbf{F}^T = [p(y)\ q(y)\ r(y)]^T$ where the known functions $p(y)$, $q(y)$ and $r(y)$ are the components in the X, Y and Z directions respectively. We use the Gibbs–Appell method to obtain the equation of motion of the particle.

The Gibbs function is given by

$$S = \frac{1}{2}\left\{\ddot{x}^2 + \ddot{y}^2 + \ddot{z}^2\right\}, \tag{8.60}$$

and the constraint may be expressed by

$$\ddot{y} = 2z\dot{z}\dot{x} + z^2\ddot{x}. \tag{8.61}$$

Eliminating $\ddot{y}$ from S, we get

$$S = \frac{1}{2}\left\{\ddot{x}^2 + \left(2z\dot{z}\dot{x} + z^2\ddot{x}\right)^2 + \ddot{z}^2\right\}. \tag{8.62}$$

Furthermore, if υ_x, υ_y and υ_z are the three components of the virtual displacement vector $\mathbf{v}$, they must satisfy the relation

$$\upsilon_y = z^2\upsilon_x. \tag{8.63}$$

The virtual work done by the impressed forces then becomes

$$\begin{aligned}\mathbf{F}^T\mathbf{v} &= p(y)\upsilon_x + q(y)\upsilon_y + r(y)\upsilon_z = \{p(y) + z^2q(y)\}\upsilon_x + r(y)\upsilon_z \\ &= P_x\upsilon_x + P_z\upsilon_z.\end{aligned} \tag{8.64}$$

Using the Gibbs–Appell equations we now get

$$\ddot{x} + \left(2z\dot{z}\dot{x} + z^2\ddot{x}\right)z^2 = \{p(y) + z^2 q(y)\} \tag{8.65}$$

and

$$\ddot{z} = r(y) \tag{8.66}$$

Note that we are not yet done. We need to add on to equations (8.65) and (8.66) the constraint equation (8.61) to complete the system of equations. We leave it for the reader to verify that an equivalent set of equations would be obtained by using equation (8.5) or equation (8.24).

Let us further investigate the vector **P** now. We know that the acceleration of the unconstrained system (in Cartesian coordinates) is given by the relation $M\mathbf{a} = \mathbf{F}$, or in partitioned form as,

$$\begin{bmatrix} M_e & 0 \\ 0 & M_I \end{bmatrix} \begin{bmatrix} \mathbf{a}_e \\ \mathbf{a}_I \end{bmatrix} = \begin{bmatrix} \mathbf{F}_e \\ \mathbf{F}_I \end{bmatrix} \tag{8.67}$$

so that,

$$\mathbf{P} = -R^T\mathbf{F}_e + \mathbf{F}_I = -R^T M_e \mathbf{a}_e + M_I \mathbf{a}_I \tag{8.68}$$

Furthermore, using the constraint equation in the form of equation (8.32), the quantity

$$\begin{aligned} \mathbf{a}^T M\ddot{\mathbf{x}} &= \mathbf{a}_e^T M_e \ddot{\mathbf{x}}_e + \mathbf{a}_I^T M_I \ddot{\mathbf{x}}_I \\ &= \mathbf{a}_e^T M_e(-R\ddot{\mathbf{x}}_I + A_e^{-1}\mathbf{b}) + \mathbf{a}_I^T M_I \ddot{\mathbf{x}}_I, \end{aligned} \tag{8.69}$$

where we have partitioned the vector **a** into two subvectors $\mathbf{a}_e$ and $\mathbf{a}_I$ having m and $(3n - m)$ components respectively. Hence by (8.69),

$$\frac{\partial\{\mathbf{a}^T M\ddot{\mathbf{x}}\}}{\partial \ddot{\mathbf{x}}_I} = -R^T M_e \mathbf{a}_e + M_I \mathbf{a}_I = \mathbf{P}, \tag{8.70}$$

where the last equality follows from equation (8.68). Using this result in the right hand side of equation (8.57), we see that the Gibbs–Appell equations can be expressed as

$$\frac{\partial\{S - \mathbf{a}^T M\ddot{\mathbf{x}}\}}{\partial \ddot{\mathbf{x}}_I} = \mathbf{0} \tag{8.71}$$

or alternately as

$$\frac{\partial\left\{S - \mathbf{a}^T M\ddot{\mathbf{x}} + \frac{1}{2}\mathbf{a}^T M\mathbf{a}\right\}}{\partial \ddot{\mathbf{x}}_I} = \mathbf{0}. \tag{8.72}$$

The addition of the term $\frac{1}{2}\mathbf{a}^T M\mathbf{a}$ in the above expression has no effect because at any time t, $\mathbf{x}(t)$ and $\dot{\mathbf{x}}(t)$ are assumed to be known, and therefore the acceleration of the unconstrained system $\mathbf{a}(t) = M^{-1}\mathbf{F}(\mathbf{x}, \dot{\mathbf{x}}, t)$ is known, and so not a function of $\ddot{\mathbf{x}}_I$. Noting the definition of S from equation (8.53), the quantity in the brackets on the left hand side of equation (8.72) is simply

$$\frac{1}{2}\ddot{\mathbf{x}}^T M\ddot{\mathbf{x}} - \mathbf{a}^T M\ddot{\mathbf{x}} + \frac{1}{2}\mathbf{a}^T M\mathbf{a} = \frac{1}{2}(\ddot{\mathbf{x}} - \mathbf{a})^T M(\ddot{\mathbf{x}} - \mathbf{a}), \tag{8.73}$$

which is nothing but our Gaussian, G! Hence we have shown that the Gibbs–Appell equation (8.57) is exactly the equation

$$\boxed{\frac{\partial \mathrm{G}}{\partial \ddot{\mathbf{x}}_I} = \mathbf{0},} \tag{8.74}$$

where the Gaussian G is considered to be a function of $\ddot{\mathbf{x}}_I$, the independent acceleration components of the constrained system. We have already eliminated the dependent components (contained in the subvector $\ddot{\mathbf{x}}_e$) of the acceleration vector $\ddot{\mathbf{x}}$ from G by using the constraint equation (8.32); we therefore no longer need to enforce the constraint equation, since it is identically satisfied! Equation (8.74) is then easily recognized as simply the necessary condition for extremizing this Gaussian!

We therefore see that Gauss's principle demands that the quantity

$$\mathrm{G}(\ddot{\mathbf{x}}) = \frac{1}{2}(\ddot{\mathbf{x}} - \mathbf{a})^T M(\ddot{\mathbf{x}} - \mathbf{a}) \tag{8.75}$$

be minimized over all $\ddot{\mathbf{x}}$ subject to the constraints $A\ddot{\mathbf{x}} = \mathbf{b}$. In arriving at the fundamental forms for the equation of constrained motion, we directly solved this constrained minimization problem and showed that our solution $\ddot{\mathbf{x}}$ gave a global minimum of G.

In the Gibbs–Appell formulation we first convert this constrained minimization problem into an unconstrained minimization problem by first expressing the Gaussian G in terms of the $(3n - m)$ vector $\ddot{\mathbf{x}}_I$ whose components are independent (thereby making G a function of only the vector $\ddot{\mathbf{x}}_I$). This we do through the use of the constraint equations. Since the constraints have now been taken account of, we then simply write down the necessary conditions for extremizing this function G, considered as a function of $\ddot{\mathbf{x}}_I$! The Gibbs–Appell equation (8.57) is then nothing but the necessary condition for the extremum of the Gaussian expressed in terms of the independent components of $\ddot{\mathbf{x}}$. When there are nonholonomic constraints, this elimination, in general, will work for the vector $\ddot{\mathbf{x}}_e$, as we have shown; but, in general, this elimination will *not* work, for the components of the vector $\mathbf{x}_e$. Components of this vector when present either in S or in $\mathbf{P}$, in nonholonomic systems, cannot be eliminated from equation (8.57). We are therefore forced to add on the constraint equations to complete the system of differential equations! This may be looked upon perhaps as partially defeating the purpose of the elimination process that brought us out so far, because we have to bring back the vector $\ddot{\mathbf{x}}_e$ (or $\dot{\mathbf{x}}_e$) through the constraint equation $A\ddot{\mathbf{x}} = [A_e \quad A_I][\mathbf{x}_e^T \quad \mathbf{x}_I^T]^T = \mathbf{b}$ to complete the formulation.

We have indeed returned after a long and exciting journey into the core of analytical mechanics back to where we started from, namely the simple and aesthetic Principle of Gauss. The fundamental forms we obtain yield a simple and straightforward solution to the constrained minimization problem which this Principle leads to, when the constraints are expressible in our standard form, $A\ddot{\mathbf{x}} = \mathbf{b}$.

PROBLEMS

8.1 Write out the fundamental forms analogous to equations (8.11)–(8.15) applicable to the situation where we use Lagrangian coordinates. Indicate explicitly the functional dependence of each of the quantities in the expressions.

8.2 Show that if $A\mathbf{v} = \mathbf{0}$, then $\mathbf{v}^T A^{\{1,4\}}\mathbf{z} = 0$ for any vector $\mathbf{z}$ which belongs to the range space of A.

8.3 Write the general form for the Lagrange multiplier vector, analogous to equation (8.17), when using Lagrangian coordinates.

8.4 Prove that the rank of the matrix $\mathbf{L}$ in equation (8.22) is $6n$.

8.5 By explicitly determining the inverse of the matrix, show that equation (8.24) is exactly the fundamental equation (8.5).

8.6 Write the analog of equation (8.24) in Lagrangian coordinates.

8.7 Show that the $\ddot{\mathbf{x}}_I$ that extremizes $G(\ddot{\mathbf{x}}_I)$ given by the Gibbs–Appell equations, actually minimizes $G(\ddot{\mathbf{x}}_I)$.

8.8 Obtain the Gibbs–Appell equations pertinent to Lagrangian coordinates.

8.9 Extend the Gibbs–Appell approach to the situation when the rank r of the m by $3n$ matrix A is such that $r < m < 3n$.

8.10 Solve Example 3.8 using equations (8.5), (8.24), (8.41) and (8.57) and show that all the approaches lead to the same equation of motion for the constrained particle.

8.11 To the system described in Example 8.1, add the constraint $\dot{y} = z^2\dot{x} + x$. Use the different methods described in this chapter to obtain the equation of motion for this system.

8.12 Discuss the advantages and disadvantages of using the different approaches presented in this chapter in arriving at the equation of motion for constrained mechanical systems.

For Further Reading

1. A2 The first chapter of this book gives a quick survey of the methods of analytical mechanics, though more from a mathematical viewpoint. We recommend that the interested reader take a look at this material.

2. D1 We have not presented Dirac's method for the determination of the equations of motion for constrained Hamiltonian systems in this book. Dirac uses extended Poisson brackets in his approach. He provides an algorithm which can even be applied to systems where the matrix M is singular. Though such systems do not usually arise in well-modeled mechanical systems, they play an important role in field theory.

3. F1 This paper by J. Franklin deals more thoroughly with the fact that equation (8.5) is still valid even when the constraint equation is satisfied not exactly, but in a least-squares sense. Fom the viewpoint of the philosophy of mechanics the implications of this paper have yet to be explored and understood.

4. K1 The method proposed by Kane is similar to our development of equation (8.33). Note that equation (8.33) needs to be, in general, augmented by equation (8.3) (or equation (8.31)) to obtain the complete set for describing the motion of a nonholonomically constrained system.

5. P1 The discussion in Pars related to the Gibbs–Appell equations on pages 197 to 202 is well worth reading because it has a somewhat different approach from that taken in this chapter.

References

A1 P. Appell, *Exemple de Mouvement d'un Point Assujeti a une Liason Exprimee par une Relation Non Lineaire Entre les Composantes de la Vitesse*, Comptes Rendus, 1911, pp. 48–50.

A2 V. I. Arnold, *Dynamical Systems III*, Springer Verlag, Berlin, 1980.

B1 R. Bellman, *Introduction to Matrix Analysis*, 2nd Edition, McGraw-Hill, New York, 1970.

D1 P. A. M. Dirac, *Lectures in Quantum Mechanics*, Yeshiva University Press, New York, 1964, pp. 1–24.

F1 J. Franklin, Least-Squares Solution of Equations of Motion under Inconsistent Constraints, *Linear Algebra and Its Applications*, Vol. 222, 1995, pp. 9–13.

G1 F. Graybill, *Matrices and Applications to Statistics*, 2nd Edition, Wadsworth, Belmont, Calif., 1983.

G2 T. Greville, Some Applications of the Pseudoinverse of a Matrix, *SIAM Rev.*, Vol. 2, 1960, pp. 15–22.

G3 C. F. Gauss, Uber Ein Neues Allgemeines Grundgesatz der Mechanic, *Journal fur Reine und Angewandte Mathematik*, Vol. 4, 1829, pp. 232–235.

G4 J. W. Gibbs, On the Fundamental Formulae of Dynamics, *American Journal of Mathematics*, Vol. 2, 1879, pp. 49–64.

G5 H. Goldstein, *Classical Mechanics*, Addison-Wesley, Second Printing, 1981.

HJ1 R. Horn and C. Johnson, *Matrix Analysis*, Cambridge University Press, 1985.

K1 T. R. Kane, Dynamics of Nonholonomic Systems, *Journal of Applied Mechanics*, 1961, pp. 574–578.

LH1 C. Lawson and R. Hanson, *Solving Least Squares Problems*, Prentice-Hall, Englewood Cliffs, N.J., 1974.

M1 The Math Works, *MATLAB User's Guide*, South Natick, Mass. 01760, 1992.

P1 L. Pars, *A Treatise on Analytical Mechanics*, Ox Bow Press, Conn., Second Printing, 1972.

PFTV W. Press, B. Flannery, S. Teukolsky and W. Vetterling, *Numerical Recipes: The Art of Scientific Computing*, Cambridge University Press, 1986.

R1 R. Rosenberg, *Analytical Dynamics of Discrete Systems*, Plenum Press, New York, 1977.

R2 C. R. Rao, *Linear Statistical Inference and Applications*, Wiley, New Delhi, 1973.

R3 J. W. S. Rayleigh, *The Theory of Sound*, Volume 1, Dover, 1945.

W1 R. Wolfram, *Mathematica*, Wolfram Inc., Ill., 1992.

W2 E. T. Whittaker, *A Treatise on the Analytical Dynamics of Particles and Rigid Bodies*, Fourth Edition, Cambridge University Press, 1989 (originally published in 1904).

Attention must also be called to the following works:

1. G. W. Housner and D. E. Hudson, *Dynamics*, Van Nostrand, Princeton, N.J., 1961.
2. R. Huston, *Multibody Dynamics*, Butterworth-Heinemann, Boston, 1990.
3. T. R. Kane, *Dynamics*, Holt, Reinhart and Winston, 1968.
4. J. L. Lagrange, *Mechanique Analytique*, Vols. 1 and 2, Gauthier-Villars, Paris, 1853.
5. C. Lanczos, *The Variational Principles of Mechanics*, Dover, 1970.
6. L. Meirovitch, *Methods of Analytical Mechanics*, McGraw-Hill, 1974.
7. Ju. I. Neimark and N. A. Fufaev, *Dynamics of Nonholonomic Systems*, American Mathematical Society Translations, Vol. 33, Providence, R.I., 1972.
8. E. Routh, *Dynamics of a System of Rigid Bodies*, Dover, 1955.
9. E. C. G. Sudarshan and N. Mukunda, *Classical Dynamics: A modern perspective*, Wiley, 1974.
10. J. L. Synge and B. A. Griffith, *Principles of Mechanics*, McGraw-Hill, 1959.
11. A. G. Webster, *The Dynamics of Particles and of Rigid, Elastic and Fluid Bodies: Lectures on mathematical physics*, Dover, 1959.

Afterword

This book primarily deals with developing the equations of motion for mechanical systems. It is a problem which was first posed at least as far back as Lagrange, over 200 years ago, and has been vigorously worked on since by many physicists and mathematicians. The list of scientists who have contributed and attempted this problem is truly staggering. A recent monograph on the subject by Neimark and Fufaev lists more than 500 recent references.

The first major step in the understanding of constrained motion was taken by Lagrange when he formulated and developed the technique of using, what are called today, the Lagrange multipliers. The next step took about a century in the making when Gibbs and Appell developed the Gibbs–Appell approach in the late eighteen hundreds. As mentioned in Pars's book (written in 1965) this approach is considered by most to provide the simplest and most comprehensive way of setting up the equations of motion for systems with nonintegrable constraints. In 1964, P. A. M. Dirac attempted to solve the problem anew and, for Hamiltonian systems with singular Lagrangians, developed a procedure for obtaining the equations of motion for constrained systems by ingeniously extending the concept of a Poisson bracket. Dirac considered constraints which were not explicitly dependent on time.

Dirac's approach has not been discussed in this book. It is well documented in advanced treatises on analytical mechanics and may be found in the list of references that we have provided; besides, it requires background material which would go well beyond an introductory text in mechanics. The purpose here was to provide a simple, new and fresh approach to the conceptualization of constrained motion. Besides its inherent simplicity, the approach presented in this book can handle many problems of interest in classical dynamics.

As with all theories which deal with fundamental aspects of nature, the development of the seemingly simple, central results in this book went through various phases of transformation. They have provided us with periods of unlimited excitement and doubt. It is their ultimate simplicity and aesthetic beauty which prompted us to write this short text, not as a research monograph pitched for the specialist but as a text for the undergraduate student learning mechanics. Accordingly, we have made sure to highlight the core fundamental concepts in mechanics, putting little to no emphasis on aspects such as numerical integration procedures and their accuracy, or the stability of the motions involved or on impulsive motions. These topics shall be reserved for a separate book, perhaps some time in the future.

The results developed here provide a new and exciting way of interpreting the manner in which motion occurs in mechanical systems. The way Nature emerges from these results is both simple and marvelous–aspects of its operation that we believe can be shared even with beginners in the field of mechanics. This book marks only the beginning of a new approach to understanding the motion of mechanical systems. We hope that while providing new spice to a mature field, it will help open up new horizons in our understanding of Nature.

Index